AF732522

The Power of New Urban Tourism

The Power of New Urban Tourism explores new forms of tourism in urban areas with their social, political, cultural, architectural and economic implications. By investigating various showcases of New Urban Tourism within its social and spatial frames, the book offers insights into power relations and connections between tourism and cityscapes in various socio-spatial settings around the world.

Contributors to the volume show how urban space has become a battleground between local residents and visitors, with changing perceptions of tourists as co-users of public and private urban spaces and as influencers of the local economies. This includes different roles of digital platforms as resources for access to the city and touristic opportunities as well as ways to organise and express protest or shifting representations of urban space. With contemporary cases from a wide disciplinary spectrum, the contributors investigate the power of New Urban Tourism in Africa, Asia, the Americas, Europe and Oceania. This focus allows a cross-cultural evaluation of New Urban Tourism and its dynamic, and changing conception transforming and subverting cities and tourism alike.

The Power of New Urban Tourism will be of great interest to academics, researchers and students in the fields of cultural studies, sociology, the political sciences, economics, history, human geography, urban design and planning, architecture, ethnology and anthropology.

Claudia Ba is a research fellow at the Department of Urban Sociology and Sociology of Space at Technical University of Darmstadt, Germany.

Sybille Frank is Professor of Urban Sociology and Sociology of Space at the Institute of Sociology at Technical University of Darmstadt, Germany.

Claus Müller is a research fellow and PhD candidate at the Department for Planning and Construction Economics/Real Estate at Technische Universität Berlin, Germany.

Anna Laura Raschke is a research fellow at the Department of Urban Sociology and Sociology of Space at Technical University of Darmstadt, Germany.

Kristin Wellner is Professor of Planning and Construction Economics/Real Estate at the Institute of Architecture at Technische Universität Berlin, Germany.

Annika Zecher is a research fellow at the Department of Urban Sociology and Sociology of Space at Technical University of Darmstadt, Germany.

Contemporary Geographies of Leisure, Tourism and Mobility

Series Editor: C. Michael Hall, *Professor at the Department of Management, College of Business and Economics, University of Canterbury, Christchurch, New Zealand*

The aim of this series is to explore and communicate the intersections and relationships between leisure, tourism and human mobility within the social sciences.

It will incorporate both traditional and new perspectives on leisure and tourism from contemporary geography, e.g. notions of identity, representation and culture, while also providing for perspectives from cognate areas such as anthropology, cultural studies, gastronomy and food studies, marketing, policy studies and political economy, regional and urban planning, and sociology, within the development of an integrated field of leisure and tourism studies.

Also, increasingly, tourism and leisure are regarded as steps in a continuum of human mobility. Inclusion of mobility in the series offers the prospect to examine the relationship between tourism and migration, the sojourner, educational travel, and second home and retirement travel phenomena.

The series comprises two strands:

Contemporary Geographies of Leisure, Tourism and Mobility aims to address the needs of students and academics, and the titles will be published simultaneously in hardback and paperback.

Routledge Studies in Contemporary Geographies of Leisure, Tourism and Mobility is a forum for innovative new research intended for research students and academics, and the titles will initially be available in hardback only. Titles include:

Degrowth and Tourism
New Perspectives on Tourism Entrepreneurship, Destinations and Policy
Edited by C. Michael Hall, Linda Lundmark and Jasmine Zhang

Sense of Place and Place Attachment in Tourism
Ning Chris Chen, C. Michael Hall and Girish Prayag

Tourism, Change and the Global South
Edited by Jarkko Saarinen and Jayne M. Rogerson

The Power of New Urban Tourism
Spaces, Representations and Contestations
Edited by Claudia Ba, Sybille Frank, Claus Müller, Anna Laura Raschke, Kristin Wellner and Annika Zecher

For more information about this series, please visit: www.routledge.com/Contemporary-Geographies-of-Leisure-Tourism-and-Mobility/book-series/SE0522

The Power of New Urban Tourism

Spaces, Representations and Contestations

Edited by Claudia Ba, Sybille Frank, Claus Müller, Anna Laura Raschke, Kristin Wellner and Annika Zecher

Routledge
Taylor & Francis Group
LONDON AND NEW YORK

First published 2022
by Routledge
2 Park Square, Milton Park, Abingdon, Oxon OX14 4RN

and by Routledge
605 Third Avenue, New York, NY 10158

Routledge is an imprint of the Taylor & Francis Group, an informa business

© 2022 selection and editorial matter, Claudia Ba, Sybille Frank, Claus Müller, Anna Laura Raschke, Kristin Wellner and Annika Zecher; individual chapters, the contributors

The right of Claudia Ba, Sybille Frank, Claus Müller, Anna Laura Raschke, Kristin Wellner and Annika Zecher to be identified as the authors of the editorial material, and of the authors for their individual chapters, has been asserted in accordance with sections 77 and 78 of the Copyright, Designs and Patents Act 1988.

All rights reserved. No part of this book may be reprinted or reproduced or utilised in any form or by any electronic, mechanical, or other means, now known or hereafter invented, including photocopying and recording, or in any information storage or retrieval system, without permission in writing from the publishers.

Trademark notice: Product or corporate names may be trademarks or registered trademarks, and are used only for identification and explanation without intent to infringe.

British Library Cataloguing-in-Publication Data
A catalogue record for this book is available from the British Library

Library of Congress Cataloging-in-Publication Data
A catalog record has been requested for this book

ISBN: 9780367555399 (hbk)
ISBN: 9780367555443 (pbk)
ISBN: 9781003093923 (ebk)

Typeset in Times New Roman
by Deanta Global Publishing Services Chennai India

Contents

Figures

Tables

Contributors

Alberto Amore, Dr, is Lecturer in Tourism and Air Travel Management at Solent University, United Kingdom. He holds a PhD in Management (University of Canterbury, New Zealand) and a MA cum laude in Tourism, Territory and Local Development (Università degli Studi di Milano-Bicocca, Italy). His research interests include urban planning, urban tourism and urban regeneration, with a focus on post-disaster urban governance and destination resilience.

Claudia Ba, PhD candidate, (editor) studied comparative cultural studies and is research fellow at the DFG-funded project "Neighbourhoods in the tourist trap?" at Technical University of Darmstadt. In her PhD thesis, she focused on the iconic coherence of visualisations of the intangible cultural heritage in Senegal and Gambia, West Africa.

Monica Bernardi, PhD, is research fellow at the Department of Sociology and Social Research (Milano-Bicocca University). Her research foci are the sharing economy in its urban impacts, social innovation spread, and the governance model of the sharing cities. She is a member of Dimmons Research Group (UOC Barcelona).

Henriette Bertram, Dr, is a researcher at the Institute for Urban Transformations at Kassel University. She has studied Cultural Sciences in Frankfurt (Oder) and Madrid and has a PhD from the University of Kassel. Her work focuses on urban memory cultures after conflict as well as the relationship of gender and space.

Wei-Fen Chen, Dr, is Lecturer in Marketing at the University of Leicester. Her research focuses on the consumer behaviour of individuals experiencing geographic relocation and social mobility, as well as its impact on personal well-being and market sustainability.

Guido Cimadomo, MArch, PhD, is Lecturer at the Department of Art & Architecture, Universidad de Malaga. He is an expert member of the ICOMOS' scientific committee CIPA for the Documentation of Architectonic Heritage and UN-Habitat UNI focal point at the Universidad de Malaga.

Madalena Corte-Real, PhD, has a PhD in Urban Studies and a background in Sociology. She is a researcher at CICS.NOVA—Centre for Interdisciplinary Social Sciences, Nova University Lisbon. Her main area of interest is the analysis of the perception and production of space in a context of territorial transformation.

Lena Eskilsson, PhD, is a Senior Lecturer in Human Geography at the Department of Service Management and Service Studies at Lund University. She has a specialisation in place development and destination marketing, mainly with an urban focus. She has worked with research projects focusing on bridging the film and tourism sectors as well as the influence of media on tourism and tourist behaviour. She is currently working on sustainable urban tourism development.

Sybille Frank, Prof Dr, (editor) is Professor for Urban Sociology and Sociology of Space at the Department of Sociology, Technical University of Darmstadt. Her work focuses on the sociology of space and place, on urban conflicts and social inequalities and on contested power relations in the fields of tourism and heritage-making.

C. Michael Hall, Prof Dr, is a professor in the Department of Management, Marketing and Entrepreneurship at the University of Canterbury; visiting professor at Linnaeus University, Kalmar; guest professor at Lund University, Helsingborg; and docent at University of Oulu. Co-editor of *Current Issues in Tourism*, his research interests include tourism, regional development, global environmental change, food, servicescapes, farnarkeling, environmental history, sustainability and World Heritage.

Mario Hernandez, PhD, is an Assistant Professor in the department of sociology at Mills College in Oakland, California. He is an urban sociologist who specialises in the study of gentrification. His most recent research focuses on the Bushwick neighbourhood of Brooklyn in New York City.

Annie Hikido, Prof, is an assistant professor of sociology at Colby College in Waterville, Maine. She is broadly interested in how race, class and gender shape intimate interactions and global processes. Her current project examines how Black women broker different types of authenticity in the South African township tourism market.

Eduardo Jiménez-Morales, MArch, PhD Urban Studies, is postdoctoral research fellow at the Department of Art & Architecture, Universidad de Malaga. He is member of UTOPIA (Urbanism, Tourism and Landscape Research Group) and leads the research project "Impact of the residential uses in the urban vulnerability diagnosis of Torremolinos".

Maria João Gomes, PhD, is a landscape architect with an MA in Human Ecology. She holds a PhD in Urban Studies and is a researcher at CICS.NOVA—Centre for Interdisciplinary Social Sciences, Nova University Lisbon. Her areas of interest are the social production of space, urban walking experience and urban tourism.

Emily Kelling, PhD candidate, is a researcher and lecturer at Technische Universität Berlin, where she completed her PhD in sociology on informal housing in London in 2020. Her research covers urban studies, housing and questions of space, politics and power. She has a keen interest in the combination of sociology and architectural tools such as mappings. She has been a guest researcher at Universidad de Chile, Facultad de Arquitectura y Urbanismo; at HafenCity Universität Hamburg; and at University College London's the Bartlett Development Planning Unit.

Luís Manata e Silva, PhD candidate, has a background in Sociology and is a PhD student in Urban Studies at the Nova University Lisbon in a partnership with ISCTE—Lisbon University Institute. His thesis is on territorial marketing taking as a case study a city in southern Portugal.

Marianna Monte, PhD Urban Studies, M.A. Urban Design, is a collaborating researcher at CICS.NOVA—Centre for Interdisciplinary Social Sciences, Nova University Lisbon. She has been working on projects concerning urban transformation, in-between uses and the impact and development of tourism in urban areas.

Claus Theodor Müller, Dipl.-Volksw., (editor) is a research fellow and PhD candidate at the department for Planning & Construction Economics/Real Estate at Technische Universität Berlin. Currently his research focuses on the influence of tourism on local social and retail infrastructure.

Giulia Mura, PhD, is a research fellow at the Department of Human Science for Education (Milano-Bicocca University). Currently her work on the impact of ICT in society focusses on sharing platforms and food policies/food waste at urban level, while previous research focused on ICT and education.

Jan Henrik Nilsson, PhD, holds a PhD in Social and Economic Geography from Lund University. He is a Reader in Human Geography and Associate Professor in Economic Geography, Department of Service Management and Service Studies, Lund University. His main research interests concern mobility, tourism geography and sustainable urban tourism development.

Anna Laura Raschke, Dr, (editor) is a postdoctoral research fellow at the Chair of Urban Sociology and Sociology of Space at Technical University of Darmstadt. Her work focuses on urban sociology, sociology of space, urban tourism and neighbourhood developments.

Marta Torres Ruiz, PhD candidate, has studied architecture at Technische Universität Berlin and worked as an architect and project leader at various architectural offices in India and Germany. Since 2020 she is partner at Bureau Ruiz Saad—Architecture Urbanism Research in Berlin. She is a collegiate of the DFG Graduate School "Identity and Heritage", where she is conducting a

PhD dissertation on the narratives and discoursivisation of transformations in cities related to the term of overtourism.

Anja Saretzki, MA, Dipl.-Kffr., is a part-time lecturer at the Institute of Urban and Cultural Area Research (IFSK), Leuphana University Lüneburg, Germany.

Christoph Sommer, PhD candidate, recently completed his PhD in geography at Humboldt University Berlin. Currently he is working as research associate at the Institute of Environmental Planning at Leibniz University Hannover. He co-founded the Urban Research Group: New Urban Tourism at the Georg-Simmel Center for Metropolitan Studies; his main areas of interest include research on urban policy, governance, the anthropology of policy, tourism and municipalism.

Tin-yuet Ting, Prof Dr, is Assistant Professor of Sociology in the Department of Applied Social Sciences at The Hong Kong Polytechnic University. His research focuses on the emergence of networked activism and political agency manifested in various forms of urban politics.

Ingrid C. Vargas-Díaz, MArch, PhD in Urban Planning, is associate fellow at the Research Cluster on Territorial Synergies, Universidad de Granada. She is visiting researcher at the Universidad La Gran Colombia and Ecole d'Architecture de Montpellier. Her researches are focused on informal settlements, urban segregation and housing policies.

Niklas Völkening, MSc Geography, is research fellow at the Chair of Human Geography and Transformation Research at the University of Augsburg. In his PhD thesis, he is analysing processes of commodification in tourism in Cuba and the associated effects on the identities of Cubans.

Kristin Wellner, Prof Dr, (editor) has been a professor of Planning and Construction Economics/Real Estate at the institute of architecture, Technische Universität Berlin since 2012 after studies in real estate economics and business administration. Her research focuses in the interfaces between economics and behavioural sciences in planning and deals with real estate market models, decisions and research on the housing quality.

Karlheinz Wöhler, Prof Dr, is a professor emeritus of tourism studies at the Institute of Urban and Cultural Area Research (IFSK), Leuphana University Lüneburg.

Annika Zecher, MA, (editor) is a research fellow at the department of Urban Sociology and Sociology of Space at Technical University of Darmstadt, where she also completed her master's degree in Sociology with a thesis on housing projects in Frankfurt, Germany and contemporary and historic perceptions of ideal urban living.

1 The power of New Urban Tourism

An introduction

Claudia Ba, Sybille Frank, Claus Müller and Anna Laura Raschke

"Tourist go home" is a slogan that has been written on walls and stickered to lampposts in cities like New York, Coimbra, Lisbon and Berlin (see Figure 1.1). Publicly expressing frustration and anger, residents worldwide fight the surge in tourism to their cities. Protesting against a phenomenon that has become known as New Urban Tourism, local activist groups call attention to many recent changes in their neighbourhoods that reach from clogged pavements and altered retail and gastronomy offers to rising rents and restricted use of public space. Since local politics often seem to be either powerless against or supportive of an increased presence of tourists in cities, more and more residents openly rally against the social and spatial developments they relate to the visitors' presence. New Urban Tourism therefore challenges both power relations within cities and established notions and routines of everyday urban life.

This book contributes to contemporary public and academic discourses on urban tourism by investigating the power of New Urban Tourism. With the term 'power', the editors of this volume intend to capture the complex transformations that the enormous quantitative rise and qualitative changes in, as well as the corresponding (re)negotiations of power relations, have brought about in the much-disputed socio-political and economic action field of New Urban Tourism in contemporary cities. The chapters of this book specifically provide new insights into the spatial relations, representations and contestations that this new form of presenting and experiencing cities entails. By this, the editors wish to add to the current research panorama by addressing topical research areas that have only been touched upon so far.

(New) Urban Tourism

Urban tourism appeared as a research field in the late 1980s when the enormous growth rates of tourism in cities attracted scholarly attention (Roche 1992, Judd and Fainstein 1999, Ashworth and Page 2011). At the time, urban tourism was analysed as a major driving force behind a perceived worldwide homogenisation and culturalisation of urban space characterised by a proliferation of "tourist bubbles" (Judd 1999). The resulting "spatially confined nature of tourist activities in cities" (Novy 2010, p. 29), shaped by the purposeful promotion

Figure 1.1 Tourists go home © Robert Nyman (Lisbon), Billie Ward Grace (New York), Scott Dexter (Coimbra), Die Partei (Berlin)

and marketing of urban festivals and sightseeing courses in central urban locations with designated monuments and new signature architectures, was however challenged in the 2000s. In view of increasing market saturation, city tourists, inspired by alternative guidebooks and 'insider tips' in digital media, "venture[d] in growing numbers into areas previously not visited, or less frequented by visitors" (Novy 2010, p. 29). What has later been coined 'New Urban Tourism' was first described by Maitland and Newman in 2004 in a paper on the discovery of Islington, a residential neighbourhood in the north of London that did not offer any particular tourist sights, as a "new tourism area" (Maitland and Newman 2004, p. 339).

Since then, the fast-growing social phenomenon of New Urban Tourism has attracted considerable attention in both urban studies and tourism studies (Maitland 2007, 2008, 2010, 2013, Maitland and Newman 2009, Novy 2010, Dirksmeier and Helbrecht 2015). The following four research strands dominated the past 15 years of interdisciplinary New Urban Tourism research: (1) theoretical reflections on New Urban Tourism as a social phenomenon; (2) studies on how New Urban Tourism has impacted on cities and urban quarters, and vice versa; (3) investigations of protests against and the discontents of New Urban Tourism; and (4) analyses of the ways in which cities set out to govern New Urban Tourism.

Theoretical reflections on New Urban Tourism as a new social trend

Publications dwelling on New Urban Tourism as a new social trend seek to relate it to broader social theories, recent analyses of social change and a concomitant rise in global mobilities of all kinds. When Maitland and Newman introduced the term "New Urban Tourism" in 2009, they built upon a variety of concepts that had been developed by an impressive body of postmodernity-, leisure- and performance-oriented tourism research. Since the 1990s, prominent tourism scholars, informed by postmodern theories (Urry 1990, Rojek 1993, Lash and Urry 1994), have drawn attention to a pervasive de-differentiation of tourism and everyday life (Crouch 2001, McCabe 2002, 2005, Hall 2005, Uriely 2005, Hannam 2011, White and White 2007). They spotlighted what urban tourists actually do when visiting a city, concluding that the basic binary distinction between tourism and the everyday that had informed tourism research up to then, had to be reconsidered (Coleman and Crang 2002, Larsen 2008). While some researchers showed that touristic travelling builds on a set of mundane habituated norms and routines (Edensor 2001, 2007), others concluded that differences between tourists and (cosmopolitan) city dwellers' practices—such as visiting a café or a museum—can no longer be sustained (Urry and Larsen 2011, Larsen 2017, 2019, Cohen and Cohen 2019, Stors *et al.* 2019). In an age of growing leisure-orientation and cultural consumption, some scholars saw the dawning of a new era of post-tourism (Feifer 1985, Urry 1990, Rojek 1993), while others found evidence for an overall end of tourism (Lash and Urry 1994). As a consequence, Martinotti (1993), for example, suggested speaking of 'city users' instead of 'tourists vs. residents', considering that a growing number of permanent city users set out to joyfully explore and experience their city, whereas temporary city users endeavoured to take part in mundane local life (Maitland and Newman 2009, Richards 2017, Wearing and Foley 2017). Last but not least, the "new mobilities paradigm" in tourism (Sheller and Urry 2006) called on tourism research to understand tourism as a complex crossover and intersection of mobilities of people, objects, knowledge, images and information (Bærenholdt *et al.* 2004, Frank 2016), and to study the multiplicity of human movements and moorings, including temporary mobilities beyond touristic visits and new forms of temporary residence (Frank and Meier 2018, Novy 2018).

Likewise, scholars from urban studies discovered urban tourism as an area of research in the 1990s. Building upon David Harvey's seminal work on the entrepreneurial city (Harvey 1989) that analysed the contemporary rise of cultural consumption and urban leisure industries, academic publications set out to describe the quantitative rise, qualitative change and overall promotion of (New) Urban Tourism as a driver of economic regeneration and urban development. It was noted that tourism perfectly tied in with the growth-oriented, market-driven and city-marketing-centred politics and policies typical for entrepreneurial cities against the background of globalisation, deindustrialisation and post-Fordist urban restructuring (Hubbard and Hall 1998, Maitland and Newman 2004, Novy 2010), as well as neoliberalisation, financialisation and urban austerity (Peck 2012,

Füller and Michel 2014). In this context, urban heritage and culture, drawing on products and atmospheres created by the so-called creative industries (Florida 2003), were discovered as a source of post-industrial consumption aimed to meet the leisure-oriented lifestyles and consumption preferences of the urban middle- and upper-classes (Zukin 1991, 1995, 2011, Novy 2010, Frank 2020). Scholars determined that tourists were more and more lured to previously less visited urban areas, such as former working-class districts characterised by an urban fabric of old streets and buildings, a socially and ethnically diverse population, a mix of functions, open spaces, small shops, cafés, bars and restaurants visited by locals. As a consequence, urban neighbourhoods were commodified and gentrified

> beyond cities' cores according to the demands of affluent population groups with sophisticated and cosmopolitan tastes, white collar professionals and workers in the creative industries, transient city users (temporary migrants, business visitors on short term assignment and so on) as well as tourists.
> (Novy 2010, p. 31)

Studies on the social and demographic characteristics of New Urban Tourists found that they are interested in a destination's qualities of place, including many small places to eat, drink and shop, in everyday culture and local atmosphere, in interacting with the locals, as well as in alternative 'non-touristic' ways of travelling and consuming. For these tourists, to be different from the broad mass of (urban) tourists by immersing themselves in local life is of high importance (Maitland and Newman 2004, Paulauskaite *et al.* 2017, Füller and Michel 2014, Mokras-Grabowska 2018).

Studies on the impacts of New Urban Tourism on cities and urban quarters

Another major research strand investigates the impacts of New Urban Tourism on cities and urban quarters (and vice versa) with a specific focus on holiday rentals, gentrification, touristification or overtourism. A lot of research has been conducted on the role of the sharing economy in contemporary tourism (Guttentag 2015, Stabrowski 2017). Specifically, scholars studied how digital platforms such as Airbnb have changed touristic patterns (Dredge and Gyimóthy 2015, Sans-Domínguez and Quaglieri 2016, Stors 2019), the hospitality industry (Aznar *et al.* 2017, Gutiérrez *et al.* 2017) and the housing market in the framework of ongoing financialisation and the discovery of short-term rentals of residential space as outstandingly profitable assets (Füller and Michel 2014, Guttentag 2015, Brauckmann 2017, Wachsmuth and Weisler 2018). Moreover, scholars investigated how the rise of New Urban Tourism, often initiating immense socio-spatial transformations of residential neighbourhoods that have been conceptualised as 'touristification' (Freytag and Bauder 2018, Sequera and Nofre 2018) or "tourism gentrification" (Gotham 2005, Minoia 2017, Mendes 2018, Cócola-Gant 2019) is related to gentrification processes (Zukin *et al.* 2009, Cócola Gant 2016,

Gravari-Barbas and Guinand 2017). Additionally, studies on overtourism (Dredge 2017, Koens *et al.* 2018, Dodds and Butler 2019, Seraphin 2019) have been published that deal with urban places exposed to different forms of urban tourism, such as cruise tourism as a fast-growing sector of the tourism industry that adds to the rise of New Urban Tourism in these places. While New Urban Tourism and its hoped-for or unwanted impacts on cities and neighbourhoods has up to now mostly been observed and investigated in Western cities, Novy and Colomb (2019) state that the phenomenon has recently spread into metropolitan areas, particularly in Asia and the Arab world, as many of them have adopted growth-oriented entrepreneurial and neoliberal city politics in an attempt to tap tourism as a source of income and economic development.

New Urban Tourism and its discontents: contestations and protests

> [T]ourism has become an object of mobilisation because there is more of it, in a wider range of urban destinations, spreading to previously "untouched" neighbourhoods, taking new forms, and because it is often not governed and regulated enough—or merely governed in the interest of a narrow range of actors.
>
> (Novy and Colomb 2019, p. 361)

This quote by Novy and Colomb indicates that the various unsettling transformations of urban residential neighbourhoods in recent times have not remained unchallenged (Colomb and Novy 2016). The authors note that changes in the material, economic, spatial and sociocultural fabric of neighbourhoods in areas outvisited by New Urban Tourists are often attributed to the presence of the latter. Generally, Novy and Colomb (2016) identified four sources of conflicts around (new) urban tourism: economic, physical, social and sociocultural as well as psychological conflicts. Principal areas of contention in this line of research are gentrification processes and a proliferation of short-term (holiday) rentals that is perceived to lead to rising rents and subsequent displacement of residents, changes in local retail leading to a loss of daily supplies for permanent dwellers, rising prices in shops and gastronomy, overcrowding of public spaces, land use conflicts, noise from festivals and parties, littering, high crime rates, as well as feelings of alienation and homogenisation due to a loss of social and ethnic diversity in the neighbourhood (Novy 2013, Füller and Michel 2014, Novy and Colomb 2016, Pinkster and Boterman 2017, Gravari-Barbas and Guinand 2017). Novy and Colomb argue that an overall politicisation of "*the city itself* and its uneven socio-spatial transformation" (Novy and Colomb 2019, p. 364) has made it difficult to differentiate mobilisations against tourists from other new social movements in cities that campaign against rising rents, neoliberal urban governance and austerity politics (Mayer 2009, 2013, Brenner *et al.* 2012). While many of these new urban social mobilisations are thus multi-focal, there are nonetheless quite a few that are explicitly directed against tourism or tourists, exposing the problem that

> tourism and leisure development frequently involves an unequal distribution of costs and benefits as profits are disproportionably reaped by community outsiders and local elites while commercial and residential gentrification, rising living costs as well as other adverse effects put additional burdens on lower-middle to low-income residents and other disadvantaged social groups. (Novy 2010, p. 32)

Analyses of the ways in which cities set out to govern New Urban Tourism

Lastly, in light of pervasive local conflicts around New Urban Tourism, a growing body of scholarship analysed the various ways in which cities have strived to govern New Urban Tourism and its effects on local communities. As Novy and Colomb (2019, p. 10f) have summed up, responses from policymakers and the tourism industry vary between an "'ignore and do nothing' approach" that prioritises growth, a delegitimisation of those who criticise New Urban Tourism, "[s]maller adjustments in policy and symbolic gestures" or "[m]ore substantial political actions and policy responses". Against the background of unaltered urban growth agendas many studies have found either a lack of, or a weak governance of, urban tourism and short-term rentals (Gravari-Barbas and Jacquot 2016, Sommer and Helbrecht 2017). This means that policymakers and tourism industry actors unwaveringly follow a growth-oriented approach that further promotes (all parts of) cities as destinations, thereby neglecting the demands of the local population that often articulates concerns regarding unwanted changes of neighbourhood life and fears of displacement. Many scholars argue that invitations by policymakers and city marketeers to participate in tourism-related decision-making mostly merely constitutes a symbolic act. Since the results of hearings and workshops are most commonly not transferred into political measures to manage tourism, they have been characterised as "post-political" (Swyngedouw 2007, 2010, Füller and Michel 2014, Füller *et al.* 2018). A similar research strand has demonstrated that tourism governance is increasingly shaped by privatisation, for example by the outsourcing of tourism management tasks to private destination management organisations or by the establishment of public–private partnerships for city marketing and tourism development (Hubbard and Hall 1998, Novy 2016). By this, tourism governance has been eliminated from democratic control mechanisms. As opposed to this, a rare example of substantial responses to local protests against unmanaged tourism growth is the case of Barcelona where social mobilisations succeeded in challenging the prevailing pro-tourism politics. In the 2015 municipal elections Ada Colau, a housing activist, was elected as new mayor who issued several political strategies that aim at containing urban tourism and at rededicating large parts of the touristified inner city to permanent city dwellers and their needs (Martins 2018, Goodwin 2019).

About this book

While building on the results of the aforementioned research strands, the editors of this volume intend to contribute to contemporary discourses on the latest changes in urban tourism by focusing on the contested power relations of New Urban Tourism. This book presents empirical research from different cities, academic disciplines and methodological perspectives to broaden the discussion on New Urban Tourism and to provide new insights into the power relations, spatial reconfigurations, representations and contestations that this recent form of urban tourism has brought about. Taking a more relational perspective, Barbara Kirshenblatt-Gimblett's observation that "[t]ourism and heritage are collaborative industries" (Kirshenblatt-Gimblett 1995, p. 371) may be expanded by conceptualising tourism and the marketable production of the urban, tourism and governance, as well as tourism and identity constructions, as collaborative industries. These linkages however do not necessarily lead to a commodification of everyday life. By considering power as relational and not something objective or stable—but rather ambiguous and processual (Massey 1993)—the phenomena related to New Urban Tourism can be reflected more diversely and conversely than the recent literature suggests. Some mentionable work on New Urban Tourism already seeks to reconceptualise what could be regarded as touristic in urban contexts (Frisch *et al.* 2019). This book aims to invite researchers and theorists alike to think of New Urban Tourism as embedded in changing social and societal structures of market and power relations.

The authors of this volume exhibit research from cities in Europe (Almada, West Belfast, Berlin, Hamburg, Lund, Milan, Torremolinos and Venice), Asia (Hong Kong), Africa (Cape Town), the Americas (Havana, New York) and Oceania (Christchurch). They address cases from a wide range of academic disciplines: sociology, economics, tourism management and marketing, human geography, cultural studies, art history, history, architecture, landscape architecture, urban design and planning. Chapter authors specifically look at the effects of New Urban Tourism on urban development and housing markets in urban centres and peripheries, on tourism-related conflicts and contested spaces in cities, on changes in communal life and urban lifestyles in residential neighbourhoods, and on questions of representation and identity constructions linked to the rise of New Urban Tourism. Gaining a broader understanding of New Urban Tourism, the book is subdivided into three parts, dealing with selected research gaps in New Urban Tourism.

First, the consumability of urban spaces via peer-to-peer platforms and the sharing economy are associated with powerful spatial (re)configurations. Taking spatial relations into account, consuming the city traditionally exacerbates the tensions between centre and periphery. Looking at and scrutinising the consumability of the city from different angles, social and spatiotemporal reorderings and rearrangements come into view, as well as new spatial orders, socioeconomic relations and claims to power and status. Periphery and centre form a complex dynamic that goes beyond processes of suburbanisation, traceable through the rise of artistic industries and investments in the sharing economy and housing quality in (former) peripheries, while former urban cores empty out and are at the same time rediscovered by specific groups of permanent and temporary city users.

Second, New Urban Tourism showcases conflicts and contestations that vary between cities and different actors in advocacy for or against New Urban Tourism. Their activities cannot be perceived as one-sided or destructive—but as co-constituting factors in the field of New Urban Tourism. The analytical strength of such a perspective lies in the empirical investigation of local power relations and processual dependencies. The answer to the question of who produces frictions and contestations, when, how and why, cannot be pre-empted but depends on the political, social and cultural conditions of the investigated city.

Third, the relation of identity constructions and representations in New Urban Tourism offers new insights into the (self-)promotion of local neighbourhoods, hosts and guests. Instead of being overrun by the New Urban Tourists, local actors draw from historical, ethnic and spatial narratives to portray their neighbourhoods as cohesive, thereby constructing stabilising new identities. This makes local facilitators their own marketers, displaying their constructions of the Self and the Other. Former power relations between locals and tourists are challenged and reinterpreted, given the embeddedness of hosts, guests and city users more generally in global mobilities. Representations on websites show how local actors stage themselves as 'authentic' by pointing to specific choices and lifestyles. Traditional forms of sight-seeing, life-seeing and New Urban Tourism are combined to tie the production of unique selling points and identification claims together.

Part I: Consuming the city: New Urban Tourism in urban centres and metropolitan peripheries

The first part of this volume, "Consuming the city: New Urban Tourism in urban centres and metropolitan peripheries", investigates the power of New Urban Tourism to transform both urban centres and metropolitan peripheries and to reshape their traditional spatial relations. New Urban Tourists' growing demand for consuming the city in a more 'authentic' way, off the beaten tracks (Cohen 1987), created a wide range of 'insider tips' and new local supplies in coveted areas—from peer-to-peer platforms like Airbnb offering overnight stays in residential apartments to the provision of new goods, services and spaces in local trade and public areas. While such economic and spatial transformations of residential quarters in or close to urban centres have been going on for quite a while, these neighbourhoods increasingly find themselves in competition with more remote urban quarters or cities that have just been discovered by New Urban Tourists, local traders and city marketers alike because of their undetected local charm. We thus witness a still underexplored massive spatial expansion of New Urban Tourism areas that turns attractive small centres in urban peripheries into new leisure areas, and less appealing ones into 'safe havens' for residents who can no longer afford the rents in the city centre or long for a life undisturbed by tourism.

In his contribution, "Bohemia and the New Urban Tourism", Mario Hernandez investigates the transformation of the Bushwick neighbourhood in Brooklyn, New York—an area that was long characterised by neglect and decay—into a widely known symbol of gentrification and a popular destination for New Urban

Tourists. The author illuminates how New Urban Tourists' consumption patterns feed into and help to sustain the local hipster subculture, but at the same time reshape the urban landscape according to standards set by marketing and corporate actors. By analysing the aesthetic features of the newly established cafés, bars and shops, Hernandez traces the local bohemian avant-garde's role in Bushwick's development from an economically neglected safe space for creativity and the arts into an urban brand that is supported by multinational corporations. While the local art scene has played a major role in the transformation of this formerly ethnically diverse and socially underprivileged working-class neighbourhood, it is itself dependent on the various bohemian networks that still cluster in Bushwick but are threatened to be driven out by further investment. Therefore, Hernandez argues that local bohemia is contributing to and profiting from the advancement of gentrification and touristification processes that in the end endanger the uniqueness of the area.

In their case study, "Tourist platformisation: New Urban Tourism in Milan", Monica Bernardi and Guilia Mura explore peer-to-peer as well as business-to-customer service providers, both being part of the platform economy, and their role in shaping the growth of urban tourism. By studying Milan, the authors showcase a city that has experienced a steady growth in tourism in the past, spiking in 2015 when Milan hosted the Expo World Fair. This event led to the proliferation of the sharing economy, while the city had generally been welcoming towards online-based sharing platforms that were regarded as social innovations catering to New Urban Tourism. Using a mixed-methods approach, the authors identify several problems that emerged with the spread of platform-based business models, especially in the tourism and hospitality sector. They include regulatory and administrative issues raised by suppliers' anonymity, as well as increased competition both on the housing market and on the market for tourist accommodation. However, interviews with representatives of the traditional hospitality industries indicate that while the platform economy had a strong impact on Milan's growing tourist sector, only a subsection of local providers experienced increased competition through informal private short-term rentals.

The influence of peer-to-peer tourist accommodation on Berlin's housing market is investigated by Claus Theodor Müller and Kristin Wellner in their chapter, "Peer-to-peer tourist accommodation and its impact on the local housing market in Berlin". Examining the market for rental flats, they show evidence that a unique situation is to be found in Berlin, where urban development is still heavily influenced by the former political division of the city after World War II and the reunification of East and West Berlin in the scope of German unification in 1990. The authors draw on different data sets to estimate the impact of the city-wide conversion of rental units into short-term rentals, and more specifically on the centrally located borough of Friedrichshain-Kreuzberg, a former working-class area that has recently become a hotspot of New Urban Tourism. In order to differentiate the impact on the neighbourhood level, the authors examine four neighbourhoods with different densities of both short-term rentals and formal tourist accommodation like hotels and hostels. By comparing the rental market in these neighbourhoods in light of the specific numbers of peer-to-peer accommodation, the authors do not find sufficient evidence for the common myth of tourism being the booster for locally increasing rents.

In "Redefining a mature destination as a low-cost neighbourhood: relations between socio-spatial segregation in Torremolinos and urban tourism in Malaga, Spain", Eduardo Jiménez-Morales, Ingrid Vargas-Diaz and Guido Cimadomo investigate the changing spatial relations between the city of Malaga and its sub-urban community of Torremolinos. They identify a major shift that is traced back to a changing demand structure in the tourist sector initiated by the advent of New Urban Tourism. Having served as an important destination for sun-and-beach tourism to Spain for decades, Torremolinos was hit hard by the oil crisis of the 1970s but recovered as a site for second homes of mainly Central Europeans in the following years. As nearby Malaga recently transformed into a destination for New Urban Tourism, Torremolinos changed its role again. Displaced by gentrification and the conversion of housing units into tourist accommodation, underprivileged former residents of Malaga, many of them employed in the hospitality industry, turned to the empty holiday flats and resorts of Torremolinos and converted them into permanent residences. This process not only further segregated both Malaga and Torremolinos; it also led to a diminished residential quality, since the buildings in Torremolinos were not intended or planned to be permanent dwellings.

The last chapter of Part I, "Tourism in a peripheral territory in the metropolitan area of Lisbon: the case study of Almada", also addresses the relationship between a major urban tourist destination and a community at its periphery. Looking at Almada, located across the river Tagus opposite Lisbon city centre, the authors Madalena Corte-Real, Marianna Monte, Maria João Gomes and Louis Manata e Silva describe a town dominated by devastated industrial structures and working-class environments that has recently attracted the interest of New Urban Tourists. Almada is not yet promoted by most of the official travel guides but has received attention as an insider tip in travel blogs and other social media outlets. Particularly the promise to immerse oneself in the authentic local everyday life, to avoid mass tourism and to experience unique restaurants are highlighted as attractions both in (social) media reports and in street surveys with visitors. Focus groups with administration officials and local entrepreneurs uncovered a generally positive view of the newly gained attention from New Urban Tourists. Locals see this development as an opportunity for investment into public infrastructure and a renewal of local commerce, but are also wary of the pitfalls of tourism gentrification and large-scale investments in the urban fabric.

Part II: Protest and frictions: contesting New Urban Tourism

The book's second part, "Protest and frictions: contesting New Urban Tourism", gives voice to studies on conflicts arising through or alongside New Urban Tourism. Showcasing research from Europe (Hamburg, Venice and Berlin) as well as Cuba and Hong Kong, it deepens our understanding of how increases in tourism are not in themselves a problem but mirror locally different effects and conflict-laden situations. While overtourism is at the centre of the debate in the Global North, the cases of Cuba and Hong Kong point to different political and power-related areas of friction when it comes to planning, regulating and engaging in urban tourism.

In the first chapter of this part, titled "New Urban Tourism in Cuba and its socio-spatial consequences", Niklas Völkening analyses the effects of New Urban Tourism in Cuba. The author shows how financial and legal reforms in Cuba paved the way for privately run tourist accommodation and the subsequent influx of much-needed hard currency to preserve socialist rule through capitalist measures since 1994. In an assessment of relevant literature, Völkening discusses New Urban Tourism as an economically, socially and spatially intriguing phenomenon that blurs the lines between locals and non-locals, gives way to potentially socially and politically destabilising processes of capitalist exploitation and leads to conflicts of interests in cities of the Global North. His investigation of Havana Vieja and Trinidad offers a Global South perspective that broadens the scope of the discussion while also exemplifying the effects New Urban Tourism has had in Cuba. The author uses interview data and mappings to show how increased tourism led to shifts in local communities, where those with access to the market have been put in significantly more powerful positions than those without. Völkening illuminates the fact that top-down approaches to tourism have left little opportunity for influence for the local population. The ongoing friction between the socialist state and the capitalist measures it implements to ensure its existence illustrates how the challenges stemming from New Urban Tourism affect Cuba in a unique way.

Highlighting the effects of, as well as local reactions to, shopping tourism to Hong Kong, Tin-yuet Ting and Wei-Fen Chen give insight into how digital media can shape protest against the effects of tourism in their chapter "Embattled consumption-space of tourism: urban contention against Chinese tourist shoppers in Hong Kong". Using netnography and media analyses, the authors show how grassroots activism in Hong Kong voices discontent with an urban development that is geared towards 'outsiders' while at the same time organising communities to form stronger bonds through local support of businesses. In Hong Kong, public urban space is economically developed to cater to incoming visitors from mainland China pouring into malls and shopping areas. Local shops are either displaced by skyrocketing rents that can only be paid by large corporations, or overrun by tourists that buy out local supplies of everyday goods. Locals turn to social media to voice their discontent through memes and public discussion but also organise protests and interventions. In contrast to other social movements, this protest is not led by elite actors and does not foster a certain political ideology of a common good. It is aimed at reclaiming urban space for local dwellers that are being pushed out or ignored.

In their case study on Hamburg's Schanzenviertel, Anja Saretzki and Karlheinz Wöhler discuss the relationship between tourism gentrification and political protest in one of Germany's urban tourism hotspots under the title "Between political protest and tourism gentrification: impacts of New Urban Tourism in Hamburg's Schanzenviertel". The authors trace the processes that unintendedly transformed the neighbourhood from a primarily residential into an entertainment and leisure district in the second decade of the 21st century. They identify four distinct discourses pertaining to the role of the Rote Flora alternative cultural centre in the local political scene: (1) the interconnectedness of overtourism and touristification with an apprehension of the sociocultural composition of the neighbourhood including disputable constructions of locals and non-locals; (2) the traditional gentrification discussion that is in this case revolving around displacement in the public space; (3) and a tourism marketing discourse that morphs the contested space of the Rote Flora into an emblem for the neighbourhoods' unique flair reinforcing a problematic relationship of becoming attractive for tourism by fighting it; (4) the authors conclude that resistance in its varying forms has and will prove unable to stop urban tourism to Schanzenviertel.

Venice is a well-established venue for overtourism research and protest against tourism. In her chapter "The empty boxes of Venice: Overtourism—conflict, politicisation and activism around the vacant houses of Venice" Marta Torres Ruiz shines a new light on the situation by focusing on the discourse around the vacant houses in the city. Based on interviews with members of local civic and resistance groups, Torres Ruiz shows how the combination of privatisation of public housing, a lack of regulation of short-term rentals and an ever-increasing tourism demand resulted not only in vacant houses in Venice, but also in debates on ownership over the city. The so-called "empty boxes" are thus more than vacant houses; they have become a metaphor for the struggle for political power in the city where touristic demands have been placed before residents' needs for quite some time. Torres Ruiz places her research at the crossing of overtourism, tourism gentrification and political activism, proving that Venice may be well-researched but still offers new insights. Her focus on local political activism exemplifies how complexly impactful the New Urban Tourism phenomenon can be on an urban community.

In his chapter "Powerful ways of (not) knowing New Urban Tourism conflicts: thin problematisation as limitation for tourism governance in Berlin" Christoph Sommer reconstructs how institutional actors in Berlin know, interpret and address frictions regarding tourism in the city. By analysing policy documents, Sommer shows how the involved parties in Berlin's tourism governance frame problems and manoeuvre through public discussion and demands for action related to tourism conflicts. Destination officers publicly frame the immensely increased tourism to Berlin as a successful comeback story and conversely address protest against it as small-minded, and problems stemming from it as minor and unavoidable side-effects. Conceptualising tourism and its effects this way negates the need for a debate on the principles of tourism governance in Berlin. Sommer further shows how governance actors in Berlin use survey data to push their tourism-friendly agenda by using survey tools to prove a positive position on tourism in the city. By applying these strategies, the governance actors construct problematic tourism as non-problematic or inevitable; in both cases allowing non-action and dismissal of opposing views. Sommer concludes with a research agenda for multi-disciplinary tourism studies to better understand, address and thereby change the approaches to (not) knowing tourism conflicts within tourism governance.

Part III: Representations and identities: hopes and challenges for New Urban Tourism

The third part, "Representations and identities: hopes and challenges for New Urban Tourism", focuses on different ways in which local identity constructions are co-constitutive and affected by New Urban Tourism. The commodification of urban life, concurrent with a decline in diversity in what is marketed as "unique" identities to urban tourists, subjects cities and neighbourhoods to image debates and to discussions about how to appeal to others and about the sources of change in the social fabric. The section sheds light on different approaches to tourists and tourism that are exploited by global investors utilising tragedy for economic development (Christchurch), drawn upon by locals to modify their collective identity in order to appeal to visitors looking for a dark, yet consumable urban past (West Belfast), regarded as a valuable resource by marginalised residents (Cape Town), blamed as scapegoats for diverse local developments (Berlin) or used to reflect on the effect of a specific group of visitors (Lund).

In the first chapter, "Shock of the new: the rhetoric of global urban tourism in the rebuild of Christchurch, New Zealand", Michael Hall and Alberto Amore focus on the relation between New Urban Tourism and global capitalism. Based on longitudinal research on the rebuilding of Christchurch after a series of earthquakes in the 2010s and on the involvement of international stakeholders in the relaunch of the city as a creative hub, they argue that a fourth neoliberal wave entered Christchurch under the guise of something "new". Triangulating data from the city council, the Canterbury Earthquake Recovery Authority (CERA) and interviews with grassroots networks, they criticise the rhetoric and interests of international consultancy companies that play on the hopes of the locals for a successful re-establishing of the city as a tourism destination. Challenging academic culture, the authors further investigate how researchers, representing the phenomenon as something new, strived to invent a new field of study in an already well-researched area.

In "New Urban Tourism in the post-conflict city: sharing experiences of violence and peace in West Belfast", Henriette Bertram reviews the role of heritage tourism and phoenix tourism as new formations of New Urban Tourism in Northern Ireland. Through an analysis of the strategies of local district marketing and ethnographic explorations in the respective neighbourhoods, she illustrates and compares actors' various intentions and strategies in the post-conflict urban space. She shows that tourism has become a new battleground for the hegemony over identity-construction in West Belfast where former literal combatants use guided visitor tours as a means to push their version of local history, while beautifying and pacifying the narrative by omitting and sugar-coating violence and despair, thus engaging in quite common capitalist marketing patterns.

Annie Hikido looks at "Race, class and gender of websites: the marketing and mythologising of urban Africa online". The author analyses the presentation of holiday rentals in Cape Town, South Africa, through image and content analysis. In doing so, she deconstructs the narrative of collective identification and rather demonstrates how desires to appeal to the worldviews of white middle-class tourists lead to racialised and gendered representations of a commodified everyday life in the Townships. By stressing that their places are safe, clean and embedded in a community of hard-working, friendly folk, the hostesses counter the common myths about black poverty, aggressive black masculinity and a failing social fabric with a counter-myth that enables affluent white tourists to experience a safe and whitewashed version of an authentic African home. The concept of the counter-myth is shown to be an integral part of New Urban Tourism that allows and invites further reflection, particularly on racialised notions of place. The study also opens up to the African continent, which has so far been neglected in research on New Urban Tourism.

In "New Urban Tourism and the right to complain: tourism as a catchall for urban problems", Emily Kelling and Annika Zecher analyse the dominant public narrative pertaining to tourism in Berlin, as contained in over 900 newspaper articles from 2008 to 2018. They find that the notion of New Urban Tourism has changed drastically from life-seeing to life-experiencing. They approach the local discourses and reported changes through the lens of relational space constitution and reconstruct the manifold disruptions of tourism narratives in the city within just a decade. The terms 'life-seeing' and 'life-experiencing' become a methodological tool to reflect on identity constructions and processes of othering in the city. By this, the authors challenge traditional assumptions about the relation between New Urban Tourism and local residents in Berlin.

Part III closes out with "Science-driven mobility as an aspect of New Urban Tourism", where Lena Eskilsson and Jan Henrik Nilsson introduce the concept of science-driven mobility (SDM) as a driver of New Urban Tourism in the city of Lund. In an effort to further the theoretical debate on the concept of urban tourism, SDM is implemented to show how Lund sees developments that are usually ascribed to New Urban Tourism but can in this case more accurately be described as effects of student internationalisation, newly built infrastructure and recently established academic institutions. Combining the analysis of interviews and marketing material with their own insights as long-time resident researchers of Lund University, Eskilsson and Nilsson see parallels between the effects of SDM and New Urban Tourism—such as the proliferation of the English language in local shops and the establishment of housing and hotels for the incoming (temporal) inhabitants of Lund. SDM is presented as a challenge to the classical notion of tourism and embedded within a theorisation of New Urban Tourism.

Key insights

In her concluding remarks on this volume Maria Gravari-Barbas asks "So, what is new about new urban tourism?" As urban destinations are not new in tourism, the author poses this question as a way to address changes in the reception and effects of urban tourism against the background of changes in the tourism economy that is part of a dynamic capitalist society. Moreover, Gravari-Barbas situates New Urban Tourism within academic discussions on a post-tourist era, in which the lines between touristic and everyday experiences and practices are blurred such that a new research agenda, as proposed in this book, offers a promising perspective.

In sum, several apparent shortcomings of the debate on urban tourism can be concluded from this volume. For one, the term 'New Urban Tourism' is at least problematic and should be reconsidered as it describes neither an entirely new phenomenon nor a veritably new form of tourism, but rather a further diversification of the urban tourism market. Furthermore, it is undoubtedly true that it is not the everyday urban life, such as working, paying bills or raising kids, that visitors seek to find but rather urban leisure experiences. Thus, they look for the extraordinary with a quite familiar gaze (Urry 1990). A more differentiated discussion about the characteristics and types of city visitors and their practices in connection with changing power relations, urban economic, political and cultural landscapes, spatial reconfigurations and communal lives is thus required to appropriately address what lurks beneath the fancy surface of New Urban Tourism.

This book offers a starting point for further theoretical reflections on the established terms of urban tourism research, as well as a collection of intriguing theoretic perspectives and original empiric studies. The cases presented show that the phenomena subsumed under the term 'New Urban Tourism' may lead to quite different outcomes and reactions in different cities. In order to adequately address and grasp this complexity, it is therefore vital to theorise New Urban Tourism from a local perspective. An understanding of the city as a socio-spatial formation with a specific history, symbolic order and set of routinised practices (Berking 2012) could be a promising and fruitful contribution to the ongoing spatial, cultural and practice-oriented reconceptualisation of urban tourism research.

For the moment though, the cycle of growing urban tourism and repeated protests against it has been disrupted by the global COVID-19 pandemic of 2020 and 2021. In attempts to contain the spread of the coronavirus, academic studies of tourism have largely come to a halt, as conducting research is as restricted as touristic travelling. While it is indisputable that tourism-related industries have suffered a lot during the ongoing pandemic, its long-term effects (not only) on tourism and cities will definitely need to be researched in the years to come.

Acknowledgements

This book is the result of a joint effort of all authors and editors. The editors wish to thank the authors of this volume for their valuable contributions. Special thanks go to Lena Roeder, Marie Duchene and Aylin Akyildiz for their competent, flexible and reliable assistance in copy editing and manuscript preparation, as well as to Emily Kelling for her support in reviewing submissions.

References

Ashworth, G. and Page, S.J., 2011. Urban tourism research. Recent progress and current paradoxes. *Tourism Management*, 32 (1), 1–15.

Aznar, J., *et al.*, 2017. The irruption of AirBNB and its effects on hotels' profitability: an analysis of Barcelona's hotel sector. *Intangible Capital*, 13 (1), 147–159.

Bærenholdt, J.O., *et al.*, 2004. *Performing tourist places. New directions in tourism analysis*. London: Ashgate.

Berking, H., 2012. The distinctiveness of cities outline of a research programme. *Urban Research & Practice*, 5 (3), 316–324.

Brauckmann, S., 2017. City tourism and the sharing economy – Potential effects of online peer-to-peer marketplaces on urban property markets. *Journal of Tourism Futures*, 3 (2), 114–126.

Brenner, N., Marcuse, P., and Mayer, M., 2012. *Cities for people, not for profit. Critical urban theory and the right to the city*. London: Routledge.

Cócola Gant, A., 2016. Holiday rentals. The new gentrification battlefront. *Sociological Research Online*, 21 (3), 10.

Cócola-Gant, A., 2019. Tourism gentrification. *In*: L. Lees and M. Phillips, eds. *Handbook of gentrification studies*. Cheltenham: Edward Elgar Publishing, 281–293.

Cohen, E., 1987. "Alternative tourism"—a critique. Informa UK Limited. *Tourism Recreation Research*, 2 (12), 13–18.

Cohen, S.A. and Cohen, E., 2019. New directions in the sociology of tourism. *Current Issues in Tourism*, 22 (2), 153–172.

Coleman, S. and Crang, M., eds., 2002. *Tourism. Between place and performance*. New York: Berghahn Books.

Colomb, C. and Novy, J., eds., 2016. *Protest and resistance in the tourist city*. London: Routledge Taylor & Francis Group.

Crouch, D., ed., 2001. *Leisure/tourism geographies. Practices and geographical knowledge*. London: Routledge.

Dirksmeier, P. and Helbrecht, I., 2015. Resident perceptions of new urban tourism: a neglected geography of prejudice. *Geography Compass*, 9 (5), 276–285.

Dodds, R. and Butler, R., 2019. *Overtourism. Issues, realities and solutions*. Berlin: De Gruyter.

Dredge, D., 2017. *"Overtourism" Old wine in new bottles?* [online], 13 September. Available from: https://www.linkedin.com/pulse/overtourism-old-wine-new-bottles-d i [Accessed 23 December 2020].

Dredge, D. and Gyimóthy, S., 2015. The collaborative economy and tourism: critical perspectives, questionable claims and silenced voices. *Tourism Recreation Research*, 40 (3), 286–302.

Edensor, T., 2001. Performing tourism, staging tourism: (Re)producing tourist space and practice. *Tourist Studies*, 1 (1), 59–81.

Edensor, T., 2007. Mundane mobilities, performances and spaces of tourism. *Social & Cultural Geography*, 8 (2), 199–215.

Feifer, M. and Lynch, J., 1985. *Going places. The ways of the tourist from Imperial Rome to the present day*. London: Macmillan.

Florida, R.L., 2003. *The rise of the creative class. And how it's transforming work, leisure, community and everyday life*. North Melbourne, VIC: Pluto Press.

Frank, S., 2016. Dwelling-in-motion: Indian Bollywood tourists and their hosts in the Swiss Alps. *Cultural Studies*, 30 (3), 506–531.

Frank, S., 2020. Entrepreneurial heritage-making in post-wall Berlin: the case of new potsdamer platz. *In*: M. Ristic and S. Frank, eds. *Urban heritage in divided cities. Contested pasts*. Abingdon: Routledge, 190–207.

Frank, S. and Meier, L., 2018. *Dwelling in mobile times. Places, practices and contestations*. London: Routledge.

Freytag, T. and Bauder, M., 2018. Bottom-up touristification and urban transformations in Paris. *Tourism Geographies*, 20 (3), 443–460.

Frisch, T., *et al.*, eds., 2019. *Tourism and everyday life in the contemporary city*. London: Routledge.

Füller, H. and Michel, B., 2014. 'Stop being a tourist!' New dynamics of urban tourism in Berlin-Kreuzberg. *International Journal of Urban and Regional Research*, 38 (4), 1304–1318.

Füller, H., *et al.*, 2018. Manufacturing marginality. (Un-)governing the night in Berlin. *Geoforum*, 94, 24–32.

Goodwin, H., 2019. Barcelona—crowding out the locals: a model for tourism management? *In*: R. Dodds and R. Butler, eds. *Overtourism. Issues, realities and solutions*. Berlin: De Gruyter, 125–138.

Gotham, K.F., 2005. Tourism gentrification. The case of New Orleans' Vieux Carre (French Quarter). *Urban Studies*, 42 (7), 1099–1121.

Gravari-Barbas, M. and Guinand, S., eds., 2017. *Tourism and gentrification in contemporary metropolises. International perspectives*. London: Routledge Taylor & Francis Group.

Gravari-Barbas, M. and Jacquot, S., 2016. No conflict? Discourses and management of tourism-related tensions in Paris. *In*: C. Colomb and J. Novy, eds. *Protest and resistance in the tourist city*. London: Routledge Taylor & Francis Group, 31–51.

Gutiérrez, J., *et al.*, 2017. The eruption of Airbnb in tourist cities: comparing spatial patterns of hotels and peer-to-peer accommodation in Barcelona. *Tourism Management*, 62, 278–291.

Guttentag, D., 2015. Airbnb: disruptive innovation and the rise of an informal tourism accommodation sector. *Current Issues in Tourism*, 18 (12), 1192–1217.

Hall, C.M., 2005. Reconsidering the geography of tourism and contemporary mobility. *Geographical Research*, 43 (2), 125–139.

Hannam, K., 2011. Chapter 6. The end of tourism? Nomadology and the mobilities paradigm. *In*: J. Tribe, ed. *Philosophical issues in tourism*. Bristol: Channel View Publications, 101–114.

Harvey, D., 1989. From Managerialism to Entrepreneurialism: The Transformation in Urban Governance in Late Capitalism. *Geografiska Annaler. Series B, Human Geography*, 71 (1), 3.

Hubbard, P. and Hall, T., 1998. The entrepreneurial city and the 'new urban politics'. *In*: T. Hall and P. Hubbard, eds. *The entrepreneurial city. Geographies of politics, regime and representation*. Chichester: Wiley, 1–23.

Judd, D.R., 1999. Constructing the tourist bubble. *In*: D.R. Judd and S.S. Fainstein, eds. *The tourist city*. New Haven, CT: Yale University Press, 35–53.

Judd, D.R. and Fainstein, S.S., eds., 1999. *The tourist city*. New Haven, CT: Yale University Press.

Kirshenblatt-Gimblett, B., 1995. Theorizing heritage. *Ethnomusicology*, 39 (3), 367–380.

Koens, K., Postma, A. and Papp, B., 2018. Is overtourism overused? Understanding the impact of tourism in a city context. *Sustainability*, 10 (12), 4384.

Larsen, J., 2008. De-exoticizing tourist travel: everyday life and sociality on the move. *Leisure Studies*, 27 (1), 21–34.

Larsen, J., 2017. Performance, space and tourism. *In*: J. Wilson, ed. *The Routledge handbook of tourism geographies*. London: Routledge, 67–73.

Larsen, J., 2019. Ordinary tourism and extraordinary everyday life. *In*: T. Frisch, *et al.*, eds. *Tourism and everyday life in the contemporary city*. London: Routledge, 24–41.

Lash, S. and Urry, J., 1994. *Economies of signs and space*. London: SAGE.

Maitland, R., 2007. Tourists, the creative class and distinctive areas in major cities. The roles of visitors and residents in developing new tourism areas. *In*: G. Richards and J. Wilson, eds. *Tourism, creativity and development*. London: Routledge, 73–86.

Maitland, R., 2008. Conviviality and everyday life: the appeal of new areas of London for visitors. *International Journal of Tourism Research*, 10 (1), 15–25.

Maitland, R., 2010. Everyday life as a creative experience in cities. *International Journal of Culture, Tourism and Hospitality Research*, 4 (3), 176–185.

Maitland, R., 2013. Backstage behaviour in the global city: tourists and the search for the 'Real London'. *Procedia - Social and Behavioral Sciences*, 105, 12–19.

Maitland, R. and Newman, P., 2004. Developing metropolitan tourism on the fringe of central London. *International Journal of Tourism Research*, 6 (5), 339–348.

Maitland, R. and Newman, P., 2009. Developing world tourism cities. *In*: R. Maitland, ed. *World tourism cities. Developing tourism off the beaten track*. London: Routledge, 1–21.

Martinotti, G., 1993. *Metropoli. La nuova morfologia sociale della città*. Bologna: Il Mulino.

Martins, M., 2018. Tourism planning and tourismphobia: an analysis of the strategic tourism plan of Barcelona 2010–2015. *Journal of Tourism, Heritage & Services Marketing*, 4 (1), 3–7.

Massey, D., 1993. Power geometry and a progressive sense of place. *In*: J. Bird, ed. *Mapping the futures. Local cultures, global change*. London: Routledge, 59–69.

Mayer, M., 2009. The 'Right to the City' in the context of shifting mottos of urban social movements. *City*, 13 (2–3), 362–374.

Mayer, M., 2013. First world urban activism. *City*, 17 (1), 5–19.

McCabe, S., 2002. The tourist experience and everyday life. *In*: G.M.S. Dann, ed. *The tourist as a metaphor of the social world. Selection of papers ... presented at the World Congress of the International Sociological Association, Brisbane, Australia, July 2002*. Wallingford: CABI, 61–77.

McCabe, S., 2005. 'Who is a tourist?': a critical review. *Tourist Studies*, 5 (1), 85–106.

Mendes, L., 2018. Tourism gentrification in Lisbon: the panacea of touristification as a scenario of a post-capitalist crisis. *In*: I. David, ed. *Crisis, austerity, and transformation. How disciplinary neoliberalism is changing Portugal*. Lanham, MD: Lexington Books, 25–46.

Miller, B. and Nicholls, W., 2013. Social movements in urban society: the city as a space of politicization. *Urban Geography*, 34 (4), 452–473.

Minoia, P., 2017. Venice reshaped? Tourist gentrification and sense of place. *In*: N. Bellini and C. Pasquinelli, eds. *Tourism in the city. Towards an integrative agenda on urban tourism*. Cham: Springer International Publishing, 261–274.

Mokras-Grabowska, J., 2018. New urban recreational spaces. Attractiveness, infrastructure arrangements, identity. The example of the city of Łódź. *Miscellanea Geographica*, 22 (4), 219–224.

Novy, J., 2010. What's new about new urban tourism? And what do recent changes in travel imply for the "Tourist City Berlin"? *In*: J. Richter, ed. *The tourist city Berlin. Tourism and architecture*. Salenstein: Braun, 190–199.

Novy, J., 2013. "Berlin does not love you". Notes on Berlin's "Tourism Controversy" and its discontents. *In*: M. Bernt, B. Grell, and A. Holm, eds. *The Berlin reader. A compendium on urban change and activism*. Bielefeld: Transcript, 223–237.

Novy, J., 2016. The selling (out) of Berlin and the de- and re-politicization of urban tourism in Europe's 'capital if cool'. *In*: C. Colomb and J. Novy, eds. *Protest and resistance in the tourist city*. London: Routledge Taylor & Francis Group, 32–52.

Novy, J., 2018. 'Destination' Berlin revisited. From (new) tourism towards a pentagon of mobility and place consumption. *Tourism Geographies*, 20 (3), 418–442.

Novy, J. and Colomb, C., 2016. Urban tourism and its discontents. An introduction. *In*: C. Colomb and J. Novy, eds. *Protest and resistance in the tourist city*. London: Routledge Taylor & Francis Group, 1–30.

Novy, J. and Colomb, C., 2019. Urban tourism as a source of contention and social mobilisations: a critical review. *Tourism Planning & Development*, 16 (4), 358–375.

Paulauskaite, D., *et al.*, 2017. Living like a local: authentic tourism experiences and the sharing economy. *International Journal of Tourism Research*, 19 (6), 619–628.

Peck, J., 2012. Austerity urbanism. *City*, 16 (6), 626–655.

Pinkster, F.M. and Boterman, W.R., 2017. When the spell is broken: gentrification, urban tourism and privileged discontent in the Amsterdam canal district. *Cultural Geographies*, 24 (3), 457–472.

Richards, G., 2017. Tourists in their own city – considering the growth of a phenomenon. *Tourism Today*, 8–16.

Roche, M., 1992. Mega-events and micro-modernization: on the sociology of the new urban tourism. *The British Journal of Sociology*, 43 (4), 563.

Rojek, C., 1993. *Ways of escape. Modern transformations in leisure and travel*. London: Macmillan.

Sans-Domínguez, A.A. and Quaglieri, A., 2016. Unravelling airbnb: urban perspectives from Barcelona. *In*: A.P. Russo and G. Richards, eds. *Reinventing the local in tourism. Producing, consuming and negotiating place*. Buffalo, NY: Channel View Publications, 209–228.

Sequera, J. and Nofre, J., 2018. Shaken, not stirred. *City*, 22 (5–6), 843–855.

Seraphin, H., 2019. Overtourism: excesses, discontents and measures in travel and tourism. *Journal of Tourism Futures*, 5 (3), 295–296.

Sheller, M. and Urry, J., 2006. The new mobilities paradigm. *Environment and Planning. part A*, 38 (2), 207–226.

Sommer, C. and Helbrecht, I., 2017. Seeing like a tourist city: how administrative constructions of conflictive urban tourism shape its future. *Journal of Tourism Futures*, 3 (2), 157–170.

Stabrowski, F., 2017. 'People as businesses': airbnb and urban micro-entrepreneurialism in New York City. *Cambridge Journal of Regions, Economy and Society*, 10 (2), 327–347.

Stors, N., 2019. Living with guests. Understanding the reasons for hosting via Airbnb in a mobile society. *In*: T. Frisch, *et al.*, eds. *Tourism and everyday life in the contemporary city*. London; New York: Routledge, 112–139.

Stors, N., *et al.*, 2019. Tourism and everyday life in the contemporary city. An introduction. *In*: T. Frisch, *et al.*, eds. *Tourism and everyday life in the contemporary city*. London: Routledge, 1–23.

Swyngedouw, E., 2007. The post-political city. *In*: G. Baeten, ed. *Urban politics now: re-imagining democracy in the neo-liberal city*. Rotterdam: NAI Publishers, 58–76.

Swyngedouw, E., 2010. Apocalypse forever? *Theory, Culture & Society*, 27 (2–3), 213–232.

Uriely, N., 2005. The tourist experience. *Annals of Tourism Research*, 32 (1), 199–216.
Urry, J., 1990. *The tourist gaze. Leisure and travel in contemporary societies*. London: SAGE.
Urry, J. and Larsen, J., 2011. *The tourist gaze 3.0*. 3rd ed. Los Angeles, CA: SAGE.
Wachsmuth, D. and Weisler, A., 2018. Airbnb and the rent gap: gentrification through the sharing economy. *Environment and Planning. Part A*, 50 (6), 1147–1170.
Wearing, S.L. and Foley, C., 2017. Understanding the tourist experience of cities. *Annals of Tourism Research*, 65, 97–107.
White, N.R. and White, P.B., 2007. Home and away. *Annals of Tourism Research*, 34 (1), 88–104.
Zukin, S., 1991. *Landscapes of power. From Detroit to Disney World*. Berkeley, CA: University of California Press.
Zukin, S., 1995. *The cultures of cities*. Cambridge, MA: Blackwell.
Zukin, S., 2011. *Naked city. The death and life of authentic urban places*. Oxford: Oxford University Press.
Zukin, S., *et al.*, 2009. New retail capital and neighborhood change. Boutiques and gentrification in New York City. *City & Community*, 8 (1), 47–64.

Part I

Consuming the city

New Urban Tourism in urban centres and metropolitan peripheries

2 Bohemia and the New Urban Tourism

Mario Hernandez

Introduction: back to the city, tourism race and gentrification

Much of the literature in urban studies in recent years has centred on cities emerging once again as the centre of economic and cultural life for the majority of Americans (Sassen 2010, Hyra 2014). Brought on by what is often referred to as the "back-to-the-city movement", census data has shown that for the first time in half a century, more people are moving to cities than away from them (Gallagher 2013, p. 14). Scholars have suggested a variety of reasons for this shift, from global market flows in the services and information sectors that nestle in cities to massive infrastructure investment on the part of city planners (Sassen 2010, Castells 2009, Hyra 2014). Efforts have historically focused on the revitalisation of Central Business Districts (CBDs) though this process has spread to other parts of the city and is frequently co-organised, if not completely outsourced to market-oriented developers (Gottdiener and Hutchison 2011, Sites 1994). Such practices have increasingly been geared towards private-sector oriented growth, focused on individualised consumption practices, attracting innovative industries and gaining a competitive edge in global tourism. Developing a unique city image or 'brand' is a necessity among such cities. Given such contemporary trends in urban revitalisation, cultural production has increasingly become the raw material and a primary commodity driving urban growth (Zukin 1998, Gottdiener and Hutchison 2011, Currid 2009).

The arts industry, from large-scale museums, arts programming, music venues and art festivals, to more grassroots artists' collectives or avant-garde art galleries at the neighbourhood level, has become an important resource in the development of the cultural identities of cities. As bohemian art scenes are developing with increasing regularity in cities around the country, New York City, with one of the most established in the country, can offer critical insights into how artists fuel the growth of city as a kind of "entertainment machine" (Clark 2003).[1] In particular, with the largest concentration of artists, galleries and artist studios anywhere in the city, the Bushwick neighbourhood of Brooklyn is arguably the centre of art production and the bohemian scene in New York City (Forman and Chaban 2017). Though still serving as the largest Latino and Latina population in Brooklyn, as it has for over a half-century—along with a sizable African American population—the neighbourhood's association with the avant-garde art scene has led to a flood of tourists, new residents and capital into the neighbourhood in recent years. The neighbourhood's newfound association with innovative cultural production has led to a spike in new bars, restaurants and cultural amenities such as outdoor food vendors and cultural arts festivals. Further, this increasing "bohemian mix" (Mele 2000, p. 3) has attracted a new wave of urban tourism as the neighbourhood buzz has attracted cultural consumers to the neighbourhood from around the world in search of an alternative to traditional tourist destinations. Apart from the novel cultural experiences associated with the bohemian lifestyle, the location of such practices in historically poor and working-class communities offer visiting tourists the promise of a more 'authentic' version of the city. But for many of these residents, this shift has ultimately led to an increasing cost of living as more middle- and upper-class whites are moving in, exacerbating the process of gentrification.

By focusing on Bushwick as a case study, this chapter describes how and why bohemian neighbourhood enclaves,[2] once relegated to a few cities at their inception in the 20th century, are an increasingly common feature of revitalising urban economies. In particular, these zones are increasingly becoming a common fixture of orienting to and consuming cities through globally established patterns of New Urban Tourism (Maitland 2007). Driven by neoliberal policies centred around market-led growth, such enclaves accentuate urban strategies of cultural production and consumption including city branding and catering to the development of creative industries. These sites function as both material and symbolic engines of urban growth by, on the one hand, serving as creative centres and network hubs, initially for artists but ultimately for an increasing array of creative industries in a variety of fields from design, marketing and technology sectors (see Hybenova 2015). Further, apart from attracting such industries, the specific set of lifestyles, tastes and cultural practices that proliferate within such enclaves have contributed to an industry of alternative lifestyle tourism. But that neighbourhoods like Bushwick, which until recently were some of the poorest and most isolated in the city, would come to serve as vital engines of such new forms of tourism and urban growth, makes this type of New Urban Tourism a particularly important area of urban research ranging from the local level to the global. I hope to highlight the intersection of the two (i.e. the "glocal") specifically with regard to the question of who has the "right to the city" (Harvey 2008).

Thus, by considering the spatial effect of the bohemian art scene in generating new industries such as alternative forms of tourism, I argue that many of the transitions in the neighbourhood are indicative of larger patterns of economic restructuring of cities more broadly. Such changes signify a shifting economic mode of production from the Fordist-era urban landscape towards a service-sector–oriented growth focused on creative industries and cultural production and consumption. In drawing out this larger structural analysis, I accentuate three separate and interrelated areas that are important in considering these trends. Those include: (1) the increasing importance of culture in postmodern economies; (2) translating culture into a commodity through consumption and lifestyle practices, primarily expressed through commercial spaces that attract a specific form of tourism consumption. The packaging and marketing of neighbourhoods and entire cities through 'city branding' becomes particularly salient in this regard. Finally (3), the gentrification and displacement of long-time residents. To this end, I begin the next section by situating Bushwick's current revitalisation within the context of its historical association with urban blight. By framing these trends around New Urban Tourism, this chapter shows how artists, as cultural producers, provide a vital role in creating an alternative and more 'authentic' version of urban tourism while standardising the once liminal zones of the city in the process.

Bushwick bohemia: revitalisation in historical context

While Brooklyn's Bushwick neighbourhood today is the centre of the New York art scene and a prime example of what Peck (2005, p. 741) calls a "buzzing, trendy neighbourhood", it was until recently a particularly pronounced example of the endemic problems felt around the city in the years following World War II. From the 1960s to the 1990s, New York as a whole had lost over 800,000 jobs (Bourgois 1995, p. 114). In Bushwick, the effect was particularly acute with the neighbourhood's unemployment rate reaching as high as 40% (Gottlieb 1993). As white residents fled to the suburbs in search of better housing, jobs and services (a product of deliberate federal policy[3]) an influx of African American and Latino residents, who had only recently migrated to the city during the war in search of work, found themselves increasing isolated and impoverished as dwindling job prospects were compounded by predatory loan practices on a dilapidated housing stock (Rausher and Momtaz 2014, Sanchez 1988, Merlis and Gomes 2012). Subsequent decades of economic recession, fiscal austerity and the criminalisation of communities of colour through such policies as the War on Drugs took its toll on Bushwick as it did in many inner-city neighbourhoods around the country. By the early 1990s, the *New York Times* had described Bushwick as home to "the most notorious drug bazaar in Brooklyn and one of the toughest in New York City" (Maher and Daly 1996, p. 469). As a result, and even with the flood of resources and new residents into the neighbourhood, Bushwick still trails behind most other neighbourhoods across the city in infrastructure and quality-of-life indicators, such as education, economic security and health services. In other words, the legacy of this history of structural inequality and racism is still clearly visible today.

During and throughout this time, the residents of Bushwick worked tirelessly to sustain Bushwick as a viable neighbourhood during its bleakest periods. Bushwick's community board director at the time, John Dereszewski (2007), notes that community residents organised and maintained the stability of the community "by, among other things, establishing a significant number of block associations".[4] Such members

> participated in anti-crime initiatives, getting trees planted, and generally holding residents of their blocks together as the area around them descended

> into chaos. And they would play a key role in the recovery that would come following the devastation of the fires and blackout-spawned looting of the 1970s.
>
> (Ibid.)

This community resilience has preserved the community's unique identity and spans generations to the current day, partially lending to the "authenticity" that tourists come to the neighbourhood in search of today.

City branding: culture as an increasing commodity

Starting as early as the 1970s, a combination of factors slowly began reviving the city and the neighbourhood of Bushwick in the process. A variety of scholars have accentuated the reasons that have contributed to this shift, including larger global market flows that have led to investment in cities such as financial service, and information sectors have increasingly centred in cities (Sassen 2010, Castells 2009, Smith 1979, Brenner and Theodore 2002). Others have described the increasing importance of culture as a commodity in cities, as tourism and the rise of the creative class has driven a demand in quality-of-life amenities, as well as leisure and entertainment services (Hannigan 1998, Gottdiener and Hutchison 2011). Such demand has fuelled a global distribution of retail brands increasingly clustered in urban centers, leading to the development of 'commercial utopias' in cities (Evans 2003, Klein 1999, p. 143). City planners have been instrumental in encouraging and fostering these kinds of market-led redevelopment strategies, either through direct large-scale infrastructure revitalisation or through indirect market incentives (Hyra 2012, Smith 1979, Sites 1994). From the redevelopment of downtown commercial shopping and business districts, to the revitalisation of waterfront areas, historic districts or the development of downtown sports arenas, such projects are seen as investments to promote consumption and entertainment (often referred to as "festival marketplaces"), to public–private partnerships in the redevelopment of commercial and residential housing sector (Gottdiener and Hutchison 2011, p. 172, Chatterton and Hollands 2003, Sites 1994). All of this has contributed to a growing industry in global tourism and cultural consumption. To facilitate this growing 'experience economy' (Richards 2013), urban planners have increasingly adopted strategies of 'city branding' (Dinnie 2011) in order to increase their competitive advantage among cities for what Zukin (2014) calls the "three Ts" of talent, trade and tourism. To this end, the art industry has been a particularly vital resource to personify and increase a city's unique 'brand' identity.

A variety of authors have described the use of art industries in fostering city branding from film festivals (Ooi and Strangaard Pederson 2010) to the music industry (Grazian 2003), literary tourism (Herbert 2001) to graffiti and street art (Dickens 2009). New York City's historical legacy as a centre in the art world meant that improving conditions also meant a resurgence in the city's art industry. The creative sector constitutes the largest industry in the city overall, as well as the largest concentration of creative sector employment anywhere in the US

(Forman 2016). In the decade from 2003 to 2013, New York City saw an increase in every creative sector from employment in film and television (53%), "architecture (33 percent), performing arts (26 percent), advertising (24 percent), visual arts (24 percent) and applied design (17 percent)" (Forman 2016, p. 3), outpacing the city's overall employment growth (12%) (ibid.). The number of artists in the city rose as well, increasing 17.4% from 2000 to 2015, which marks an all-time high (Forman and Chaban 2017).

In terms of their spatial distribution across the city, with the cost of living skyrocketing in the East Village and Manhattan overall, artists began moving to the outer boroughs and Brooklyn in particular in the late 1980s and early 1990s. Brooklyn has seen the greatest upsurge, and Voon (2017) writes, "[a]t the start of this century [...] 150 artists lived in what was then a very industrial area; 15 years later, 1,824 called the neighbourhood home", an increase of 1,116% and the single largest of any neighbourhood in the city (Voon 2017). Though particularly striking for Bushwick, this reflects the growing significance of art culture industries for New York City as a whole. This is particularly the case since as Forman (2016, p. 7) notes, among other creative sectors, 28% of the country's fashion designers live in New York City, 12% of its art directors, 11% of the country's jobs in film and 15% in television among other industries are located there.

With art and entertainment industries providing such an important role to the resurgence of the city as a whole, neighbourhoods like Bushwick serve as a central location for early career artists and recent college graduates looking for cheap rent while seeking to gain entry into these industries. As will be shown, by clustering in bohemian centres such as Bushwick, young people draw on, and establish their own networks for everything from job prospects to fashion and style choices. In this way, living in such neighbourhoods becomes as critical for gaining entry into such fields as the necessity for finding relatively cheap rent while working for low wages in entry-level positions in an increasingly unaffordable city. These areas also double as sites of cultural production, as established culture industry producers draw on areas like Bushwick for cutting-edge art, fashion and design from such bohemian youth subculture enclaves. Finally, and within this context, these areas serve as central destinations of New Urban Tourism for a younger generation of savvy travellers in search of alternative ways of experiencing cities.

The creative influx has contributed to Brooklyn emerging into what the *New York Times* calls the "epicenter of cool" (Zukin 2009). Recognised and marketed globally as a centre of hip urban style and cutting-edge culture, the borough is referenced in music, television and films. In recent years, its mixology bars, commercial design, fashion trends and branding (e.g. 'Made in Brooklyn') are recognised worldwide. Williamsburg's bohemia was central in generating the cultural "brand" of Brooklyn as a whole, as the neighbourhood's artists have been both celebrated and derided for developing the hipster subculture in the US. Indeed, when various media outlets refer to cities with emerging bohemian scenes as the 'newest Brooklyn', it is usually with reference to Williamsburg in mind (DeVito 2015). Given this image, many corporate brands have opened concept stores in the area to test and market product design strategies. Such examples include Adidas'

Figure 2.1 Bushwick Block Party performance art Hernandez (2019)

"Brooklyn Creator Farm" and BMW's creative studio "A/D/O" for its Mini line of cars (Rothstein 2017). Others have created event spaces that function as cultural centres such as Van's "House of Vans" and Red Bull's contemporary art space. Red Bull's recent exhibition on the avant-garde artist and hip-hop pioneer Rammellzee offers a poignant example of the use and juxtaposition of corporate brands and underground artists (Montes 2018). In these ways, artists have helped shape the symbolic image of Brooklyn and New York City as a whole, driving a global industry of New Urban Tourism in the process (see Figure 2.1). Given the context of city branding as a tool for growth, the rest of this chapter will focus on the role of artists in driving tourism, as well as the consequences of Bushwick's revitalisation.

Bushwick's bohemia and urban growth

Artists began to arrive in Bushwick in large numbers by the late 1990s and early 2000s. Priced out of Williamsburg, the large industrial spaces that would serve as live-in studio spaces in the neighbourhood provided an additional pull factor. These were the facilities where the brewing and garment industries once flourished during Bushwick's heyday. In 2007 artists had established four galleries in the neighbourhood, and by 2017, that number would swell to over 70. The growth and expansion of Bushwick Open Studios, a yearly art festival where artists open their studio spaces to the general public, can give a further indication of the proliferation of artists in the neighbourhood. Starting with just 85 artist spaces in 2006, the venue had, by its height in 2015, over 600 registered spaces and over 3,000 participating artists. Additionally, a variety of tour companies have emerged in recent years to take visitors around the neighbourhood to highlight the numerous murals painted by artists from around the world. The murals are coordinated by the Bushwick Collective, who also host a yearly summer concert festival called the "Bushwick Collective Block Party" (Complex 2016). By 2010, The *New York Times Magazine* called Bushwick, Brooklyn "the coolest place on the planet" (Chavez 2017); four years later *Vogue* named it one of the coolest neighbourhoods in the world (Remsen 2014). This image has become self-perpetuating, as trendy bars and restaurants from around the country have flocked to Bushwick to be associated with New York City's latest "buzzing trendy neighbourhood" (Peck 2005).

Figure 2.2 Tourists being taken around the neighbourhood to see the local murals. Hernandez (2016)

The establishment of the art scene has attracted a variety of creative spaces that have made Bushwick a popular destination for younger residents across the city and tourists alike: music venues such as "House of Yes", theatre and dance performance spaces such as "Bushwick Starr", experimental spaces like the exotic dance venue "Heck", "Bizarre Bushwick" or drag events like the "Bushwig Festival" and even an urban opera to Bushwick (Brennan 2016, O'Regan 2014, see also Hybenova 2016). Underground music venues such as "Trans-Pecos", "Silent Barn" (which closed in 2018), "Bohemian Grove" and warehouse party promoters have also been attracted to Bushwick, as much for the avant-garde scene for the remote location of the industrial area (where they do not have to worry about neighbours complaining about the noise). Today, Bushwick's music venues constitute an established stop for indie bands that tour the country, joining a network of neighbourhoods in cities around the country and globe where such music scenes incubate.

The proliferation of these creative entertainment industries has led to a "multiplier effect", as a wide range of commercial establishments including coffee shops, cocktail bars, boutiques, bookstores and clothing stores have followed these industries to capitalise on the neighbourhood's popularity (Zukin and Braslow 2011, p. 133). Though initially centred around locally sourced goods and producers, an increasing array of commercial entrepreneurs that cater to hipster tastes have established outposts in neighbourhoods like Bushwick.

San-Francisco–based "Blue Bottle", "Mission Chinese" and "Tartine Bakery", for example, now have bi-coastal locations. "Beauty Bar", the self-described "hipster chain" with locations around the country, opened a location in Bushwick in the initial wave of such bar openings. The Danish brewers "Mikkeller", have locations in 37 countries, including San Francisco as well as the Norrebro neighbourhood of Copenhagen. This readily identifiable array of brands facilitates New Urban Tourism by orienting the hipster tourist to an otherwise unfamiliar urban landscape. And though brand recognition is an important phase of standardising of such neighbourhoods towards global consumer tastes, the process is more nuanced.

Bohemian commercial spaces: design, taste and lifestyle practices

Commercial establishments such as coffee shops, bars and restaurants have historically functioned as the primary venues for fostering the bohemian ethos. These spaces have been important to the development of bohemian enclaves not just for providing a space for artistic networks and intellectual production, but also for the development of a specific set of taste and lifestyle practices. Thus, while individual artists may develop their craft in isolation, as Wilson (2000, p. 34) writes, it was in commercial settings such as coffee shops that individual artists "became" bohemian. In a similar way, the development of globally identifiable consumptions patterns has led to the development and proliferation of bohemian scenes like Bushwick's that have engendered an entire industry of such cultural production and consumption.

Aesthetically, the bars and restaurants share a similar set of patterns in their look and 'feel', which cater to a New Urban Tourism centred on the hipster subculture. In their physical construction, for example, such spaces accentuate a 'rustic', make-shift, post-industrial feel of the neighbourhood in their design, and valorise repurposed materials and minimalism in their aesthetics. Common patterns such as exposed brick, untreated wood, industrial fixtures and turn-of-the-century kitsch comprise the dominant themes in their decor. Such design choices are not unique to Bushwick and have been collectively referred to as "AirSpace" (Sloane 2016). Campanella (2017), for example, mapped the use of the popular Edison-style light bulb across Brooklyn and proposed it as a measure to trace the proliferation of neighbourhood gentrification. Design choices ranging from materials, colour palates, menu options and music selections, all function to signal a specific system of meaning. Embedded within this design aesthetic is a sign system of 'authenticity' that speaks to the 'target audience' or consumer group (see Figure 2.3) (Zukin 2008, Rössler 2008). Increasingly, a global industry in the production and consumption of such taste practices pervades such scenes, making such lifestyle practices a readily identifiable and consumable urban fixture. Such practices have developed a global industry of New Urban Tourism, as travellers cultivated in such taste practices seek out the newest cocktail bar, micro-brewery or 'local' dive bar. But, by signifying the "appropriate" patterns of consumption and performance patterns in such places, such commercial spaces work to reinforce who "belongs" and who does not (see Centner's, 2008, discussion on 'spatial capital' as well as Sloane, 2016, on the politics of such aesthetics and design).

Figure 2.3 Blue Bottle Coffee Shop in Bushwick (Image by Nicolas Doyle)

Gentrification and the "buzzing, trendy neighbourhood"

The persistent theme of this chapter has been in noting that the economic impact artists have on urban economies is not relegated to art-related industries alone, but influences other spheres of the economy as well, such as the real estate, commercial sectors and tourism industry in particular. As Bushwick's popularity has taken off, investors, developers and building owners have sought to profit off the rising real estate demand. Quoting the developer-forecasting agency Terra CRG, Short (2017, p. 2) writes:

> Bushwick remains one of the most sought-after neighbourhoods for development for all types of properties. Last year, there were 72 multifamily building sales in Bushwick for a total of $136.6 million (second highest among all neighbourhoods in Brooklyn), 38 mixed-use building sales for a total of $74.4 million, 23 residential development sites for a total of $39.4 million, three retail building sites for a total of $14.6 million, and four industrial or office buildings at a total of $18.2 million.

But the report also notes,

> that housing in the city's creative neighbourhoods also reflects skyrocketing rents. In the ten community districts 'most closely associated with art and creativity', rents rose by at least 32% from 2000 to 2012; the median increase across all districts for the same period was 23%.
>
> (Voon 2015)

The average rent for a studio increased just one year later, from $1,700 in 2013 to $2,300. Equally significant is the increase in the number of Airbnb rentals that cater largely to tourists. In a recent report on the impact of Airbnb on rent in New York City, the city's Comptroller, Scott Stringer, noted that between 2009 and 2016, Airbnb listings skyrocketed while rents increased by an average of 25% citywide during the same period (Stringer 2018). Bushwick was listed among the neighbourhoods where Airbnb listing were most heavily concentrated and where rent had increased by 39.5% above the borough average. As Wachsmuth and Weisler (2018) note, the cultural desirability of neighbourhoods like Bushwick has led to an "extra-local" tourist demand on sites like Airbnb. This new source of revenue flows is heightening the process of gentrification as short-term stays create a "new rent gap" by artificially inflating the value (or "potential use") of the housing market.

Bushwick's recent growth has had an immense effect on the racial demographic of the neighbourhood. Bushwick's total population stands at about 136,000 residents. Latinos and Latinas still make up the majority of the population of Bushwick at 62%, followed by African Americans at 17.5% and Whites who make up 13% (The Furman Center 2016). But, although the Latino population has actually increased in New York City as a whole from the 2000 census to 2010 (from 27% to 29%), in Bushwick their numbers have declined (to 59%).

Additionally, and consistent with an alarming trend occurring in many major cities in the US, the total number of African Americans decreased in the city as a whole (from 25% to 22%) and decreased in Bushwick from 23% to 17%. Whites on the other hand, while decreased in the overall population of the city (35% to 32%), increased in the population of Bushwick from just 3% to 17% (ibid.). The increase of Whites in Bushwick from 2000 to 2014 constitutes the single greatest percentage growth in either direction by any racial group in the neighbourhood.

Finally, it should be noted that arguments vary on the normative assessment of gentrification, ranging from those that focus on the negative effects of gentrification, to others that see it as a potential benefit for long-time residents (The Economist 2015, Buntin 2015). Such arguments accentuate the benefits of an increasing standard of living in the form of increasing quality of city services and commercial amenities (Florida 2002, Hyra 2014). The increase in safety is seen as a primary factor, along with potential economic opportunity and increasing diversity of once hyper-segregated areas. These benefits are particularly pronounced for those residents that own their homes and are thus able to see their property values appreciate with the rising real estate market in the neighbourhood while also enjoying the benefits of the increasing number of amenities in the area. But in poor neighbourhoods, home ownership tends to be far lower than the city average, such as in Bushwick, where home ownership was 10.9% in 2014, compared to 28.5% in Brooklyn as a whole and 31.2% citywide (The Furman Center 2016). Indeed, the lack of a stable housing market is what makes such neighbourhoods susceptible to the destabilising effect of gentrification in the first place.

Summary: Bohemia and New Urban Tourism

The aestheticisation of once neglected neighbourhoods through the bohemian ethos has reframed the cultural image of Bushwick, making an area once synonymous with urban blight a central destination for new forms of urban tourism. Increasingly, neighbourhoods such as Bushwick provide an alternative to popular tourist destinations by offering the promise of a more "authentic" version of the city through a glimpse into the everyday lived reality of poor and working-class inhabitants. The contradiction of this form of tourism is that it also relies on an identifiable and homogenous set of consumption patterns and lifestyle tastes that constitute the hipster subculture. These consumption patterns serve to orient the savvy urban traveller and mediate hyperlocal experiences. Further, with the increasing importance of such cultural practices in standardising neighbourhoods, Bushwick has been shown to be a product of neoliberal urban growth in cities around the world. By conceptualising the development of this subculture as a lifestyle category of consumption taste practices, this chapter shows how the once outsider aesthetic of bohemians has gone from the fringe to reshaping the urban landscape of advanced capitalist cities.

At the local level, this chapter has shown how a market-oriented approach to growth on the part of the city planners since the 1970s has led to a wave of "reverse" migration of middle- and upper-class, largely white residents, back to the city. This process has led to skyrocketing real estate prices and a steadily increasing cost of living in the neighbourhoods like Bushwick, which in turn has led to a familiar process of gentrification and displacement of many of the neighbourhood's long-time poor and working-class African American and Latino residents. Though beyond the scope of this chapter, long-time residents have mobilised to resist this displacement, most notably around the Bushwick Community Plan proposed in 2018 to counter the city's proposed rezoning (see Bushwick Community Plan 2018). Given this familiar historical pattern, it would follow that the reintegration of neighbourhoods like Bushwick would be fraught. It is for this reason that gentrification constitutes a central focus of race and class relations in many contemporary cities. The argument here shows how such divisions become normalised through lifestyle tastes and consumption practices, which ultimately serves to mask racial and class conflict.

Notes

1 See also Silver *et al.* (2010) for a definition and detailed discussion on the development of 'scenes'.
2 These have intermittently been referred to as bohemian, neo-bohemian or 'hipster' enclaves.
3 Because suburbanisation was largely tied to federal policies and programs like the GI Bill and agencies like the Federal Housing Association, Bushwick accentuates the dramatic and lasting consequences to contemporary race relations (Anderson 2012, Mahoney 1995, Brodkin Sacks 1997, Kirp 1997, Jackson 1987, Katznelson 2005).
4 DeMause, *How Bushwick Fought to Rebuild itself After The 1977 Blackout*. https://gothamist.com/news/how-bushwick-fought-to-rebuild-itself-after-the-1977-blackout

References

Anderson, E., 2012. The iconic ghetto. *The Annals of the American Academy of Political and Social Science*, 642 (1), 8–24.

Bourgois, P., 1995. *In search of respect: selling crack in El Barrio*. Cambridge: Cambridge University Press.

Brennan, C., 2016. Underground strip club run by dancers operates out of Bushwick apartment [online]. *The Daily News*, 23 September. Available from: http://www.nydailynews.com/new-york/brooklyn/underground-strip-club-operates-bushwick-apartment-article-1.2802532 [Accessed 16 December 2020].

Brenner, N. and Theodore, N., 2002. *Spaces of neoliberalism: urban restructuring in North America and Western Europe*. Malden, MA: Blackwell Publishing.

Brodkin Sacks, K., 1997. GI bill: whites only need apply. *In*: R. Delgado and J. Stefancic, eds. *Critical white studies: looking behind the mirror*. Philadelphia, PA, Baltimore, BD: Temple University Press, 310–313.

Buntin, J., 2015. The gentrification myth [online]. *Slate*, 14 January. Available from: https://slate.com/news-and-politics/2015/01/the-gentrification-myth-its-rare-and-not-as-bad-for-the-poor-as-people-think.html [Accessed 10 October 2020].

Bushwick Community Plan, 2018. [online]. Available from: http://www.bushwickcommunityplan.org/welcome [Accessed 16 December 2020].

Campanella, T., 2017. Mapping the Edison bulbs of Brooklyn [online]. *Citylab*, 24 October. Available from: https://www.bloomberg.com/news/articles/2017-10-24/mapping-brooklyn-s-gentrification-via-edison-bulbs [Accessed 10 December 2020].

Castells, M., 2009. *The rise of the network society. The information age: economy, society and culture volume I.* 2nd ed. Hoboken, NJ: Wiley-Blackwell.

Centner, R., 2008. Places of privileged consumption practices: spatial capital, the dot-com habitus, and San Francisco's internet boom. *City and Community*, 7 (3), 193–223.

Chatterton, P. and Hollands, R., 2003. *Urban nightscapes: youth cultures, pleasure spaces, and corporate power*. New York: Routledge.

Chavez, D., 2017. Burning Bushwick [online]. *NYC Archive*, 22 November. Available from: http://www.archives.nyc/blog/2017/11/22/burning-bushwick [Accessed 10 December 2020].

Clark, T., ed., 2003. *The city as an entertainment machine*. Lanham, MD: Lexington Books.

Complex, 2016. *No free walls: art and gentrification collide in Bushwick*. YouTube Video [online], 14 September. Available from: https://www.youtube.com/watch?v=7n5pDoKAcgw [Accessed 1 December 2020].

Currid, E., 2009. Bohemia as subculture; "Bohemia" as industry: art, culture and economic development. *Journal of Planning Literature*, 23 (4), 368–382.

Dereszewski, J.A., 2007. Bushwick notes: from the 70s to today. *Up From Flames*. Brooklyn Historical Society, New York, 1–13. Available from: http://upfromflames.brooklynhistorical.org/uff_resources [Accessed 1 December 2020].

DeVito, L., 2015. 'New York Times' continues to perpetuate the myth that Detroit is an awesome place for artists [online]. *Detroit Metro Times*, 13 July. Available from: https://www.metrotimes.com/the-scene/archives/2015/07/13/new-york-times-continues-to-perpetuate-the-myth-that-detroit-is-an-awesome-place-for-artists [Accessed: 1 December 2020].

Dickens, L., 2009. *The geographies of post-graffiti: art worlds, cultural economy and the city*. Thesis (PhD). University of London.

Dinnie, K., ed., 2011. *City branding: theory and cases*. New York: Palgrave Macmillan.

The Economist, 2015. Bring on the hipster: gentrification is good for the poor [online]. *The Economist*, 19 February. Available from: https://www.economist.com/news/united-states/21644164-gentrification-good-poor-bring-hipsters [Accessed 1 December 2020].

Evans, G., 2003. Hard-branding the cultural city: from Prado to Prada. *International Journal of Urban and Regional Research*, 27 (2), 417–440.

Florida, R., 2002. *The rise of the creative class: and how it's transforming work, leisure, community and everyday life*. New York: Basic Books.

Forman, A., 2016. Creative New York [online]. *Center for an Urban Future*, June. Available from: https://nycfuture.org/research/creative-new-york-2015 [Accessed 17 December 2020].

Forman, A. and Chaban, M., 2017. Artists in schools: a creative solution to New York's affordable space crunch. *Center for an Urban Future* [online], July Report. Available from: https://nycfuture.org/research/more-NYC-artists-fewer-studios-schools [Accessed 1 August 2020].

The Furman Center, 2016. *State of New York City's housing & neighborhoods –2015 report* [online]. Available from: http://furmancenter.org/research/sonychan/2015-report [Accessed 1 August 2020].

Gallagher, L., 2013. *The end of the suburbs: where the American dream is moving*. London: Portfolio.

Gottdiener, M. and Hutchison, R., 2011. *The new urban sociology*. Boulder, CO: Westview Press.

Gottlieb, M., 1993. Bushwick's hope is a public project [online]. *New York Times*, 15 August. Available from: http://www.nytimes.com/1993/08/15/nyregion/bushwick-s-hope-is-a-public-project.html?pagewanted=all&pagewanted=print [Accessed 15 November 2020].

Grazian, D., 2003. *Blue Chicago: the search for authenticity in urban blues clubs*. Chicago, IL: University of Chicago Press.

Hannigan, J., 1998. *Fantasy city: pleasure and profit in the postmodern metropolis*. New York: Routledge.

Harvey, D., 2008. The right to the city. *New Left Review*, 53 (Sept–Oct), 23–40.

Herbert, D., 2001. Literary places, tourism and the heritage experience. *Annals of Tourism Research*, 28 (2), 312–333.

Hybenova, K., 2015. Boston and Dallas developers will turn a Bushwick warehouse into "Creative Offices" [online]. *Bushwick Daily*, 29 September. Available from: https://bushwickdaily.com/bushwick/categories/bushwick/3336-boston-and-dallas-developers-will-turn-a-bushwick-warehouse-into-creative-offices [Accessed 1 December 2020].

Hybenova, K., 2016. NSFW photos: a shy woman comes to lust, Bushwick's erotic party. *Bushwick Daily*, 18 November. Available from: https://bushwickdaily.com/bushwick/categories/arts-and-culture/4322-shy-woman-lust-party-bushwick-lot-54 [Accessed 1 December 2020].

Hyra, D., 2012. Conceptualizing the new urban renewal: comparing the past to the present. *Urban Affairs Review*, 48 (4), 498–527.

Hyra, D., 2014. The back-to-the-city movement: neighborhood redevelopment and processes of political and cultural displacement. *Urban Studies*, 52 (10), 1753–1773.

Jackson, K., 1987. *Crabgrass frontier: the suburbanization of the United States*. Oxford: University Press.

Katznelson, I., 2005. *When affirmative action was white: an untold history of racial inequality in twentieth century America*. New York: W.W. Norton & Company.

Klein, N., 1999. *No logo: taking aim at the brand bullies*. Toronto: Knopf Canada and Picador.

Kirp, D., 1997. *Our town: race, housing, and the soul of suburbia*. Rutgers: University Press.

Maher, L. and Daly, K., 1996. Women in the street-level drug economy: continuity or change? *Criminology*, 34 (4), 465–492.

Mahoney, M.R., 1995. Segregation, whiteness and transformation. *University of Pennsylvania Law Review*, 143 (1659), 1659–1684.

Maitland, R., 2007. Tourists, the creative class and distinctive areas in major cities. The roles of visitors and residents in developing new tourism areas. *In*: G. Richards and J. Wilson, eds. *Tourism, creativity and development*. London: Routledge, 73–86.

Mele, C., 2000. *Selling the lower east side: culture, real estate and resistance in New York City*. Minneapolis, MN: University of Minnesota Press.

Merlis B. and Gomes, R., 2012. *Brooklyn's Bushwick and east Williamsburg communities*. Wyckoff, NJ: Gomerl Publishing.

Montes, P., 2018. New Rammellzee exhibit is coming to New York City. *Hypebeast*, 21 February. Available from: https://hypebeast.com/2018/2/rammellzee-exhibit-red-bull-arts-new-york [Accessed 21 July 2020].

Ooi, C.-S. and Srandgaard Petersen, J., 2010. City branding and film festival: re-evaluating stakeholder's relations. *Place Branding and Public Diplomacy*, 6 (4), 316–332.

O'Regan, K., 2014. Bushwick's drag festival is back– hide the kids! (OR, Heck, bring 'Em) [online]. *Bedford and Bowery*, 3 September. Available from: http://bedfordandbowery.com/2014/09/bushwig-festival-of-drag/ [Accessed 16 December 2020].

Peck, J., 2005. Struggling with the creative class. *International Journal of Urban and Regional Research*, 29 (4), 740–770.

Rausher, C. and Momtaz, S., 2014. *Brooklyn's Bushwick - urban renewal in New York, USA. Community, planning and sustainable environments*. Cham: Springer International Publishing.

Remsen, N., 2014. Global street style report: mapping out the 15 coolest neighborhoods in the world. *Vogue* [online], 5 September. Available from: https://www.vogue.com/slideshow/fifteen-coolest-street-style-neighborhoods [Accessed 15 November 2020].

Richards, G., 2013. Creativity and tourism in the city. *Current Issues in Tourism*, 17 (2), 119–144.

Rothstein, E., 2017. Brooklyn developers choosing coolness over credit for their retail spaces [online]. *Bisnow*, 16 October. Available from: https://www.bisnow.com/new-york/news/retail/brooklyn-developers-choosing-coolness-over-credit-for-retailers-80339 [Accessed 15 December 2020].

Rössler, M., 2008. Applying authenticity to the cultural landscape. *Journal of Preservation Technology*, 39 (2/3), 47–52.

Sanchez, T., 1988. *Bushwick: neighborhood profile*. New York: Brooklyn In Touch Information Center.

Sassen, S., 2010. The city: its return as a lens for social theory. *City, Culture and Society*, 1 (1), 2–11.

Short, A., 2017. What does the next wave of development mean for Bushwick artists? *Hyperallergic*, 22 September. Available from: https://hyperallergic.com/401969/bushwick-commercial-real-estate-artists/ [Accessed 10 October 2020].

Silver, D., Nichols Clark, T. and Yanez, C.J.N., 2010. Scenes: social context in an age of contingency. *Social Forces*, 88 (5), 2293–2324.

Sites, W., 1994. Public action: New York City policy and the gentrification of the lower east side. *In*: J.L. Abu-Lughod, ed. *From urban village to east village: the battle for New York's lower east side*. Oxford: Blackwell Publishing, 189–212.

Sloane, M., 2016. Inequality by design? Why we need to start talking about aesthetics, design and politics. *The London School of Economics and Political Science* [online], 12 September. Available from: https://blogs.lse.ac.uk/researchingsociology/2016/09/12/inequality-by-design-why-we-need-to-start-talking-about-aesthetics-design-and-politics/ [Accessed 15 December 2020].

Smith, N., 1979. Toward a theory of gentrification: a back to the city by capital, not people. *Journal of the American Planning Association*, 45 (4), 538–548.

Stringer, S.M., 2018. The impact of airbnb on NYC rents [online]. *Bureau of Budgets Reports*, 3 May. Available from: https://comptroller.nyc.gov/reports/the-impact-of-airbnb-on-nyc-rents/ [Accessed 1 December 2020].

Voon, C., 2015. Report finds New York City's creative sector is thriving, for now [online]. *Hyperallergic*, 9 July. Available from: https://hyperallergic.com/220200/report-finds-new-york-citys-creative-sector-is-thriving-for-now/ [Accessed 20 July 2020].

Voon, C., 2017. Not just Bushwick: new study charts where NYC artists live [online]. *Hyperallergic*, 13 July. Available from: https://hyperallergic.com/390330/not-just-bushwick-new-study-charts-where-nyc-artists-live/ [Accessed 20 July 2020].

Wachsmuth, D. and Weisler, A., 2018. Airbnb and the rent gap: gentrification through the sharing economy. *Environmental and Planning A: Economy and Space*, 50 (6), 1147–1170.

Wilson, E., 2000. *Bohemians: the glamorous outcasts*. New Brunswick, NJ: Rutgers University Press.

Zukin, S., 1998. Urban lifestyles: diversity and standardization in spaces of consumption. *Urban Studies*, 35 (5–6), 825–839.

Zukin, S., 2008. Consuming authenticity. *Cultural Studies*, 22 (5), 724–748.

Zukin, S., 2009. Retail capital and neighborhood change: boutiques and gentrification in New York City. *City and Community* 8 (1), 47–64.

Zukin, S., 2014. Post-card perfect: the big business of city branding [online]. *The Guardian*, 6 May. Available from: https://www.theguardian.com/cities/2014/may/06/postcard-perfect-the-big-business-of-city-branding [Accessed 20 July 2020].

Zukin, S. and Braslow, L., 2011. The life-cycle of New York's creative districts: reflections on the unanticipated consequences of unplanned cultural zones. *City, Culture and Society*, 2 (3), 131–140.

3 "Tourist platformisation"

New Urban Tourism in Milan

Monica Bernardi and Giulia Mura

Introduction

The chapter starts from the assumption that the sharing economy and digital platforms are opening opportunities and challenges for cities, as well as shaping a new kind of urban tourism. Contemporary tourists (Gavinelli and Zanolin 2019) gain new chances in terms of hospitality, mobility and experience (European Union 2015, Dredge and Gymothy 2015); nevertheless, the massive spread of short-term rental (STR) platforms is activating/strengthening processes of gentrification (Guttentag 2013), hotelisation and Disneyfication (Bryman 2004, Lee 2016) for some cities, exacerbating the already difficult situation of affordable housing (Brauckmann 2017, Wachsmuth and Weisler 2018) (see section "Risks and challenges of the sharing economy for urban tourism").

The study adopts a mixed-method approach (see section "Methodology"): a focus group with experts has deepened the specific case of short-term rentals (STR) in relation to urban tourism (section "Focus group: the vision of experts"). The case study chosen is Milan, a city that, according to the Global Destination Cities Index[1], scores fifth place in Europe, surpassing Rome (eighth) and Venice (tenth), and fifteenth place around the world. Milan has adopted specific guidelines to promote and manage the 'sharing economy', signed the Sharing Cities Declaration and launched several local initiatives such as the civic crowdfunding campaign, the coworking vouchers and the city food policy. The section "The case study: Milan as a tourist destination" frames the case study relying on institutional and scientific materials for the 'policy part', while the section "An outline of the local tourist sharing offer in Milan" outlines the main local sharing services for the 'offer part', showing a variety of active tourist platforms in the city. Lastly, a set of semi-structured interviews (section "The perception of traditional hospitality services") with local owners of hostels and low-budget hotels collect the perception of the traditional local hospitality sector towards 'tourist platformisation'.

The conclusions propose preliminary considerations about the emergence of a New Urban Tourism based on sharing economy services and platforms. The analysis shows different levels of awareness: the voice of experts contextualises the case study in the bigger picture of risks and challenges posed by digital platforms, while the opinions of traditional business operators testify to a perception of the phenomenon still confused and only partially aware of its impacts. The contribution paves the way for further analysis on the topic.

A last remark is related to the global pandemic that has completely changed the tourist landscape by zeroing out tourism and emptying our cities, reversing also the STR scenario and probably setting the ground for a "new New Urban Tourism" (Bernardi 2020, Gainsforth 2020).

Methodology

The adoption of a mixed-method approach, based on the integration of qualitative and quantitative data, was chosen to obtain a better understanding of the phenomenon and test the starting assumptions (Pearce 2012). The research builds on the literature on the topic, flanking it using different methodological tools.

A focus group was conducted with experts on STR, to deepen the understanding of risks and challenges at a general level and collect information about the policy interventions required by cities to manage the phenomenon. The focus group was carried out during the Sharing Cities Encounter organised within the Smart City Expo World Congress held in Barcelona in November 2019. One of the authors led and moderated the panel: "City sovereignty and Data policies: Special focus on Tourism Platforms" with experts from various fields.

We analysed institutional materials about the case of Milan, retrieved from the official website of the Municipality and the main literature on the topic. The local tourist sharing services in the city were outlined in four categories: hospitality with sharing accommodation (STR), sharing transport/ mobility, tours and experience sharing, and share of food and belongings.

Semi-structured interviews were held with operators of the traditional hospitality sector in the city. The sample has been selected relying on the Open Data Portal of the Lombardy Region, considering the hotel and extra-hotel accommodations active in Milan. Among them, the local owners of hostels and low-budget hotels have been selected because, according to the literature, they are the more affected by the presence of STR solutions (among the others: Koh and King 2017, Zervas *et al.* 2017). The sample was stratified in order to cover the different administrative areas of the city and the participation was voluntary.

The impact of COVID-19 on STR has been dramatic and has promoted new reflections that are represented in the final remarks of this chapter.

Risks and challenges of the sharing economy for urban tourism

The spread of digital platforms has undeniably changed the tourist panorama, opening a new era of "tourism platformisation" and shaping a new kind of urban tourism (WTO 2017). Thanks to digital platforms—which provide access to a wide range of services in a faster and more affordable way compared to the traditional tourism businesses—the overall supply of travel options has considerably increased (OECD 2016). By recirculating idle resources, "dead capital" (Koopman *et al.* 2015) and latent expertise, digital platforms provide new monetisation opportunities and open new entrepreneurialism chances; they allow more flexibility, personalisation and connection with locals, increasing trust and visitor satisfaction through systems of peer-to-peer feedback, direct contact, interactive communication and transparent transactions (European Union 2015, Dredge and Gymothy 2015). This is the type of platform we refer to in this chapter (with the awareness that other types exist); referring to Botsman and Roger (2010), we consider business-to-consumer (B2C) platforms—such as BikeMe, in which a company provides the assets (bikes in this case) and users can access them as a service—and consumer-to-consumer (C2C) or peer-to-peer (P2P) platforms—such as Airbnb and many more, in which the company provides the digital tool (a mobile app/platform) to connect seekers and providers. The argument that we make is that, besides the opportunities, digital platforms are also opening new issues or exacerbating old urban problems.

In legal terms, digital platforms often move in a normative grey zone (Codagnone *et al.* 2016) outside the traditional regulative system (Smorto 2015) and raise ethical issues due to information asymmetries, tax and privacy issues and the difficulty of distinguishing occasional transactions from more professional services provision (Arcidiacono 2017). In some cases, platforms (defined as "extractive" by Bauwens and Kostakis 2014) have adopted an aggressive profit-oriented business model, opening an era of "platform capitalism" (Srnicek 2016, Kenney and Zysman 2016) with negative effects on different fields such as labour exploitation (platforms of food delivery, ride-sharing, short-term contracts or freelance work provision) or exacerbation of the housing crisis. This last aspect is particularly critical in the tourist sector; the literature (Picascia *et al.* 2017) is denouncing the impact that home-sharing platforms are having on house shortage: the increase in STR supply is eroding housing units from the traditional housing market, adding them to the tourist hospitality market. This trend leads to the relocation of residents that (have to) move to other districts and city areas, thereby activating or strengthening processes of gentrification (Wachsmuth and Weisler 2018) and segregation (Guttentag 2013, Ball *et al.* 2014). City centres and entire neighbourhoods lose their dwellers, replacing them with new residents, thus changing their identity, becoming a place of tourist services marked by social desertification (Semi 2015) and fuelling the process of Disneyfication (Bryman 2004) and hotelisation (Lee 2016).[2]

Focus group: the vision of the experts

The focus group hosted representatives of the cities of Amsterdam, Vienna and Barcelona, all of which are heavily impacted by the STR platforms spread; as well as the European Network for Short-Term Rentals, an advocacy group of more than 20 cities (including the three mentioned above) aiming at lobbying at EU level in order to ensure their sovereignty in the regulation of the short-term rental sector and protect the citizens' right to housing; the founder of Inside Airbnb, an investigatory/watchdog website with an independent, non-commercial set of tools and data that allows an exploration of how Airbnb is growing in cities around the world. Also taking part were representatives of CIDOB, the Barcelona Centre for International Affairs, which has recently published "Cities versus short-term rental platforms: the European Union battle" (CIDOB 2020) reporting how STR platforms are lobbying the ecosystem in Brussels (see Haar 2019, Tansey and Haar 2019) through the European Holiday Home Association (EHHA) in order to have a centralised top-down position in the EU from which to overcome all the hurdles eventually posed by single cities; the last participant was an expert in data visualisation, digital commons and participation technologies from Estudio Montera34.

The discussion, while recognising that new digital platforms facilitate tourists' access to a range of services with substantial time and cost savings, confirmed what is discussed above: the increase in STR impacts the housing market and housing costs, intensifying the loss of residential housing supply, and strengthening the process of gentrification for many neighbourhoods and city centres. Experts mentioned also the worsening of overtourism, especially for cities already impacted by this phenomenon, and the commercialisation and disruption of residential commons as a further issue.

Four important findings mark the contribution of this chapter to the discussion: first, many operators, while offering normal services, 'hide' behind profiles in STR platforms that are officially dedicated to non-professional hosts, introducing a competition considered unfair, as they manage to avoid the standard regulation that applies to traditional operators like hotels, hostels and bed and breakfasts.

At the same time, cities are struggling to respond adequately to the spread of STR platforms. However, the fact that listings are 'anonymised' by platforms makes it difficult to know exactly on which platform a facility is being offered and who is responsible for it. Additionally, platforms refuse to share their data, which would be extremely useful to the local administration to observe and regulate the sector, levy taxes and enforce laws if necessary.

Third, the legislations affecting short-term rental services and platforms are legion, and at different levels: national, regional and even international legislations do not always move in a coherent way, and leave the administration at city level with limited power. Some cities have serious difficulties in defending their sovereignty.

This is amplified by the fact that these platforms are trying to lobby at EU level through the above-mentioned EHHA, to avoid the need of direct confrontation between platforms and single cities, in order to limit cities' ability to impose regulations on a local basis. For this reason, it is increasingly important for cities to collaborate and network in order to exercise a power, albeit soft, to shape the transnational agenda (Foster and Swiney 2019). There are two good examples in this sense: the "Cities for Adequate Housing Declaration"—presented at the High-Level Political Forum at the United Nations in New York in July 2018, and the "Sharing Cities Declaration"—presented at the Sharing Cities Summit in Barcelona in November 2018.

The case study: Milan as a tourist destination

Milan is known mostly as a city of business and an international capital of fashion, design and finance, an identity gained during the deindustrialisation process of the 1980s. In 2016, Costa *et al.* described Milan as a city "lost in the transition" from the paradigm of the growth machine (Molotch 1976) to that of social innovation (Andreotti and Le Galès 2019). Since 2014, more space has been dedicated to reflections related to the sharing economy and its most critical aspects, and the city has adopted specific guidelines to promote and manage a fair integration of platform-based services for city users (Milan Municipality 2014, Bernardi and Diamantini 2018, Bernardi and Mura 2018).

In this transition, and due to the pull of EXPO 2015, the tourist sector has grown, in a process of city rebranding that has repositioned it in the international tourist panorama (Ferrari and Guala 2017, Sainaghi *et al.* 2019). From the 2015 record with its (unmatchable) peak of 22.2 million of overnight stays, Milan has kept a growing tendency in terms of presence: in 2019, for example, the metropolitan area has registered almost 11 million tourists (overnight stays), 9.2 percentage points more than in 2018 (Municipality of Milan—OpenData 2020).

EXPO 2015, with its 22.2 million tourists (Expo 2015, 2018), was also the starting point for a reflection on the sharing economy. ShareExpo, a path aimed at identifying proposals and initiatives for the experimentation of the sharing economy during EXPO 2015, pushed the Municipality to seriously consider the phenomenon (Sharexpo 2015), adopting specific guidelines (Municipality of Milan 2014) to encourage sharing experimentations in the city.

Note that in 2014 the perception of the sharing economy at a global level was more enthusiastic than today, since its most critical aspects were not yet evident. Scholars such as Diamantini *et al.* (2014), remark that Milan has always had a *human* approach in planning and policy-making, evident also in its promotion of social innovation (Vitale and Polizzi 2017) and sharing economy: both the city's Guidelines and its 'subscription' of the Sharing Cities Declaration testify to it. The Declaration, signed in Barcelona in 2018 during the Sharing Cities Summit by nearly 50 cities, introduced a set of principles and commitments for city sovereignty regarding the platform economy, to promote a fair and respectful collaboration between city governments and sharing platforms. The Municipality of Milan has developed a variety of initiatives inspired by the Guidelines and the Declaration, such as vouchers for the use of coworking and fablabs/makerspaces, a Register of Experts and Operators of the sharing economy in the city, and more. The openness of the administration has favoured the emergence of different types of digitally mediated services, some oriented to the promotion of social cohesion and social inclusion, linked to the local dimension by a collaborative/bottom-up approach—such as the social streets (Pasqualini 2018) or the neighbourhood concierge to connect citizens of the same district (see Bernardi 2019); others with a more profit-oriented business model. At the same time, the city has attracted start-ups and platforms already active abroad, in particular big players such as Uber, Airbnb, BlaBlaCar.

Looking to STR platforms, Airbnb is the bigger player: according to Inside Airbnb data (Inside Airbnb n.d.), in February 2020 it counted 20,354 listings in the city (while in November 2018 there were 17,659) (Inside Airbnb 2018). The Municipality has tried to maintain good relations with the platform and, according to scholars such as Amore *et al.* (2020, p. 9), it has adopted a *laissez-faire* approach favouring the growth of listings. In the authors' opinion, this approach reflects a market-driven attitude oriented to a quick return on investment. The fact that 70% of EXPO 2015 revenues were earned by Airbnb local households (Statista 2016) seems to confirm the authors' assumption. On this regard, Aguilera *et al.* (2019) highlighted that the presence of Airbnb in Milan was initially welcomed to mitigate the effects of the financial crisis on the housing market, as a winning solution to support "new public and private sharing services and products in a more sustainable and inclusive economy" (ibid. p. 12) and to reposition the city of Milan as a leading tourist destination. Airbnb was seen as one of the players of the wider sharing economy panorama. The phenomenon was seen as a tool to support the city in its goal to combine economic growth with social inclusion through technologies (Gasco *et al.* 2016) and to promote public participation and service-coproduction (Pais *et al.* 2019). In 2015, the Municipality and Airbnb signed a memorandum of understanding including four types of actions: support of digital literacy, impact measurement, increase in the accommodation offer during big events and collaboration in terms of legal and tax rules. Nevertheless, the negative impacts on housing stocks and early signs of tourism gentrification did not take long to become noticeable.

In order to contain the overflow of the Airbnb offer, or at least to have a minimum of control over it, in 2018 the Municipality of Milan and the platform signed an agreement on the collection of tourist taxes and to reduce the shadow market. Thanks to the agreement, Airbnb now pays the taxes collected from guests who book their stays on the platform directly to the Municipality.

However, the expansion of STR and its impact on the city is still strong. Looking at the Inside Airbnb data analysis of November 2018, when 17,659 listings where registered, 72.5% were entire houses or apartments, 25.5% were private rooms and only 2% were shared rooms. This data shows how the platform has changed its nature since its birth in 2007, and how strong the presence of hosts that offer entire accommodations, namely unused property houses that were previously rented to students and families and are now made available mostly to tourists, is. Another study by the University of Siena (Picascia *et al.* 2017) shows also the relevance of multi-owner (agencies or real estates that own several listings on the platform) with considerable effects in terms of redistribution of revenues and economic inequalities: in Milan, a single owner accumulated more than 520,000 euros in 2016 alone, while 75% of hosts earned less than 5,000 euros in a year. The same situation is present in other big Italian cities as Rome, Venice or Florence and confirms the 80/20 Pareto rule: 20% of the Airbnb listings produce more than 80% of the total revenue, and these 20% are generally located in the old town centre of a city. Hence, there is a double inequality: an interpersonal and a spatial one (ibid).

Recently, the city councillors for Urban Planning, Green and Agriculture and those for Tourism have denounced how the STR market is still eroding the real estate market by taking away availability for long-term rentals, and how it is important for Milan to limit the maximum number of rental days during the year and the number of multi-owners, as other European capitals are already doing.

An outline of the local tourist sharing offer in Milan

The previous section mainly focused on Airbnb and the hospitality sectors; nevertheless, there are other sharing services with a tourist appeal in the city.[3]

A total of 78 platforms have been identified and divided into four subsectors, labelled as follows: hospitality, tours and experience, transport and mobility, sharing of food and belongings. This collection shows the expansion of opportunities for a tourist that in the city chose to rely on sharing economy platforms and services.

Hospitality

The hospitality sector counts 22 platforms offering different services (see Table 3.1).

Eleven platforms provide P2P sharing accommodation, while four are home-exchange platforms and three are based on work in exchange for hospitality. Four have a respective non-monetary approach, with two offering completely free hospitality and two in exchange for specific services. Five out of 21 also offer a B2C service. In eight cases, the transaction/exchange is of a non-monetary type. Only two platforms are Italian companies: CasaVacanza (P2P and B2C home sharing) and HomeLinkItalia (home exchange). Airbnb is the biggest player of this market.

Table 3.1 Platforms in the hospitality sectors

11 platforms for P2P sharing accommodation	Airbnb; VRBO; HomeAway; Casa Vacanza; Wimdu; Spot at Home; Nestpick; HomeStay; BedyCasa; WayToStay; Couchsurfing
4 home-exchange platforms	Home Exchange; HomeLinkItalia; Love Home; HospitalityClub
3 platforms based on work exchange for hospitality	WorldPackers, WorkingTraveller and StayDu
1 offers free hospitality specifically to bike tourists	Warmshowers
1 is based on a non-monetary exchange asking in-home pet care for a free place to stay in return	TrustedHouses Sitters
1 provides hospitality on a non-monetary basis in exchange for teaching a language	GoCambio
1 has a charitable approach to hospitality	Servas

Tours and experience

In this sector, there are 22 active platforms offering tourist services, to visit and explore the city or to live local experiences (see Table 3.2). The main distinction relates to the national/international level.

Table 3.2 Platforms in the tours and experience sector

11 international companies operating in Milan	Airbnb with its "experiences", ContextTravelTours; ToursByLocals; GetYourGuide; WebGeoService; Like a Local; Rent a Local Friend; Eat With; Withlocals; Showaround; Come Home
11 Italian platforms	Curioseety; GuideMeRight; Zes Trip; Italian Stories; Affitto Giardino; What a Space; Cuccagna; SharryLand; Esperia Aviator; Guido Tour Sharing; Cesarine

Eleven are international companies operating in Milan, eleven are Italian platforms. Five platforms mainly offer tours of the city while 13 have a more varied offer providing also the so-called 'experiences', for a total of 18 platforms active in this subsector. Added to this are two platforms based on food experience: Eat With offers 'food & wine tours' while Cesarine proposes cooking classes. Another two platforms are related to the sharing of specific locations: AffittoGiardino offers the opportunity to rent a private garden for an event, while What a Space allows one to rent a private space (commercial or non-commercial) to organise an event.

Transport

There are 19 active platforms offering sharing services, and 17 offer rental services: cars, vans, scooters, electric scooters, kick scooters and campers. The emerging segment of kick scooters is gaining momentum and its market is expanding also in the city, counting 5 services out of 17. The municipal bike sharing service (BikeMi) has been active since 2008 and offers 3,650 traditional bikes and 1,150 e-bikes to dwellers, tourists and city users. There is a platform offering a bus sharing solution (Flixbus) and a peer-to-peer car sharing service (BlaBlaCar). Five out 19 are Italian platforms, while the remaining are international companies.

Food and belongings

There are 15 active platforms in the city that can be divided into three categories (see Table 3.3).

About the seven platforms of food delivery, note that this category falls into the so-called 'platform capitalism' (Srnicek 2016, Kenney and Zysman 2016), embodying what has been termed the 'gig economy',[4] far from the original idea

Table 3.3 Platforms in the food and belongings sector

7 platforms dedicated to food delivery	Uber Eats, Just Eat, Glovo, Foodora, Deliveroo, MyMenu, Quomi, Diet To Go
4 platforms for social eating	Gnammo, SoLunch, Babette, Bon Appetour
4 platforms to share food and reduce food waste	Toogoodtogo, Regusto, MyFoody, BringTheFood

of sharing; while the social eating platforms allow users to offer or search for a shared meal, prepared by the host; and the last category is mainly focused on reducing food waste.

Half of the group is composed of food delivery platforms (50%), confirming the analysis of the last B2C eCommerce Observatory Report of the Milan Polytechnic (2019), according to which the food delivery sector in 2019 registered a +59% growth compared to 2018. Nevertheless, there is a tough debate about riders' rights and working conditions, and some initial informal union experiences (among others: Aloisi 2018, Marrone and Finotto 2019).

It should be highlighted that other tourism-oriented services are available through platforms in the city. They mainly concern logistics aspects, such as BagBnb, the first luggage storage network; or ShareWood for sports equipment rental.

The perception of traditional hospitality services

This part of the fieldwork started in January 2020 with a plan of 40 face-to-face interviews, 20 to hostel owners and 20 to low-budget hotel owners; unfortunately, it had to be interrupted due to the outbreak of the COVID-19 pandemic; therefore, the analysis can only count on 15 interviews and the results should be taken as just an indication of a general trend.

The interviews touched on four main points: sharing platforms used for promotion; opinion on the impacts of these platforms on the tourist sectors (advantages and disadvantages perceived) and in particular in the city of Milan; opinion on the habits of their regular customers; opinion on the role of the Municipality concerning STR platforms. Our main five findings concluded as follows.

Almost all of the interviewees (14 out of 15) use online channels to promote and advertise their business and to gain more visibility among customers (1), in particular Booking and Expedia, but none use sharing platforms. The data shows that interviewees are still confused when asked about sharing economy platforms but are anyway familiar with the online tools. On average, hoteliers have been using these online booking services for three to five years, considering them effective, as many of the customers book through them; in addition, the official website of the hotels/hostels is still used by tourists to book a room, especially in the case of so-called 'repeaters' (long-standing customers). In the interviewees' opinion, online services (with no distinction between platforms and other services) increase advertising and visibility and facilitate the finding of information for customers. Nevertheless, they also increase competition (price dumping), fees and commissions are high and managing them is not easy due to various booking methods.

The interviewees do not appear to be well informed on the topic of STR, but they all state that, since the spread of this type of platforms, they have witnessed an increase in the number of tourists (2). Among the owners of low-budget hotels, it is a common opinion that STR platforms, Airbnb in particular, create price competition, and that tourists prefer to rent an entire apartment rather than a room. For the owners of hostels, the presence of Airbnb and its competitors does not have this effect, as customers who choose hostels have different needs (they look for low prices and are willing to share the room). Both believe that the presence of STR platforms is destined to grow.

The interviewees did not observe significant changes in the profile of their customers (3). For the hotels, customers are still mainly young people, couples, families, groups with different travel motivations; looking for a clean comfortable room in central areas with a reasonable price and supported by kind staff. The typical customers of the hostels are mainly students, occasional travellers, foreigners, solo travellers and solo senior travellers; they look for safe and central areas, a friendly environment and entertainment services.

When asked to assess the impacts of these platforms on the city, the interviewees are divided on two positions (4). In general, they all believe that digital platforms have contributed to changing the face of the city, but while one group has a less positive vision, the other one gives a more positive interpretation. For one group, STR platforms have opened new entrepreneurial opportunities for private citizens and subtracted potential customers from hotels and hostels, moving fluxes of tourists into new areas. Nevertheless, they also think that the growing tourists' flux brought an increase of traditional accommodation facilities too, especially new bed and breakfast and small hotels. In their opinion, STR platforms have not pushed local tourism, since the city has always attracted millions of tourists. On the contrary, the growing tourist appeal of the city has brought about new accommodations (both traditional and through platforms). For them, the sharing platforms in Milan are not very influential in economic terms. Another group, while believing that digital platforms have changed the face of the city, also believe that the platforms have had a positive economic impact, providing more services for tourists and residents, more employment and more business development. Hoteliers and hostels owners appear to be not so well informed or at least confused about the normative regulating STR platforms (5). They believe that there are not enough official inspections to control the compliance of normative requests among businesses, but they mainly refer to traditional accommodation solutions and not to the STR offer. Moreover, they think that the tourist tax introduced by the Municipality should be managed differently; in their opinion, the price is too high, and customers often ask for a discount on the room price to compensate for the city tax. In addition, they believe the Municipality should give more protection to traditional accommodation facilities considering the jobs they create, while, in their opinion, Airbnb and similar platforms do not have the same economic potential. Also, they would impose greater economic constraints on these sharing economy platforms in order for them to be on an equal footing.

Conclusions and future research

The assets of Milan, both in terms of attractiveness and offer, outline new trends that are shaping a New Urban Tourism. The city has widened its tourist brand image by going beyond fashion and finance, reimagining and structuring a new international profile based on tourism too (Ferrari and Guala 2017). The diffusion of digital platforms has also played a role in this restructuring process.

Just as the tourism relaunch of the city went through EXPO 2015, in the same way the spread of digital platforms has benefited from the mega event and from a public administration willing to stimulate forms of social innovation. The Municipality has enthusiastically welcomed the phenomenon of the sharing economy since 2014, considering it a harbinger of new opportunities, in line with its own programmatic lines. The approach adopted and the role played—as facilitator and connector—has made the city particularly attractive and welcoming for the spread of services mediated by platforms.

The research has highlighted the richness of the tourist offer mediated by platforms, showing four different categories of services: hospitality through home sharing, mobility through sharing or renting solutions, tours and experience through the engagement of residents, and food sharing plus some services of logistic or technical support for tourists. As underlined by the literature, the expansion of the tourist's offer brings new experiential and economic opportunities.

Nevertheless, 'tourist platformisation' carries risks and challenges with it. In particular, home sharing through STR platforms has had a heavy impact at the urban level. As is happening in many other European cities, these platforms are also opening new or exacerbating old urban issues. What was initially perceived as a winning solution to support sharing services/goods in a sustainable and inclusive way proved to be a double-edged sword, reinforcing gentrification, strengthening touristification and eroding housing units from the traditional housing market (Lee 2016, Picascia *et al.* 2017, Barron *et al.* 2018, Clancy 2020). Experts have confirmed that the issues opened by these platforms are shared by the main European (but also American) cities, and have highlighted the need for a collective, cities-based action to limit the lobbying power of these platforms at EU level.

In the city of Milan, the perception of risk comes not only from local government: in 2017, ATR, the Hoteliers' Association member of Confesercenti Milano, created the website www.hotelvsairbnb.it to highlight and denounce the unfair competition that STR platforms constitute for the traditional sector; nevertheless, this initiative seems to have disappeared in the net: there are no recent updates and the site is no longer active. The initiative OspitaMi, an association of volunteers that promote responsible home sharing in Milan, is also interesting.

If we consider the risk perception of hoteliers and hostel owners, the interviews allow the emergence of a different level of awareness compared to that of experts: they recognised that the phenomenon is related to the increase of tourist fluxes and that Airbnb seems destined to grow, but they do not reveal an excessive concern. However, they ask for more support from the Municipality in order to stabilise their position in the market.

These results could suffer from the limited number of interviews realised in the field and they can only suggest some preliminary reflection. Nevertheless, the lack of recent initiatives coming from official agencies representing hoteliers can be indicative of a general perception that STR platforms are not a real threat to business; or, on the contrary, it can be the expression of the market-oriented approach that characterised the city till EXPO 2015, when the STR sector experienced a surge.

The contribution paves the way for further analysis on the topic, based on an improved dataset, and in light of the implications that the recent COVID-19 outbreak will have in the tourist sector. The global pandemic has completely changed the tourism scenario, by zeroing out tourism and emptying the city of Milan. Analysts such as Nomisma[5] suppose that the Italian cities that have focused more on attracting foreign corporate investment, such as Milan, will pay the most for the coming crisis. The city has attracted 40% of foreign investments in the hotel and logistics sectors (3.3 and 1.4 billion euros in 2019); and its centrality—so close to the epicentre of the Italian viral pandemic—represents an additional factor of concern in terms of perspective (Gainsforth 2020). The booking resetting of STR solutions, together with the zeroing of customers for traditional operators, will completely change the image of tourism, setting the ground for a 'new New Urban Tourism' (Bernardi 2020).

Notes

1 The Index ranks cities in terms of the number of their total international overnight visitor arrivals and the cross-border spending by these same visitors in the destination cities, and gives international overnight visitor growth forecasts.
2 The terms describe the commercial transformation of a city centre to resemble the Walt Disney Parks and Resorts to host tourists, based upon rapid Western-style globalisation and consumerist lifestyles.
3 The collection of data was carried out in November 2019 during the laboratory on "Sharing economy and tourism" of the bachelor degree in Tourism Science and Local Community of the Milano-Bicocca University.
4 The "gig economy" refers to small tasks or partial jobs (the "gigs") that individuals carry out through a contract with platform companies.
5 Nomisma is an independent company that carries out economic research and consulting for businesses, associations and public administrations. See: https://www.nomisma.it/ and https://leasenews.it/news/dati-di-settore/nomisma-il-settore-immobiliare-ai-tempi-del-coronavirus.

References

Aguilera, T., Artioli, F., and Colomb, C., 2019. Explaining the diversity of policy responses to platform-mediated short-term rentals in European cities: a comparison of Barcelona, Paris and Milan. *Environment and Planning A: Economy and Space*. Available from: https://doi.org/10.1177/0308518X19862286.

Aloisi, A., 2018. 'With great power comes virtual freedom'. A review of the first Italian case holding that (food-delivery) platform workers are not employees. *Comparative Labor Law & Policy Journal* [online]. Available from: http://dx.doi.org/10.2139/ssrn.3260669.

Amore, A., de Bernardi, C., and Arvanitis, P., 2020. The impacts of Airbnb in Athens, Lisbon and Milan: a rent gap theory perspective. *Current Issues in Tourism* [online]. Available from: https://doi.org/10.1080/13683500.2020.1742674.

Andreotti, A. and Le Galès, P., 2019. Introduzione. Governare Milano nel nuovo millennio. *In*: A. Andreotti, ed. *Governare Milano nel nuovo millennio*. Bologna: Il Mulino, 7–28.

Arcidiacono, D., 2017. Economia collaborativa e startup: forme alternative di scambio economico o mito della disintermediazione? *Quaderni di Sociologia*, 73, 29–47.

Ball, J., Arnett, G., and Franklin, W., 2014. London's buy-to-let landlords look to move in on spare room website Airbnb. *The Guardian*, 20 June [online]. Available from: https://www.theguardian.com/technology/2014/jun/20/buy-to-let-landlords-leasing-properties-airbnb-uk [Accessed May 2020].

Barron, K., Kung, E., and Proserpio, D., 2018. The sharing economy and housing affordability: evidence from Airbnb. In *EC 2018: proceedings of the 2018 ACM conference on economics and computation*. Ithaca, NY: Cornell University, 18–22.

Bauwens, M. and Kostakis, V., 2014. *Network society and future scenarios for a collaborative economy*. New York: Palgrave Pivot Macmillan.

Bernardi, M., 2019. Portinerie di Quartiere: innovazione sociale tra digitale e locale. *In*: G. Nuvolati, ed. *Enciclopedia Sociologica dei Luoghi, Volume 1*. Milano: Ledizioni, 335–351.

Bernardi, M., 2020. Short-term rental trends and the global pandemic: "emancipatory catastrophism" in a metamorphosing world. *LabGov*, 30 April [online]. Available from: https://labgov.city/theurbanmedialab/short-term-rental-trends-and-the-global-pandemic-emancipatory-catastrophism-in-a-metamorphosing-world/ [Accessed May 2020].

Bernardi, M. and Diamantini, D., 2018. Shaping the sharing city: an exploratory study on Seoul and Milan. *Journal of Cleaner Production*, 203, 30–42.

Bernardi, M. and Mura, G., 2018. Sharing economy e istituzioni pubbliche: l'innovazione sociale nei contesti urbani. *Sociologia Italiana–Ais Journal of Sociology*, 11, 51–75.

Botsman, R. and Rogers, R., 2010. *What's mine is yours: the rise of collaborative consumption*. New York: HarperBusiness.

Brauckmann, S., 2017. City tourism and the sharing economy–potential effects of online peer to-peer marketplaces on urban property markets. *Journal of Tourism Futures*, 3 (2), 114–126.

Bryman, A., 2004. *The disneyization of society*. London: SAGE.

CIDOB, 2020. *CIDOB - Cities versus short-term rental platforms: the European Union battle* [online]. Available from: https://www.cidob.org/en/publications/publication_series/notes_internacionals/n1_222/cities_versus_short_term_rental_platforms_the_european_union_battle [Accessed June 2020].

Clancy, M., 2020. Tourism, financialization, and short-term rentals: the political economy of Dublin's housing crisis. *Current Issues in Tourism*. https://doi.org/10.1080/13683500.2020.1786027.

Codagnone, C., Biagi, F., and Abadie, F., 2016. The passions and the interests: unpacking the 'sharing economy' [online]. Institute for Prospective Technological Studies, JRC Science for Policy Report EUR 27914 EN. Available from: http://dx.doi.org/10.2139/ssrn.2793901

Costa, C., Cucca, R., and Torri, R., 2016. Milan: a city lost in the transition from the growth machine paradigm towards a social innovation approach. *In*: T. Brandsen et al., eds. *Social innovations in the urban context. Nonprofit and civil society studies (An International Multidisciplinary Series)*. Cham: Springer, 125–142.

Diamantini, D., Borrelli, N., and Bernardi, M., 2014. Smart and slow city. The case study of Milan EXPO 2015. In *Arte-Polis 5 International Conference and Workshop, 8-9 August 2014*. Indonesia: Bandung Technoloy Institute.

Dredge, D. and Gyimóthy, S., 2015. The collaborative economy and tourism: critical perspectives, questionable claims and silenced voices. *Tourism Recreation Research*, 40 (3), 286–302.

European Union, 2015. *The sharing economy and tourism: tourist accommodation*. EPRS-European Parliamentary Research Service. Bruxelles. Available from: http://www.europarl.europa.eu/RegData/etudes/BRIE/2015/568345/EPRSBRI(2015)568345_EN.pdf [Accessed June 2020].

Expo2015, 2018. *Expo Milano 2015*. Official Report [online]. Available from: https://issuu.com/expomilano2015/docs/expo-milano-2015_official-report_en [Accessed May 2020].

Ferrari, S. and Guala, C., 2017. Mega-events and their legacy: image and tourism in Genoa, Turin and Milan. *Leisure Studies*, 36 (1), 119–137.

Foster, S.R. and Swiney, C.F., 2019. City power and powerlessness on the global stage. *In*: E. García-Chueca and L. Vidal, eds. *Urban futures* [online]. Barcelona: CIDOB, 19–28.

Gainsforth, S., 2020. L'Italia è piena di case vuote, abbiamo bisogno di un piano per riabitarle. *CheFare*, 7 April [online]. Available from: https://www.che-fare.com/gainsforth-italia-abitare-case/?utm_content=bufferd955b&utm_medium=social&utm_source=facebook.com&utm_campaign=buffer&fbclid=IwAR30VKf8Wxj4Q8toJWYKvXdhMHCpyLD7kVPGd6A6zHoc3ESaV5XpVMwsdP4 [Accessed May 2020].

Gascó, M., Trivellato, B., and Cavenago, D., 2016. How do southern European cities foster innovation? Lessons from the experience of the smart city approaches of Barcelona and Milan. *In*: J.R. Gil-Garcia, T.A. Pardo and T. Nam, eds. *Smarter as the new urban agenda. A comprehensive view of the 21st century city*. Cham: Springer, 191–206.

Gavinelli, D. and Zanolin, G., 2019. *Geografia del turismo contemporaneo: pratiche, narrazione e luoghi*. Roma: Carocci.

Guttentag, D., 2013. Airbnb: disruptive innovation and the rise of an informal tourism accommodation sector. *Current Issue on Tourism*, 18 (12), 1192–1217.

Haar, K., 2019. *UnFairbnb*. Brussels: Corporate Europe Observatory.

Inside Airbnb, 2018. *Milan* [online]. Available from: http://insideairbnb.com/milan/ [Accessed 19 December 2020].

Inside Airbnb, n.d. *Get the data* [online]. Available from: http://insideairbnb.com/milan/ [Accessed 19 December 2020].

Kenney, M. and Zysman, J., 2016. The rise of the platform economy. *Issues in Science and Technology*, 32 (3), 61–69.

Koh, E. and King, B., 2017. Accommodating the sharing revolution: a qualitative evaluation of the impact of Airbnb on Singapore's budget hotels. *Tourism Recreation Research*, 42 (4), 409–421.

Koopman, C., Mitchell, M., and Thierer, A., 2015. The sharing economy and consumer protection regulation: the case for policy change. *Journal of Business, Entrepreneurship & the Law*, 8 (2), 529–545.

Lee, D., 2016. How Airbnb short-term rentals exacerbate Los Angeles's affordable housing crisis: analysis and policy recommendations. *L. Pol.'y Rev*, 10, 229–255.

Marrone, M. and Finotto, V., 2019. Challenging Goliath. Informal unionism and digital platforms in the food delivery sector. Case of Riders Union Bologna. *Partecipazione e Conflitto*, 12 (3), 691–716.

Milan Polytechnic, 2019. *Osservatorio eCommerce B2c. L'eCommerce B2c: il motore di crescita e innovazione del Retail!* [online]. Available from: https://blog.osservatori.net/it_it/e-commerce-in-italia?utm_source=(direct)&utm_medium=comunicatinew&utm_campaign=ecommerce [Accessed January 2020].

Molotch, H., 1976. The city as a growth machine: toward a political economy of place. *American Journal of Sociology*, 82 (2), 309–332.

Municipality of Milan, 2014. *Milano sharing city* [online]. Available from: https://economiaelavoro.comune.milano.it/sites/default/files/2019-02/milano%20sharing%20city_finale.pdf [Accessed April 2020].

Municipality of Milan - OpenData, 2020. *Tourism* [online]. Available from: https://dati.comune.milano.it/it/organization/comunedimilano?q=turismo&sort=score+desc%2C+metadata_modified+desc [Accessed April 2020].

OECD, 2016. *OECD tourism trends and policies 2016*. Paris: OECD.

Pais, I., Polizzi, E., and Vitale, T., 2019. Governare l'economia collaborativa per produrre inclusione: attori, strumenti, stili di relazione e problemi di implementazione. *In*: A. Andreotti, ed. *Governare Milano nel nuovo millennio*. Bologna: Il Mulino, 215–37.

Pasqualini, C., 2018. *Vicini e connessi. Rapporto sulle Social Street a Milano*. Milano: Fondazione Giangiacomo Feltrinelli.

Pearce, L.D., 2012. Mixed Methods Inquiry in Sociology. *American Behavioral Scientist*, 56 (6), 829–848.

Picascia, S., Romano, A., and Teobaldi, M., 2017. The airification of cities: making sense of the impact of peer to peer short term letting on urban functions and economy. In *proceedings of the annual congress of the association of European schools of planning*, 11–14 July, Lisbon.

Sainaghi, R., et al., 2019. Mega events and seasonality: the case of the Milan World Expo 2015. *International Journal of Contemporary Hospitality Management*, 31 (1), 61–86.

Semi, G., 2015. *Tutte le città come Disneyland?* Bologna: Il Mulino.

Sharexpo, 2015. *Milano città condivisa per Expo 2015. Documento di indirizzo* [online]. Available from: https://issuu.com/sharexpo/docs/documento_d_indirizzo_sharexpo/7 [Accessed April 2020].

Smorto, G., 2015. Verso la disciplina giuridica della sharing economy. *Mercato concorrenza regole*, 27 (2), 245–277.

Srnicek, N., 2016. *Platform capitalism (theory redux)*. New York: Wiley.

Statista, 2016. Number of Airbnb hosts in the Italian city of Milan in 2015 and during the Milan Expo 2015. *Statista Research Department*. [online]. Available from: https://www.statista.com/statistics/629800/airbnb-number-of-hosts-in-milan-and-during-milan-expo-italy/#statisticContainer [Accessed April 2020].

Tansey, R. and Haar, K., 2019. *Über-influential?* Brussels: Corporate Europe Observatory/AK Europa.

Vitale, T. and Polizzi, E., 2017. Governo collaborativo e catene relazionali di innovazione. Spunti a partire dal caso di Milano. *Quaderni di Rassegna Sindacale*, 18 (2), 129–147.

Wachsmuth, D. and Weisler, A., 2018. Airbnb and the rent gap: gentrification through the sharing economy. *Environ Plan A*, 50 (6), 1147–1170.

WTO-World Tourism Organization, 2017. *New platform tourism services (or the So-called sharing economy): understand, rethink and adapt.* Madrid: World Tourism Organization.

Zervas, G., Proserpio, D., and Byers, J.W., 2017. The rise of the sharing economy: estimating the impact of Airbnb on the hotel industry. *Journal of Marketing Research*, 54 (5), 687–705.

4 Peer-to-peer tourist accommodation and its impact on the local housing market in Berlin neighbourhoods

Claus Müller and Kristin Wellner

Investigating tourism and housing in Berlin

Platforms like Airbnb cater to New Urban Tourists' demand for 'authentic' experiences. While this opens up new areas of the city to tourists, it also bears the risk of conflict and intensified competition for residential space and infrastructure. For this chapter, we draw on findings from an ongoing research project on neighbourhood tourism in Berlin.[1] Using public data from market reports as well as rental offerings from a real estate online portal, we investigate the residential rental market in four Berlin neighbourhoods with different concentrations and types of tourist accommodation.

Berlin has undergone fundamental changes since the fall of the Berlin Wall in 1989. Two cities of politically and economically antagonistic systems were (re)unified and became Germany's capital, home of the parliament, federal ministries and many other government agencies. History did not simply change Berlin's position within Germany and Europe, but also the city's urban structure. Neighbourhoods that used to be at the periphery of either West or East Berlin, near the Berlin Wall with its armed guard towers and no-man's land, became transforming parts of the new inner city (Brandt *et al.* 2019): rents are on the rise, new inhabitants move in, buildings, streets and places change their appearance. For several years now, one of the most controversial topics in Berlin politics and media has been the local housing market, which is in a state of rapid change from rent levels that have been comparatively low for a long time to a sharp increase over the last decade.

At the same time, Berlin has gained a place among the most visited cities in Europe (European Cities Marketing 2019), with an ever-increasing number of tourists coming to Berlin. The city attracts a wide variety of visitors interested in Berlin's unique history, in its (sub)cultural music scene and famous clubs, but also in immersing themselves in local everyday life. Especially since the emergence of peer-to-peer holiday accommodation platforms such as Airbnb, tourism, while being a source of income and commerce for the city and its inhabitants, is portrayed as the agent of rapid and often unwanted change (Kritische Geographie Berlin 2014, Novy 2013, Novy and Huning 2014). Especially in neighbourhoods in the inner city that are known to be somewhat hip, young and alternative, a connection between rising rents, scarcity of available and affordable housing and the increase of visitors to these neighbourhoods is being made both by the local media (analysed in detail by Kelling and Zecher in Chapter 15 of this issue) and politics (Loy 2013). Especially the conversion of rental units into more profitable tourist accommodation has been controversial and led to regulatory interventions, namely the "Zweckentfremdungsverbot-Gesetz" (ZwVbG), the law against the misappropriation of housing (Duso *et al.* 2020). In this chapter, we investigate the relationship between the proliferation of different types of tourist accommodation and its influence on the housing market.

Our analysis of the housing market is concentrated on the development of rental prices for flats on different spatial levels. Ranging from examinations of the residential market of Berlin to an analysis of the small-scale effects on the neighbourhood level, using public data from the state-level statistics office, market reports by real estate research agencies and banks as well as ads on a real estate web portal, we cast a wide light on Berlin's housing market, overall and on a neighbourhood level, showcasing the development of asking rents in Berlin's residential market and in specific neighbourhoods between 2008 and 2018.

We study the development of tourism for all of Berlin, as well as the specific emergence of Airbnb accommodations in selected neighbourhoods since the founding of Airbnb in 2008. After giving an overview of the developments on the local real estate market in the aftermath of Berlin's division and subsequent reunification, we use publicly available data to examine the potential impact of Airbnb on the Berlin housing market. We then analyse the impacts at a smaller level by concentrating on four specific neighbourhoods with different patterns of tourist lodging, and we use apartment listings to investigate small-scale effects like shifts in rental prices on the local market for rental flats from 2008 to 2018.

The research presented is part of the ongoing research project "Neighbourhood in the tourist trap? An examination on the changing residential quality through tourist accommodation in selected Berlin residential neighbourhoods", in which we investigate the presence of tourists and their respective accommodations' impact on the daily life and housing quality of local residents using qualitative as well as quantitative methods.

The discussion around the impact of tourism on city life is not specific to Berlin. Multiple studies highlighting different aspects of this phenomenon in very different locations such as New York City (Wachsmuth and Weisler 2018), Vienna (Seidl *et al.* 2018) and Barcelona (Garcia-López *et al.* 2020) among many others, or even at country level (Campbell *et al.* 2019), have been published. The case of Berlin has been studied by Schäfer and Braun (2016) as well as Duso *et al.* (2020). Investigating the Berlin housing market at district level (an administrative subdivision between borough and neighbourhood), Schäfer and Braun (2016) find the steepest rent increases in central districts with high shares of offerings on Airbnb. Duso *et al.* (2020) analyse the regulation of home sharing by the Berlin state government which restricted the use of housing space as tourist accommodation in 2016 and even further in 2018. They find that both regulatory interventions reduced the number of entire homes on Airbnb significantly. Concerning the effect on the housing market, they treat these policy changes as natural experiments. With this, they find that any given apartment on Airbnb within a 250 meter distance of a rental flat leads to a rent increase of seven cents per square meter on average, differentiated by size and rental period of the apartments (ibid., p. 41).

Berlin's housing market since 1990

Berlin's housing market is largely dominated by rental flats. The share of owner-occupied units is very low, at about 15% (Investitionsbank Berlin 2019), compared to the German average of 51.4%. Germany already has the lowest ownership rate within the European Union (Eurostat 2019), making the rate for Berlin even more striking.

The rental market in Berlin has experienced a dramatic development since the German reunification in 1990. The predominantly state-run housing industry of East Berlin was transformed in favour of a more market-driven housing supply in line with the evolution of the market system in the former socialist part of Germany. Large parts of the public housing supply were privatised, and previously expropriated estates were restituted to their former owners (Krätke and Borst 2000). At the time of the reunification, the East Berlin housing stock of the Wilhelminian era was mostly in a run-down state. The housing policy of the socialist German Democratic Republic (GDR) was concentrated on the construction of pre-fabricated housing blocks outside the centres, which were intended to replace the older buildings in the long run (Hannemann 1996). So, for the eastern part of Berlin the change in ownership structure, combined with long overdue investments into the Wilhelminian housing stock, laid the foundation to a gentrification process ongoing to this day.

The fall of the Berlin Wall caused an overnight change of Berlin's urban structure in the western as well as in the eastern part. Previously wall-adjacent, and therefore undesirable areas of West Berlin, such as Kreuzberg, became attractive locations due to their newly gained centrality. Although with a different causal background, these parts of West Berlin found themselves in a comparable situation to their eastern counterparts: with a run-down housing stock of mostly Wilhelminian-style buildings but in a newly promising, now central location.

In the decade after the German reunification, the development of the Berlin housing market did not live up to the expectations of economic growth and prosperity associated with the fall of the Berlin Wall and Berlin's new role as the capital of Germany. After a brief sharp rise in rents due to the lack of supply in the immediate aftermath of reunification, high unemployment rates, a slow economic development and increasing suburbanisation (Krätke and Borst 2000) led to a decreasing population in the inner city and therefore to a declining market for rental flats with—compared to other European, especially capital, cities—low prices and high vacancy rates. So, even after the reunification of Germany, investments in the Berlin housing stock were not lucrative and were often suspended. On the other hand, this left many buildings in a dilapidated and underused state, creating affordable spaces for alternative culture and lifestyles. Thus, Berlin became increasingly attractive to young and creative people from other parts of Germany and from abroad. This contributed to Berlin gaining a reputation as "Capital of Cool" (Frary 2018), resulting in a stronger demand for living space, decreasing vacancy rates and finally in rising rental prices after the economic crisis of 2007 due to a high immigration rate (Investitionsbank Berlin 2018). Since Berlin had not experienced urban growth for many decades, neither the public administration nor the construction industry were prepared to provide for the demand for housing as well as office and industrial space (Becher 2016, Bauindustrieverband Ost 2019).

Peer-to-peer home sharing and the housing market

Simultaneously with Berlin's renaissance and the concomitant population growth, home sharing gained relevance as an urban phenomenon with the emergence of platforms like Airbnb or Wimdu. While tourist accommodation has been dominated by the hotel industry for decades, those platforms enable private households to rent out (parts of) their home to tourists. Airbnb highlights the aspects of sharing in their business model and assures that the majority of their hosts rent out their own homes while they are not using them, which would contribute to a more efficient use of living space, since homes are used that otherwise would lie empty. But the notion that Airbnb is predominantly about sharing underused space is challenged on many accounts (Seidl *et al.* 2018, Cócola Gant 2016, Cócola Gant and Gago 2019, Wachsmuth *et al.* 2018). Not only is Airbnb's promise of a more efficient usage of living space questionable, its emergence as an easy-to-use rental platform is even suspected to result in the removal of rental units from the housing market as their conversion into tourist accommodation seems a more profitable type of use (Wachsmuth and Weisler 2018). With constant or, in the case of Berlin, growing demand for rental flats, this reduction of supply in areas with a high density of Airbnb units would be expected to drive up local asking rents.

The data provided by Airbnb on their listings is scarce and selective. For our investigation, we therefore used information from Airbnb investment analyst AirDNA, as well as from the data scraping and visualisation projects InsideAirbnb and Dwarshuis to investigate the spatial and temporal distribution of Airbnb offerings. For Berlin, research by AirDNA shows that 46%[2] of the properties on Airbnb are available for three months per year or more, suggesting that these might be used as secondary residence at most. AirDNA also lists 23% as full-time investment properties (AirDNA 2019); these units are only available for tourist accommodation and are therefore completely removed from the housing market. The actual number of Airbnb listings in the city is somewhat disputed, partly since it varies quite significantly over time, but also, since the regulation of rental properties is a controversial political issue, the different parties involved try to highlight data that supports their respective agendas. AirDNA denotes 8,552 entire home rentals (58% of 14,758 active rentals), while the independent data project Inside Airbnb denotes 10,722 entire homes or apartments (47.5% of 22,552 listings), out of which 8,935 have been reviewed within the last six months or are booked for

Table 4.1 Comparison of the number of listings on Airbnb for Berlin by data source

	AirDNA	*Inside Airbnb*
Listings	14,758	22,552
Entire home	8,552	10,227
% of housing stock	0.44%	0.53%
Recently/frequently booked		4,216
% of housing stock		0.22%
Highly available (>90 days)		3,624
% of housing stock		0.19%
Recently/frequently booked and highly available		2,221
% of housing stock		0.11%

more than 90 days per year (see Table 4.1). The difference here is most likely due to a difference in definition of what constitutes an "active rental" between Inside Airbnb and AirDNA (James 2018). Comparing these values with Berlin housing stock, we find that all "entire home" listings on Inside Airbnb make up for 0.55% (0.38% using AirDNA data) of the overall housing stock of the city. If we consider only recently and frequently booked listings, the share is reduced further to 0.22%. When only including apartments, which are available for at least 90 days, the share accounts for only 0.19% of the housing stock. Combining the latter constraints in order to limit the examination to apartments which are both active and highly available, and therefore are most likely to be removed from the housing market entirely, reduces the share of presumably misused housing units to 2,221 or 0.11%. Compared to 135,000 apartments that are reported to be calculatively missing in Berlin (Investitionsbank Berlin 2018), the impact of misused Airbnb apartments seems negligible.

Airbnb listings are, however, far from being equally distributed within the city. The most affected areas in Berlin are the centrally located boroughs Mitte, Friedrichshain-Kreuzberg and the northern part of Neukölln. Both Mitte and Friedrichshain-Kreuzberg are also the boroughs with the highest share of expensive rentals of 14€/m² or higher (38% each) (Investitionsbank Berlin 2018). Concentrating the analysis on Friedrichshain-Kreuzberg, 1.63% of the housing stock is listed on Airbnb as "entire home"; this share is reduced if the analysis is focused on listings with frequent activity (0.57%) or high availability (0.45%, or 0.28% if for both restrictions combined) (see Table 4.2). So, even when

Table 4.2 Share of housing stock in Friedrichshain-Kreuzberg that is listed on Airbnb

Entire home	*Entire home and frequent activity*	*Entire home and high availability*	*Entire home and frequent activity and high availability*
1,63%	0,57%	0,45%	0,28%

concentrating on a certain submarket with a high density of Airbnb listings the potential influence of the use of rental flats as tourist accommodation on rental prices is still fairly low.

Investigating housing market developments on a local level

The German market for rental housing is highly regulated: contracts are usually open-ended and while they can simply be terminated by the tenant on a three months' notice, a termination by the landlord is limited to very specific situations like gross misconduct by the tenant, failure of payment or the right to repossession for personal usage by the landlord. The possibility of increasing the rent during an existing rental contract is also limited by law; therefore, rents are usually increased with new contracts, which is restricted as well, but less so than during a contract. For a growing market, this means that the rental prices that are stated in apartment listings are expected to be higher than the average paid rent, which also includes older contracts. This effect is exacerbated by the fact that in rising markets, people avoid moving in order to keep their old, usually cheaper, rental contract. Therefore, the gap between the asking rents and the comparative rents in existing contracts is increasing over time and increases the differences between established and new tenants (see Figure 4.1). Due to the limited availability of small-scale data of actual rental prices in contracts, the analysis on neighbourhood level uses asking rents. Asking rents are relevant for the tenants' decision on whether to apply for a specific rental unit, but also for the landlord's decision on whether to offer a unit on the rental market, to leave it vacant or to offer it as a tourist accommodation, e.g. on Airbnb.

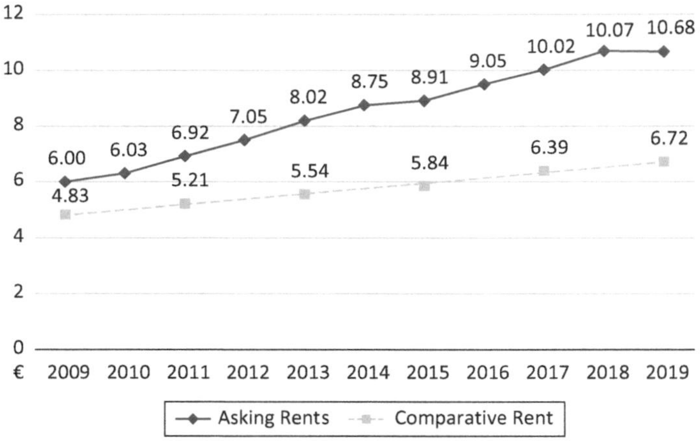

Figure 4.1 Development of asking rents and comparative rents in Berlin overall © Claus Müller and Kristin Wellner

Selection of investigation areas

Doing research on the neighbourhood level is often made difficult because data is not available on such a small scale, or the geographic limits of statistical data do not concur with perceived neighbourhoods in everyday life. Berlin is a different case, as the city's administration has divided the city into 448 so-called "Lebensweltlich orientierte Räume" (LORs)—neighbourhoods that share a life-world ("LOR" and "neighbourhood" are used synonymously in the following). Of the 448 LORs, four were chosen as investigation areas according to the local availability of different kinds of tourist accommodations (as shown in Figure 4.2).

We selected four areas along the lines of an assumed impact of tourism on the neighbourhoods based on offerings of accommodation: Boxhagener Platz was picked due to the high density of hotels and hostels as well as peer-to-peer tourist accommodation, for which the number of Airbnb offerings is used as a proxy. At Askanischer Platz, many hotels and hostels are located, but only few accommodations on Airbnb are offered. The opposite situation is found in Reuterkiez, where there are just a handful of formal tourist accommodations and a high density of listings on Airbnb. In the LOR Scharnweberstraße, no noteworthy accommodation industry could be found; hardly any hostels and barely any listings on Airbnb are traceable. This neighbourhood was chosen to control our results for tourism impact.

Three of the four selected neighbourhoods are located within the inner city (see Figure 4.3), marked by the circular railway line (Ringbahn). Only LOR Scharnweberstraße lies outside the circle-line (see Figure 4.3). It has become

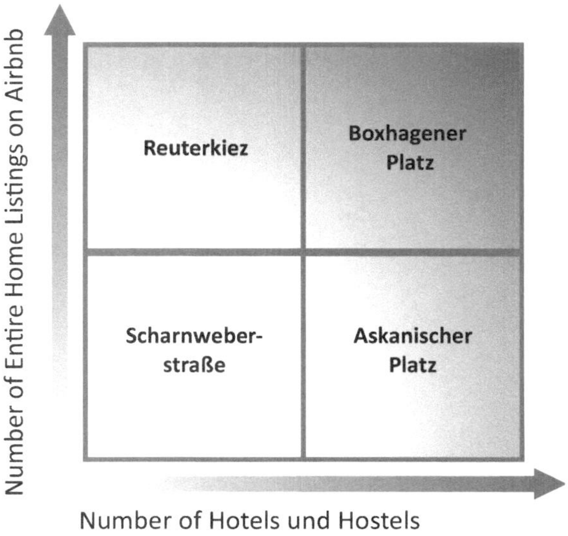

Figure 4.2 Selected LORs according to accommodation density © Claus Müller and Kristin Wellner

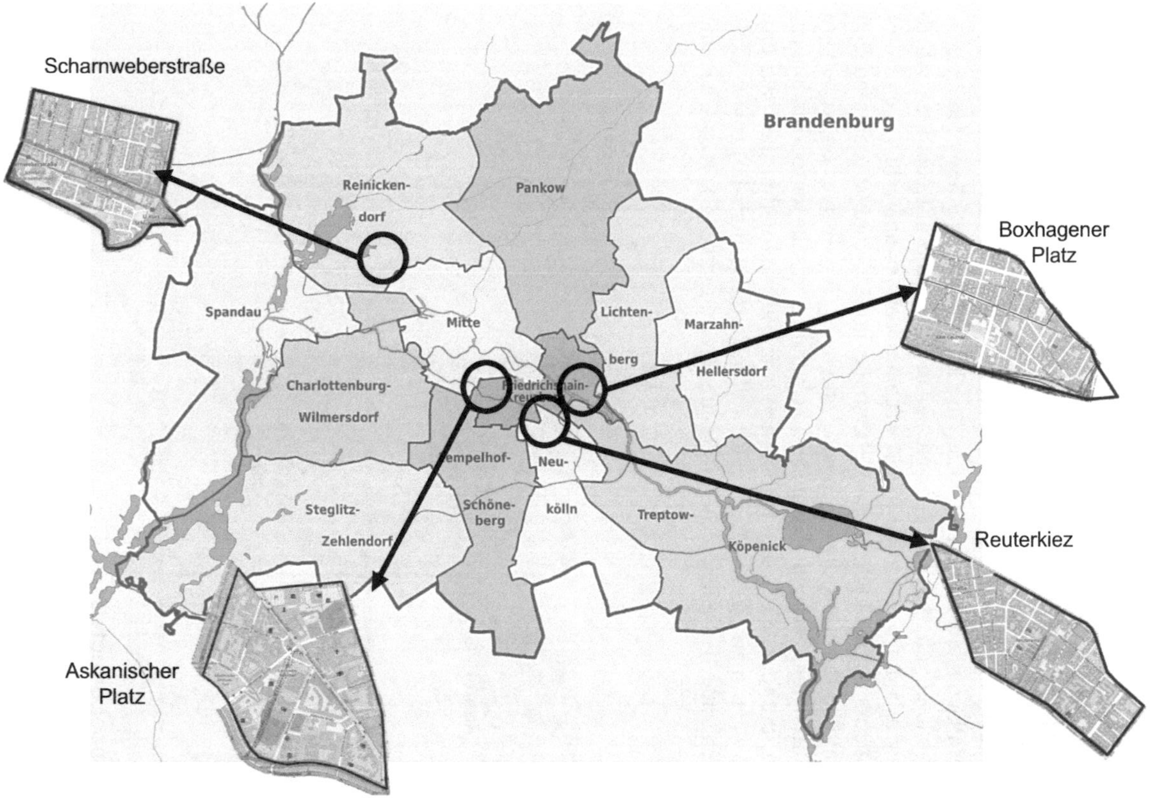

Figure 4.3 Berlin's administrative boroughs and the location of the investigation areas. © Image by the authors based on TUBS on Wikimedia Commons & data from OpenStreetMap and OpenStreetMap Foundation; licensed under CC BY-SA 3.0

necessary to exit this infrastructural borderline in order to find neighbourhoods with a low number of listings on the website, as the inner city is in its entirety heavily affected by Airbnb's proliferation.

The LOR Askanischer Platz is located close to the centre of the city and near the main touristic sights of Potsdamer Platz, Brandenburger Tor and Reichstag. The neighbourhood has 6,481 inhabitants, dwelling in 2,829 units, of which 11.5% are publicly owned. The share of public housing is below Berlin's overall average of 14.9% (Senatsverwaltung für Stadtentwicklung und Wohnen 2017), but since publicly owned properties are predominantly located in peripheral housing estates, this is to be expected for all of the four selected neighbourhoods. Due to its centrality and proximity to some of the main sights of Berlin, Askanischer Platz is a major site for hotels and hostels. The neighbourhood itself offers a number of well-known tourist attractions like the Topography of Terror documentation centre, the Berlin Story Bunker, the exhibition venue Martin-Gropius-Bau and the Tempodrom concert hall. The northern part of the LOR Askanischer Platz, adjacent to the Potsdamer Platz area, is dominated by office buildings, used mainly by government officials of the Federal Ministry of Economic Cooperation and Development and two big newspapers with their editorial offices. The housing stock in the neighbourhood is very mixed with few Wilhelminian-style buildings, student apartments and a range of model houses of the International Architectural Exhibition (IBA) of 1987, but also one large high-rise apartment building. Large plots of waste land were apparent in this area for decades; while most of them are now home to newly built luxury apartments and office developments, some sites are still unused. Tourism is quite noticeable at Askanischer Platz, but since major hotels and hostels provide the predominant accommodation, there is no significant impact on the housing market, especially the asking rents to be expected.

Boxhagener Platz is an East Berlin neighbourhood in close proximity to the river Spree which served as the border between East and West Berlin until the fall of the Berlin Wall. Being located in Friedrichshain, it is part of a traditional working-class neighbourhood consisting of mostly Wilhelminian-style buildings. The housing stock in the LOR consists of 12,255 units, with about 4% public housing. The area was a site of political and social protest, of squatting and alternative lifestyles after the fall of the Berlin Wall, and recently became infamous for gentrification and touristification (Döring and Ulbricht 2016, Glatter and Sturm 2019). It is home to nightlife, bars and restaurants with many hotels, hostels and holiday rentals. In the southern part of LOR Boxhagener Platz is an abandoned Railway Works, which is partly reused as a location for (sub)cultural activities with event spaces, art galleries and techno clubs. Especially the latter led the so-called "RAW-Gelände" to become infamous for its drug scene (Hausdorf and Goller 2015). Since Berlin developed a reputation as a major centre for electronic music, the RAW-Gelände and its surrounding area has become a destination for party-tourism over the last 20 years. Unsurprisingly, this development did not go without conflict: the area is notorious for its high levels of noise and litter pollution as well as a high crime rate. Despite these issues, Boxhagener Platz is also renowned as one of the most gentrified areas in Berlin, with considerable investments in the building stock, rising housing prices and a rapidly changing structure of inhabitants (Glatter and Sturm 2019, Amacher *et al.* 2019). With its alternative vibe and subcultural history, the LOR is in itself a major attraction for New Urban Tourists, so it comes as no surprise that hotels and hostels, but also many offerings on Airbnb, can be found here. In terms of the local housing market this would mean that the already high demand, fuelled by gentrification, is faced with a reduced supply since many apartments are used as tourist accommodation.

Reuterkiez is the northernmost neighbourhood of the West Berlin borough of Neukölln, adjacent to the Landwehrkanal waterway, which also used to serve as the border between East and West Berlin in this area. With 27,792 inhabitants, Reuterkiez is the most populous of our investigation areas. The housing stock in the LOR consists of 15,206 units, of which 3% are public housing. Neukölln has always been a working-class area with mostly Wilhelminian-style buildings, with Reuterkiez being a little bit more upscale compared to other Neukölln neighbourhoods. This did not prevent it from experiencing urban decay from the 1960s onwards. Lack of investment and the competition of large-scale modernist housing estates in the periphery made the dense, inner-city neighbourhoods like Reuterkiez seem out of date, dilapidated and undesirable. The northern, older part of Neukölln became home to many migrant workers, who were discriminated against, not least on the housing market, so they were forced into the dilapidated quarters with run-down houses. Over the decades, Neukölln became a hub for different migrant groups, with a dominant Turkish community for a long time, and an increasing Arabic community in recent years, giving the central street, Sonnenallee, the nickname "Arab Street" (Alkousa 2018). While this concentration of migrants was primarily the product of discrimination and was frequently problematised by politics and the media (Buschkowsky 2013), it also gave Neukölln an international vibe that allured new residents, such as students and expats, and attracted a vibrant night life. So, within the past decade, the north of Neukölln changed its image from a troubled area to a gentrifying hotspot of hipster culture (Holm 2013). Hotels and hostels are still quite rare at Reuterkiez, the few newly established ones seeming to cater directly towards a young and hip audience. Peer-to-peer tourist accommodation is the predominant form of hospitality in Reuterkiez, so from a housing-market perspective, a comparable development to Boxhagener Platz is to be expected.

Scharnweberstraße is the least central of the investigated areas. It is located in the southern part of the borough of Reinickendorf, directly north-east of Tegel Airport which ceased operations in late 2020 and is due to close permanently in 2021. Until then, the level of noise exposure in the neighbourhood was extremely high, especially in the area around Kurt-Schumacher-Platz where this LOR is located, which reduced local quality of life significantly. Since the airport's closure, drastic change in the area is to be expected, not only due to reduced noise exposure, but also with the planned reuse of the nearby airport grounds for scientific and industry purposes, as well as for housing developments. Scharnweberstraße is connected to the inner city via subway and buses, and borders the always up-and-coming Wedding district. The LOR surrounding the eponymous high street,

Scharnweberstraße, has 10,663 inhabitants; the housing stock consists of partly Wilhelminian style, partly post-war buildings on the through streets and residential buildings of the 1920s in the back roads. The south of the neighbourhood between the subway and the autobahn consists of an eclectic mixture of industrial sites, arbours and single-family homes. Altogether, Scharnweberstraße provides 5,685 dwelling units, of which less than 1% are in public ownership. Despite its proximity to Tegel Airport, tourism is barely noticeable, the two hostels located in the LOR are mainly used as accommodation for construction workers, and only very few accommodations are available on Airbnb—we therefore cannot expect any influence of tourism on rental prices or the housing market in general.

Housing market analysis on a neighbourhood level

In order to investigate the influence of Airbnb on the local housing market the examination compares the development in the LORs described above. Data on more than 31,000 online rental offerings between 2009 and 2018 in the selected neighbourhoods were acquired from Immobilienscout24.de, the leading web portal for real estate in Germany. All ads on the website disclose the postal code of the listing, but not every listing discloses the full address, and unfortunately the postal codes do not always concur with LOR-boundaries. Listings positively within the selected research areas (9,321 ads) and listings within the same postal code area as the selected neighbourhoods, but without the exact location (14,248 adverts), had to be distinguished. Whether ads without exact location were included in the analysis or not, did not show significant changes in the results. Ads positively located outside the research areas were excluded from the analysis.

Given the general situation on the Berlin real estate market, rising rents were to be expected in all four investigated LORs (see Figure 4.4). Upon closer inspection, the least increase both in average asking rent, as well as in asking rent per square metre, was found in Scharnweberstraße, at 183% over the period between 2009 and 2018. At Boxhagener Platz, average asking rents per flat increased by 211% and asking rent per square metre increased by 183%, which also suggests that the average apartment size grew. The opposite can be seen in LOR Reuterkiez, where the average asking rents per flat rose by 198%, while asking rents per square metre increased by 212%. The steepest incline in asking rents per square metre is found at Askanischer Platz with 275%, while average asking rents per flat increase by a mere 195%. We find steeper inclines in asking rents in both neighbourhoods, with high numbers of Airbnb listings in the neighbourhood with very few tourist accommodations, but the steepest increase in asking rents is found in the neighbourhood where tourism is dominated by the traditional hospitality industry. While the findings for Boxhagener Platz and Reuterkiez, with an almost parallel, although deferred, development of rental prices were expected, the increase of asking rents at Askanischer Platz cannot be explained by the emergence of Airbnb apartments within this research set-up and the available data set. Explanations for this outcome would need closer inspection; presumptions could be a rising appreciation of centrality within a city in order to save travel costs or the unequal

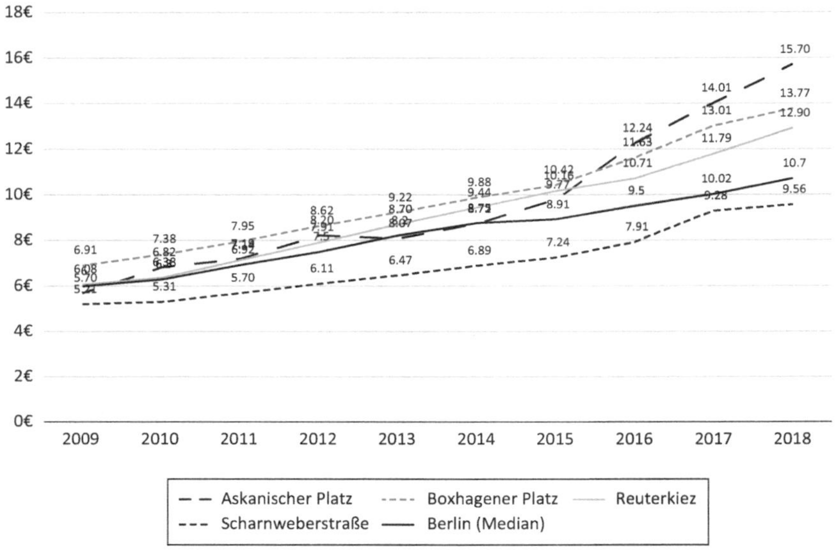

Figure 4.4 Asking rent per m² in the investigated LORs © Claus Müller and Kristin Wellner

distribution of investment into real estate like newly constructed premium condominiums, which can be found at Askanischer Platz as well as at Boxhagener Platz.

Conclusion

Peer-to-peer tourist accommodation is simultaneously an aspect of gentrification, an aspect of an ever-growing tourism industry and a contributing factor to the scarcity of housing space in a growing metropolis like Berlin. While research on New York City reports that in some areas more than 3% of the housing stock was removed from the market in favour of tourist accommodation (Wachsmuth and Weisler 2018), a case study on Vienna (Seidl *et al.* 2018), as well as our own research presented in this chapter, find considerably lower shares. Considering the impact of this reduced supply of long-term rentals on price levels, a case study in the highly visited city of Barcelona (Garcia-López *et al.* 2020) did find that short-term rentals lead to increased housing costs for locals, which is in line with the findings of Duso *et al.* (2020) for Berlin. Our analysis of the Berlin housing market does not provide simple results. Although from a theoretical standpoint, the enduring (mis)use of living space for tourist accommodation is expected to drive up rental prices by the reduction of supply, the empirical study shows that its impact is at best very small and likely outdone by other price-raising factors. Nonetheless, every single apartment on Airbnb is one less apartment for permanent dwellers, which needs to be considered in a growing city like Berlin.

Investment in the housing stock, such as in luxury apartments, and changes in the demand structure, like a higher appreciation of central parts of the city, could explain these outcomes. With our analysis of market data on a local level, our findings are in line with those of Schäfer and Braun (2016), while the results of Duso *et al.* (2020), who use policy changes as natural experiments and find increased asking rents in the vicinity of Airbnb apartments, could not be reproduced here. Our findings support the generally accepted notion that the Berlin housing market is being subject to rapid rent increases. While we cannot pinpoint peer-to-peer tourist accommodation to be the problem, we are wary about it being the solution as portrayed by Airbnb, who eagerly point out that the letting of tourist accommodation does offer additional income that might help to afford the increasing housing costs in central parts of the city.

How the current COVID-19 pandemic affects tourism in the long run, and the demand for housing in city centres and real estate in general, remains to be seen, especially since a new rent control regulation was put in place by the Berlin state government in 2020 which is expected to affect both sides of the market, and makes quantitative analysis almost impossible. Therefore, future research on the Berlin housing market will need to take various, sometimes seemingly unorthodox, factors like tourism or public health into consideration.

Notes

1 *Kiez in der Tourismusfalle? Eine Untersuchung zur Veränderung von Wohnqualität durch touristische Übernachtungsmöglichkeiten in ausgewählten Berliner Wohnquartieren*, funded by the German Research Foundation—DFG (GZ: FR 2522/5-1 & WE 5894/2-1).

2 Due to the massive impact of the COVID-19 pandemic especially on tourism and hospitality, we decided to only include data up to the year 2019 into the analysis, even if newer data was available.

References

AirDNA, 2019. *Berlin* [online]. Available from: https://www.airdna.co/vacation-rental-data/app/de/berlin/berlin/overview [Accessed 28 May 2019].

Alkousa, R., 2018. On Berlin's 'Arab Street,' two worlds collide [online]. *Reuters*, 6 February 2018. Available from: https://www.reuters.com/article/us-germany-migrants-street-insight/on-berlins-arab-street-two-worlds-collide-idUSKCN1IX48I [Accessed 30 November 2020].

Amacher, D., et al., 2019. Wem gehört(e) der Boxi? *Der Tagesspiegel Interaktiv and FH Potsdam*, 21 October. Available from: https://interaktiv.tagesspiegel.de/lab/wem-gehoert-der-boxi/ [Accessed 4 December 2020].

Bauindustrieverband Ost e.V., 2019. *Planungs- und Baukapazitäten im Land Berlin und Umland* [online]. Potsdam: Bauindustrie Ost e.V., 3 April, statement by the construction industry association of East Germany at the parliamentary commission for urban development and housing of the Berlin City Parliament, Available from: https://www.bauindustrie-ost.de/artikel-56/stellungannahme-des-bauindustrieverbandes-ost-e-v.html?file=files/statements/2019-04-03_Anh%C3%B6rung_AGH.pdf [Accessed 16 December 2020].

Becher, A.R., 2016. *Der Prüfarchitekt* [online]. Berlin: Bund Deutscher Architekten, 24 October, [press release]. Available from: https://www.bda-bund.de/wp-content/uploads/2016/11/Der-Pr%C3%BCfarchitekt-Andreas-R.-Becher.pdf [Accessed 16 December 2020].

Brandt, S., Müller, C., and Raschke, A.L., 2019. Nuisance or economic salvation - the role of new urban tourism in today's Berlin. *In*: N. Čamprag and A. Suri, eds. *Three decades of post-socialist transition. Three decades of post-socialist transition*. Darmstadt: TU Prints, 378–383.

Buschkowsky, H., 2013. *Neukölln ist überall*. Berlin: Ullstein.

Campbell, M., et al., 2019. Disrupting the regional housing market. Airbnb in New Zealand. *Regional Studies, Regional Science*, 6 (1), 139–142.

Cócola Gant, A., 2016. Holiday rentals. The new gentrification battlefront. *Sociological Research Online*, 21 (3), 112–120.

Cócola Gant, A. and Gago, A., 2019. Airbnb, buy-to-let investment and tourism-driven displacement. A case study in Lisbon. *Environment and Planning A: Economy and Space*, 21 (3), 1–18.

Döring, C. and Ulbricht, K., 2016. Gentrification-hotspots und Verdränungsprozesse in Berlin. *In:* I. Helbrecht, ed. *Gentrifizierung in Berlin. Verdrängungsprozesse und Bleibestrategien*. Bielefeld: Transcript, 17–44.

Duso, T., et al., 2020. Airbnb and rents. Evidence from Berlin. *DIW Berlin Discussion Paper* No. 1890.

Eurostat, 2019. *Verteilung der Bevölkerung nach Wohnbesitzverhältnissen, Haushaltstyp und Einkommensgruppe - EU-SILC Erhebung*. Eurostat. https://appsso.eurostat.ec.europa.eu/nui/show.do?dataset=ilc_lvho02&lang=de last accessed by the author March 24 2021.

European Cities Marketing, 2019. The European Cities Benchmarking Report. Executive Summary. *15th Offcial Edition 2018–2019*.

Frary, M., 2018. How Berlin became the capital of cool [online]. *The Times*, 14 April. Available from: https://www.thetimes.co.uk/static/how-berlin-became-capital-of-cool/ [Accessed 30 November 2020].

Garcia-López, M.-À., et al., 2020. Do short-term rental platforms affect housing markets? Evidence from Airbnb in Barcelona. *Journal of Urban Economics*, 119 (2020). https://doi.org/10.1016/j.jue.2020.103278

Glatter, J. and Sturm, C., 2019. Lokale Ökonomie der Gentrifizierung – der Wandel des lokalen Gewerbes als Baustein, Effekt und Symbol der Aufwertung von Quartieren. *In*: S. Henn, M. Behling, and S. Schäfer, eds. *Lokale Ökonomie – Konzepte, Quartierskontexte und Interventionen*. Berlin: Springer, 1–15.

Hannemann, C., 1996. Entdifferenzierung als Hypothek - Differenzierung als Aufgabe. Zur Entwicklung der ostdeutschen Großsiedlungen. *In*: H. Häußermann and R. Neef, eds. *Stadtentwicklung in Ostdeutschland*. Opladen: Westdeutscher Verlag, 87–106.

Hausdorf, N. and Goller, A., 2015. *Superstructural Berlin. A superstrictural tourist guide to Berlin for the visitor and the new resident*. Winchester: Zero.

Holm, A., 2013. Berlin's gentrification mainstream. *In*: M. Bernt, B. Grell and A. Holm, eds. *The Berlin reader. A compendium on urban change and activism*. Bielefeld: Transcript (Urban studies), 171–188.

Investitionsbank Berlin, 2018. *Wohnungsmarktbericht 2018*. Berlin: Investitionsbank Berlin.

Investitionsbank Berlin, 2019. *Wohnungsmarktbericht 2019*. Berlin: Investitionsbank Berlin.

James, L., 2018. *The AI that fuels AirDNA* [online]. Available from: https://web.archive.org/web/20191219053201/https://www.airdna.co/blog/short-term-rental-data-methodology [Accessed 30 November 2020].

Krätke, S. and Borst, R., 2000. *Berlin: Metropole zwischen Boom und Krise*. Wiesbaden: VS Verlag für Sozialwissenschaften.

Kritische Geographie Berlin, 2014. Touristification in Berlin. *sub\urban*, 2 (1), 167–179.

Loy, T., 2013. Ärger mit den Kurzzeitnachbarn [online]. *Der Tagesspiegel*, 20 January. Available from: https://www.tagesspiegel.de/berlin/ferienwohnungen-aerger-mit-den-kurzzeitnachbarn-/7660030.html [Accessed 17 December 2020].

Novy, J., 2013. "Berlin does not love you". Notes on Berlin's "tourism controversy" and its discontents. *In*: M. Bernt, B. Grell, and A. Holm, eds. *The Berlin reader. A compendium on urban change and activism*. Bielefeld: Transcript (Urban studies), 223–237.

Novy, J. and Huning, S., 2014. New tourism areas in the "New Berlin". *In*: R. Maitland, ed. *World tourism cities. Developing tourism off the beaten track*. London: Routledge, 87–108.

Schäfer, P. and Braun, N., 2016. Misuse through short-term rentals on the Berlin housing market. *International Journal of Housing Markets and Analysis*, 9 (2), 287–311.

Seidl, R., Kadi, J., and Plank, L., 2018. Tourismus in der digitalen Stadt: Das Geschäft mit Airbnb. *Forum Wohnen und Stadtentwicklung*, 2, 71–74.

Senatsverwaltung für Stadtentwicklung und Wohnen, 2017. *Monitoring Soziale Stadtentwicklung 2017. Regiokontext; Senatsverwaltung für Stadtentwicklung und Wohnen*. Berlin.

Wachsmuth, D. and Weisler, A., 2018. Airbnb and the rent gap: gentrification through the sharing economy. *Environment and Planning A: Economy and Space*, 50 (6), 1147–1170.

Wachsmuth, D., et al., 2018. *The high cost of short-term rentals in New York City, urban politica and governance research group, 30 January*. Montreal: School of Urban Planning: McGill University.

5 Redefining a mature destination as a low-cost neighbourhood

Relations between socio-spatial segregation in Torremolinos and urban tourism in Malaga, Spain[1]

Eduardo Jiménez-Morales, Ingrid C. Vargas-Díaz and Guido Cimadomo

Introduction

Malaga, with a population of 574,654 inhabitants, is a city located on the southern coast of Spain that has become a benchmark for cultural and urban tourism within the space of a few years. The city has experienced a continuous growth in the number of tourists since the beginning of the century, moving up from 700,000 visitors in 2003 to nearly 4,500,000 in 2018 (Observatorio Turístico 2018). This progression in tourist influx has not been coincidental, but the result of almost three decades implementing intense public policies linked to private interests. The priority of these public–private agents was to strengthen the role of the city in the field of culture, as well as to improve the capacity of its transport infrastructures for travellers, to increase and diversify the offer of tourist accommodation and to specialise its historic city centre in the so-called visitor economy (Cocola-Gant 2015).

Additionally, the city of Malaga was the target of urban regeneration policies implemented by the European Union from 1994. After years of neglect, its historic city centre was refurbished and its cultural heritage was enhanced, first through the URBAN Plan and then through the URBACT Programme. The marginalisation found in this urban area was the result of suburbanisation that had depopulated the historic city centre since the 1960s (Marín Cots *et al.* 2017). In those years, public–private investments also shifted from Malaga to the periphery, where the neighbourhood of Torremolinos captured all the attention after having become an international sun-and-beach tourist destination since the mid-1950s. Consequently, Torremolinos had a concentration of up to 90% of Malaga's tourist accommodation (Navarro-Jurado 2000).

However, the oil crises of 1973 and 1979, the loss of competitiveness compared to other coastal tourist destinations and several years of declining municipal investment precipitated Torremolinos' independence in 1988. This way, Malaga lost both the municipal income that sun-and-beach tourism provided, and the tourism activity itself. Therefore, and in order to compensate for the economic decline in the sector, Malaga gained interest in rehabilitating its cultural heritage. On this basis, since the 1990s, municipal policies of culture, heritage and urban intervention have been developed to strengthen the cultural tourism product in the city (Barrera-Fernández 2013).

However, this tourism model has its drawbacks, especially in those cities where the cultural tourism sector has a greater specific weight within the local economy (Ashworth and Page 2011). This economic dependence is visible in Malaga, where public policies respond earlier to the demands of tourism than to those of its citizens. Although in 2018 the economic benefit that tourism brought to the city reached three billion euros, the city centre had already lost 33% of the population that resided there in the 1990s (OMAU 2018). The massive influx of visitors and the tertiarisation of the local economy have aggravated the situation in recent years, forcing many residents to leave the city due to the thrust of tourist gentrification (Colomb and Novy 2016) and to the significant rise in rental prices.

The displaced population, with a precarious socio-economic situation in most cases, is forced to seek a housing alternative in other nearby municipalities where it is more affordable. Torremolinos meets these conditions, especially in the districts that are farthest from the coast. These areas have an extensive stock of small-sized homes with low-quality standards that are turning into low-cost residential neighbourhoods for the citizens of Malaga. As a consequence, the displaced population would be feeding the socio-economic dichotomy that exists between the coastal strip and the inner-city areas of Torremolinos. While the former concentrates the tourist activity, the other areas are disadvantaged due to their lack of attractions. This characteristic is apparent in many traditional tourist destinations in Spain (Antón Clavé 1998).

From the above, two main objectives are proposed to guide this research. On the one hand, to analyse the connections between the social exclusion suffered by the population with medium and low incomes and the harmful effects that cultural and urban tourism causes on housing prices in Malaga. On the other hand, to verify whether the relocation of this excluded population occurs in spatially segregated areas of Torremolinos, with worse conditions of habitability and urban quality. The concepts of exclusion and segregation share a clear spatial connotation in this research. The former distinguishes between the population that is included in the centre of the social structure and those that are located in its periphery (Touraine 1991). Segregation, instead, implies a physical distance between the residential locations of different social groups (Castells 1991).

To address these objectives we use the following data sources. First, the Residential Variation Statistics of the Instituto de Estadística y Cartografía de Andalucía was analysed to dimension the inter-municipal residential migration. Then, in order to know the socio-economic profile of its population, The Migrations and Municipal Register was consulted. Furthermore, the Census of Housing by the Instituto Nacional de Estadística was used to quantify the types of housing and to calculate the degree of conversion from second homes to primary residences with the revised formula of Jesus C. Montosa (2012). By the same token, the Census was used to characterise the housing stock of the municipalities under study and the average price of the rentals was extracted with the data obtained from the online platform for real estate Idealista. Finally, the reports of the Observatorio de Medio Ambiente Urbano were consulted to correlate urban tourism with the increase of rental housing prices in Malaga.

Tourism, migration, speculation and obsolescence

Nowadays Torremolinos is a city that has a total area of 20km^2, a population of 68,661 inhabitants and an urban density of 3,385 inhabitants per km^2. The municipality provides a wide range of hotel accommodation with a total of 23,515 beds (IECA 2020) and it is ranked fourth by number of beds per 1,000 inhabitants in the European Union (Eurostat 2016). These data show the dynamism of the tourism sector in Torremolinos and its prominent position among sun-and-beach destinations in Spain (Rodríguez and Such 2014). However, the origin of the city was quite modest. In fact, it was a small fishing settlement until the 1950s when its generous beaches became one of the main attractions of Malaga. Since then, Malaga and Torremolinos have developed a close spatial relationship heavily influenced by tourism (see Figure 5.1).

At this time, the commitment of the dictatorial government (1939–1975) to the promotion of mass tourism as the main source of income for the country definitively introduced the Spanish tourism sector into the organised travel industry (Lash and Urry 1998). Since 1950 and especially from the opening in 1959 of the Pez Espada, the first luxury hotel on the Malaga coast, Torremolinos consolidated itself as one of the main centres of the tourism industry in Spain due to the continuous growth of the supply and demand of its accommodation. The favourable tourist labour market promoted, in turn, an unprecedented rural exodus towards the coastline that significantly increased the population growth in Malaga, as well as the inhabitants of the seaside municipalities. The provincial capital had an increase of 82% in its population compared to 1950 and it reached a total of 503,251 inhabitants in 1980 (IECA 2020).

Figure 5.1 Geographic location of Torremolinos and Malaga in Spain © Eduardo Jiménez-Morales, Ingrid C. Vargas-Díaz and Guido Cimadomo

This double flow of both tourist and labour migration also had an impact on the construction industry in Malaga. The subsequent and growing demand for housing in a city that lacked urban regulation brought great economic benefits to the construction sector. The repeal of Malaga's General Urban Development Plan in 1964 provided a favourable context for property speculation and, consequently, led to an uncontrolled expansion of the city with two highly polarised growth fronts. On the one hand, there were the working-class neighbourhoods destined for immigrants from the countryside, who came to Malaga looking for job opportunities in the booming tertiary sector. On the other hand, there were hotels, tourist apartments and resorts for national and international tourists, who flocked to Torremolinos attracted by the climate and the low cost of living.

Available data confirms the scale of the urban expansion and transformations in the peripheral areas of Malaga. The population growth in the city between 1960 and 1980 resulted in the construction of 95,000 new dwellings in its urban extensions (OMAU 2015). In the same period of time, the urbanisation rate in Torremolinos district was 36.1 hectares per year, which meant an increase of almost 220% in the consolidated urban fabric from 75.9 to 242.4 hectares (Almeida-García *et al.* 2013). In both scenarios, private investors agreed to promote low-quality urban planning with high-building density rates (more than eight stories high) and a high spatial concentration of the population (more than 150 homes per hectare). These real estate operations were especially lucrative for the construction industry, at the expense of crowding the city.

In addition, the district of Torremolinos grew with the functional and morphological singularities typical of tourist urbanisation (Mullins 1991, Antón Clavé 1998). The popular tourist apartments, together with hotels and recreational areas, monopolised this neighbourhood and specialised it in the seasonal consumption of leisure activities. Its urban growth also took place in a fragmented manner through isolated and autonomous real estate developments, which saturated the available land. A fierce speculation defeated any attempt at urban planning and simplified the role of public initiative to facilitate private activity. In this regard, the local administration transformed the land uses of the agricultural plots with the aim of intensifying tourism activity and ensured the accessibility of the tourist resorts through investments in infrastructure.

Clearly, the interest of public–private agents in making tourist activity in Torremolinos profitable was to the detriment of urban and architectural quality. The tourist apartment block responded perfectly to the needs of tourism as it concentrated a high number of visitors in a few square meters. However, this building typology also promoted overcrowding in tiny flats (with an available area of less than 35m^2) that did not meet basic construction standards, such as thermo-acoustic insulation or elemental infrastructures (see Figure 5.2). Moreover, in its immediate surroundings, the public space and the commercial and social infrastructure were always deficient. The municipal administration did not invest in these facilities since there were too few long-term residents.

With hardly any economic activity other than tourism and few services, the district of Torremolinos became a mono-functional urban fabric of a tourist-residential

Figure 5.2 Tiny homes crammed into high-rise apartment buildings. Los Congresos Neighbourhood (Torremolinos, Spain). General view. © Ingrid C. Vargas-Díaz (2020)

character with few attractions outside the summer period. Furthermore, since the end of 1973 and due to the oil crisis, the international decline of tourist demand showed how vulnerable this type of specialised urban planning was. The tourism sector collapsed and the consequent bankruptcy of the real estate companies, which depended on the tour operators, left the Malaga coastline full of empty hotels and tourist apartments. At the end of the 1980s, once part of the tourist demand had been recovered, the surplus of real estate supply on the coast had already lost its attraction as a second home. The apartment buildings that represented Torremolinos architecture in the 1960s had become obsolete.

Instead of these flats, the middle- and upper-class tourists were looking for another way to inhabit the coast in line with a quality tourist offer and a longer holiday stay (one month or more). Many tourists, mostly Central Europeans, even decided to move their residences to Torremolinos in the search for a better climate, quality of life, leisure and a more satisfying social environment. Due to the impossibility of finding a home in the provincial capital that would respond to the emerging suburban lifestyle, part of the middle-class population in Malaga also participated in this residential mobility towards Torremolinos. These changes in trend also had an impact on the territorial model. The high-density urban growth of the 1960s changed towards a low-density urbanisation based on tourist and residential developments that had little to do with the tourist architecture that had preceded it.

Non-tourist residential use

According to the International Recommendations for Tourist Statistics (Organización Mundial del Turismo 2008), residential mobility originated two different types of residents in Torremolinos: the tourist and the non-tourist. The first type corresponded to residential tourism related to the second-home real estate market. It involved tourists, who rent or buy private housing to stay for different periods of time. The second group was related to the relocation of the primary residence to dwellings that were not intended for tourist but for new residents. In general, these new residents migrated for work reasons (economic migration) or retirement condition (migration with a non-work residential vocation) (Williams and Hall 2000). However, in Torremolinos, many of them were also displaced by the migratory dynamics of the Metropolitan Area of Malaga (Montosa Muñoz 2012).

The statistical data on non-tourist residential use in Torremolinos are conclusive. In 1998 there were 3,387 foreign residents registered in Torremolinos from the European Union. More than 80% of this population was aged 65 years or older and represented 9% of its inhabitants. Although the number of these retirement residents increased to 8,524 in 2008, it decreased by more than 17% ten years later. The booming labour market around tourism in turn attracted a large number of non-EU foreign residents. While in 1998 the number of residents coming from Africa and Latin America was 1,427, in 2008 this quota rose to 6,015 inhabitants in the municipality, which represented 9.53% of its total population. It decreased to 5.8% in 2018 (IECA 2020).

Between 1988 and 2008 a total of 29,810 residents of provincial or national origin were added to this foreign population group. In the same period, the residence relocation of many people from the city of Malaga fuelled an urban exodus to the periphery that reached 121,941 inhabitants (Montosa Muñoz 2012). The causes of this internal residential mobility were diverse, but the main motivation was the demand for a type of suburban housing that was almost non-existent in Malaga. Approximately 16% of this population had as their main destination Torremolinos, where the number of residents grew by 152% due to the juxtaposition of the different migratory flows between 1988 and 2008. This situation had an important impact on the real estate market, which had an urbanisation rate of 22.5 hectares per year until 2007 (Almeida-García *et al.* 2013).

At this point, Torremolinos became a more polarised city, at least in terms of its architecture. On the one hand, there were the real estate developments aimed at residential tourism or to satisfy the demands of new residents. In other words, quality housing for a middle-class population built in low-density urban environments. On the other hand, there were the hotels, tourist apartments and resorts of the 1960s, whose architectural features were the most suitable for a tourism model that had already become obsolete. Many of these properties lost their appeal in the tourism market and therefore increased their devaluation as real estate assets. The most affected were the tiny second homes crammed into the old apartment blocks, especially if they were far from the coast in areas with a lack of services and facilities.

As a result, much of the holiday accommodation that had these conditions gradually acquired a different function to their original tourist use. The devaluation of their purchase or rental prices made these properties affordable for low-income populations such as construction or hotel workers and economic immigrants in general. This situation favoured a change in the function of many of these properties from tourist use to residential use and allowed their consequent integration into the non-tourist housing market. This transformation also promoted a socio-economic segregation of the population in Torremolinos because residents with fewer resources ended up concentrated in the neighbourhoods with this type of housing, where habitability conditions were more deficient.

In order to estimate the conversion degree of second homes to primary residences in Torremolinos, this study takes into account the calculation of the Residential Index (RI) (Montosa Muñoz 2012) between 1988 and 2008 and also uses some data provided by the Census of Housing in the periods closest to the interests of the study (IECA 2020). Initially, we calculated the total volume of new homes in the period 1991–2011 and then estimated which of them were primary residences. The result was that in 2011 the total number of these residences was 10,610. Next, we added this figure to the number of primary residences registered in 1991. As a result, we obtained the estimated number of those residences for the period 1991–2011.

The difference between the primary residences registered in 2011 and the number estimated by the study for the period 1991–2011 showed a positive balance of 8,967 homes. In other words, the RI indicated a balance in favour of the second homes transformed into permanent housing in Torremolinos. The projection of this RI indicator between 2008 and 2018, based on the Housing Census Report for 2021, shows that this conversion was sustained over time. Conversely, in Malaga, the RI for the period 1988–2008 showed a negative balance, which indicated a conversion from primary residences to empty or holiday housing. Specifically, there was a first period, from 1991 to 2001, in which the growth of unoccupied housing was 78.27%. Although this trend was partially offset in the following decade, this transformation continues to rise today.

These results are understandable when they are related to the migratory dynamics that the city of Malaga has experienced in recent decades. At the end of the 1980s, the capital lost its attractiveness as a place of residence, which caused a first exodus towards the periphery that reached a total of 121,941 migrants in 20 years. In the same period (1988–2008), the total internal migration in Torremolinos represented 49,199 migrants, 39% of whom came from the capital. Urban regeneration policies implemented in Malaga since 1994 managed to partially mitigate this depopulation of the capital at the beginning of the 21st century. However, the migratory flow to Torremolinos increased by 24% between 2008 and 2018.

In fact, the number of migrants from the capital in this period was approximately 11,900 and represented 32% of all internal migration to Torremolinos (a total of 37,332 migrants) (IECA 2020).

This data indicates that a sector of the population has continued to leave the provincial capital to reside in the periphery. Taking into account the data from IECA, this population was mostly between 16 and 39 years old, followed by the group under 64 years old. This means that most of them were young people who needed a home but had also been hit by unemployment and job insecurity in the years of the Great Recession (2008–2014). In addition, between 2008 and 2018, approximately 19.15% of the migrants who arrived in Torremolinos from Malaga were foreigners, mainly from developing countries. However, the total number of non-tourist foreign residents (economic and retirement migrants) in Torremolinos decreased in those same ten years by 19%, while the total population of the municipality grew by 8.2%. Consequently, it has been the residential migration of Spanish origin that has contributed the most to the residentialisation of Torremolinos in the last ten years.

Undoubtedly, the unfavourable economic context experienced by the country during the Great Recession favoured this internal migration, especially when families decided to convert their second homes into primary residences for younger relatives who could not afford a home. Despite the economic recovery, residential migration has continued. In fact, between 2014 and 2018, almost 4,600 migrants left the provincial capital to reside in Torremolinos (IECA 2020). Likewise, the municipality kept a similar conversion degree of second homes to primary residences to that which obtained between 2001 and 2011 (18% of the residential stock). This situation has remained since 2014, due to the strong pressure exerted by the real estate market in Malaga with rising prices caused by the boom of cultural and urban tourism.

The housing market and socio-spatial segregation

The evolution of Malaga as a cultural and urban destination has been spectacular in the last five years, mainly as a result of the recovery of the national tourism market but also by the accompanying growth in the international tourist market. With 41,847 beds in 2018, tourist accommodation was quadrupled compared to beds offered in 2013 (Observatorio Turístico de Malaga-Costa del Sol 2018) and Malaga ranked second among national destinations with the greatest increase in the number of tourists. However, the massive influx of visitors (about 5.5 million according to the INE) and the tertiarisation of the local economy have also caused many residents to leave the city due to the thrust of tourist gentrification and the excessive increase in rental prices. Between 2014 and 2018, the historic city centre alone lost 500 inhabitants per year, approximately 11% of its population. This decline also affected other nearby neighbourhoods such as Lagunillas, San Rafael and Trinidad (OMAU 2018).

While Malaga was losing its population, the number of homes for tourism purposes grew by 1,200% between 2014 and 2017 (OMAU 2018), reaching 5,634 tourist homes one year later (Registro de Turismo de Andalucía 2018). In other words, tourists were replacing the resident population and, therefore, tourist uses were replacing residential uses in the city. This phenomenon, that was the opposite of the residentialisation process in Torremolinos, proved to be consistent with the results obtained in the calculation of the RI for Malaga. The interest aroused by short-term rental housing among owners and investors, with a profitability up to four times higher than those of long-term rental (Arias-Sans 2018), favoured the transformation of primary residences into holiday homes. This situation notably reduced the supply of rental housing for the resident population and caused the consequent rise in prices.

The strong demand for residential rentals during the economic recovery may also have led to higher prices, especially in the absence of an affordable property market. Likewise, the incursion of real estate investment trusts and the progressive deregulation of the rental market could also have encouraged a general increase in rental prices. However, the report published by Malaga City Council and the Observatorio de Medio Ambiente Urbano in 2018 found that there was a direct relationship between the tourist housing and the generalised increase in rental prices in Malaga (OMAU 2018). The Bank of Spain estimated this rise at 45.5% between 2013 and 2019 and ranked Malaga in third place on the list of Spanish cities with the highest increase in rental prices (López-Rodríguez and de los Llanos 2019).

Furthermore, in the same year the minimum house price in the provincial capital was €75,000, the highest in Andalusia and only slightly lower than prices in Madrid and Palma de Mallorca (Euroval 2019). However, the average gross income in the municipality was €25,647 in 2017 (33% less than the average income in Madrid), with unemployment figures of 39.59% and temporary employment that reached approximately 91.85% of workers (Observatorio de las Ocupaciones-SEPE 2019). Therefore, the real estate market did not match the socio-economic situation of a large part of the resident population. Since 2014, the low evolution of available household income, the upward trend in rental prices and the reduction in the size of the rental property market have reduced the possibilities for citizens to enjoy decent and affordable housing in Malaga.

Under these unfavourable conditions, part of the local low-income population has been leaving the central areas of the city to live in other nearby locations. Between 2014 and 2018, 39% of the total residential migration from Malaga moved to one of the four municipalities located in the first peri-urban belt of the Metropolitan Area. Torremolinos and Rincon de la Victoria topped the list, receiving 11% and 14% of this population respectively. However, the socio-economic profile of these migrants was different for each city. While in Torremolinos 21% of the population coming from Malaga was of foreign origin (generally from developing countries), this population represented only 5.6% in Rincon de la Victoria. In turn, Torremolinos doubled the number of migrants from Rincon who were between 19 and 39 years old (IECA 2020).

The singular features of the residential stock in Torremolinos are one of the main reasons why a young population and economic migrants have decided to

Table 5.1 Percentage distribution of rental housing by price (€/month) and floor space (m²).

	Rental price (€/month)				*Floor space (m²)*			
	< 400	*400–500*	*501–600*	*> 600*	*< 30*	*30–60*	*61–90*	*> 90*
Torremolinos	**1.15%**	**11.01%**	**15.36%**	72.45%	**1.95%**	**27.36%**	47.29%	23.36 %
Malaga	0.54%	2.82%	7.02%	89.96%	0.30%	13.13%	57.84%	28.71%
Rincon de la Victoria	0.00%	3.80%	10.86%	85.31%	0.00%	13.14%	47.78%	38.63%
Alhaurin de la Torre	0.00%	0.00%	12.96%	87.04%	0.00%	8.93%	37.10%	53.70%
Benalmadena	0.78%	5.50%	8.05%	85.64%	1.92%	20.92%	43.92%	33.21%

© Eduardo Jiménez-Morales, Ingrid C. Vargas-Díaz and Guido Cimadomo, data from real state agency IDEALISTA.com (consultation held in February 2020) and Instituto Nacional de Estadística (INE) – Census of population and housing 2011 (updated 2013)

move to this coastal destination. Due to the residentialisation process that has taken place in Torremolinos since the 1980s, its housing market presents the highest concentration of homes with a substandard floor space compared to the other municipalities. In fact, 30% of its housing stock has less than 60m² of floor space, while in the rest of the towns this type of homes represents only 16%. In contrast, the group of dwellings with more than 90m² constitute only 23% of the housing offered in Torremolinos, while this group represents almost 40% in the other municipalities evaluated (see Table 5.1).

Furthermore, a high percentage of these small houses have significant deficits in habitability and urban quality. Therefore, with a housing stock with these characteristics, Torremolinos has a significant volume of rental housing with a more affordable price than nearby cities. Specifically, 12% of the rental housing available in Torremolinos require a monthly rent of less than €500 and 28% are offered for less than €600 (see Table 5.1). This last percentage is higher than those presented in other cities such as Alhaurin de la Torre (12.96%), Rincon de la Victoria (14.6%) or Benalmadena (14.3%), and considerably higher than the figures for Malaga (10.3%). Therefore, since 2014, Torremolinos has been transformed into a low-cost residential area for those social groups that have economic difficulties in accessing housing in Malaga.

However, the existence of this low-cost residential stock in Torremolinos cannot be generalised to its entire rental housing market. In fact, the spatial distribution of these dwellings in the city has a very specific pattern. Most of them are located next to the city centre, in the areas furthest from the coastline and its promenade. This is the case of the districts of Los Congresos and El Calvario. Both areas were diagnosed as vulnerable neighbourhoods with an average level according to the urban vulnerability basic indicators (studies, unemployment and housing) of 1991, 2001 and 2011 (Observatorio de la Vulnerabilidad Urbana 2011). Consequently, the processes of social exclusion in Malaga due to the socio-economic imbalances caused by cultural and urban tourism are encouraging a socio-spatial segregation in the most disadvantaged neighbourhoods of Torremolinos.

Concluding thoughts

Since the beginning of the 20th century, and especially from 1950, Malaga and Torremolinos have developed a close spatial relationship marked by tourism. In the 1960s, when Torremolinos was an urban quarter of the city of Malaga, the strong growth in tourist demand led to a rapid increase in the supply of accommodation, and the lack of urban planning regulations promoted an uncontrolled expansion of the city. The resulting real estate speculation crowded the coastline with the construction of tourist apartments and hotels while squeezing the few tourist resources available. The gradual decline of the tourism sector due to saturation and oversupply was evident in the following decade. The oil crisis of 1973 broke investment and tourist demand on Torremolinos. This left a model of a mass tourism city predominated by a low architectural quality and tourists overcrowded in small apartments.

By the end of the 1980s, the popular tourist apartment buildings became obsolete in the face of changes in tourism and real estate consumption patterns. The devaluation of these property assets in the tourist market encouraged the subsequent transformation of many of these second homes into primary homes and their consequent integration into the residential stock of Torremolinos. In this way, the municipality currently has a high concentration of small and poor quality housing compared to nearby municipalities. Therefore, the rental prices are significantly lower and housing is much more affordable for social groups with fewer economic resources.

With a low-quality housing stock, part of Torremolinos has become a low-cost neighbourhood for many citizens of Malaga. This is demonstrated by data on residential migration and the common socio-economic profile of migrants who have arrived in the municipality. This population with medium and low incomes suffers from social exclusion due to the pernicious effects of cultural and urban tourism on housing prices in Malaga. The inner city of Torremolinos concentrates the largest percentage of this population, especially in the areas that are furthest from the coastline and its promenade. Therefore, the processes of social exclusion that occur in Malaga are, in turn, promoting the socio-economic polarisation of Torremolinos with a displaced population which is segregated in the most disadvantaged neighbourhoods of the municipality.

Based on these results, it is clear that in recent years, the socio-residential imbalances caused by cultural and urban tourism in Malaga have acquired a supra-municipal dimension. In other words, the effects of tourist gentrification in terms of social and urban cohesion transcend the local sphere and affect other nearby municipalities, such as Torremolinos. The forced displacement of the medium- and low-income population towards the periphery might be encouraging a generalised suburbanisation of poverty in the metropolitan area that would need to be studied in greater depth. Finally, as long as the institutions do not decide to promote public policies for housing (for rent or sale) at municipal and supra-municipal level or do not approve legislative measures to regulate the scale of rental prices, the social and urban segmentation will continue to increase.

Note

1 This chapter presents the results of the research project "Impact of the residential uses in the urban vulnerability diagnosis of Torremolinos" funded by Ayudas B3 (I Plan Propio)-Universidad de Málaga and Andalucia-Tech.

References

Almeida-García, F., Cortés-Macías, R., and Balbuena, A., 2013. Torremolinos, análisis de la transformación urbana y hotelera. *In*: *XXIII Congreso de Geógrafos Españoles Proceedings*, 23–25 October 2013, Palma de Mallorca. Palma de Mallorca: UIB, 505–514.

Antón Clavé, S., 1998. La urbanización turística. De la conquista del viaje a la reestructuración de la ciudad turística. *Documents D'anàlisi Geogràfica*, 32, 17–43.

Arias-Sans, A., 2018. Turismo y gentrificación. Apuntes desde Barcelona. Papers. *Regió Metropolitana de Barcelona. Territori, estratègies, planejament*, 60, 207–211.

Ashworth, G.J. and Page, S.J., 2011. Urban tourism research: recent progress and current paradoxes. *Tourism Management* [online], 32 (1). Available from: doi:10.1016/j.tourman.2010.02.002 [Accessed 21 October 2019].

Barrera-Fernández, D., 2013. Políticas incidentes en la adecuación turística de la ciudad histórica. *AGIR - Revista Interdisciplinar de Ciencias Sociais e Humanas*, 1, 94–120.

Castells, M., 1991. El auge de la Ciudad Dual. Teoría social y tendencias sociales. *Alfoz: Madrid, territorio, economía y sociedad*, 80, 89–103.

Cocola-Gant, A., 2015. Gentrificación y turismo en la ciudad contemporánea. *Turismo y Desarrollo*, 14, 1–7.

Colomb, C. and Novy, J., eds., 2016. *Protest and Resistance in the Tourist City. Contemporary Geographies of Leisure, Tourism and Mobility* [online]. Available from: doi:10.4324/9781315719306 [Accessed 10 March 2019].

Eurostat-European Commission. Statistical Office of the European Union, 2016. *Urban Europe: statistics on cities, towns and suburbs. Tourism and culture in cities* [online]. Available from: https://ec.europa.eu/eurostat/en/web/products-statistical-books/-/KS-01-16-691 [Accessed 10 October 2019].

Euroval, 2019. *Informe de Coyuntura Inmobiliaria (Diciembre 2019)* [online]. Available from: https://euroval.com/informe-de-coyuntura-inmobiliaria-de-euroval/ [Accessed 1 January 2020].

IECA-Instituto de Estadística y Cartografía de Andalucía, 2020. *Censos de población y viviendas 1991, 2001, 2011. Clase de vivienda* [online]. Available from: https://www.juntadeandalucia.es/institutodeestadisticaycartografia/censos/index.htm [Accessed 15 March 2020].

IECA-Instituto de Estadística y Cartografía de Andalucía, 2020. *Estadística de variaciones residenciales en Andalucía from 2008 to 2018* [online]. Available from: https://www.juntadeandalucia.es/institutodeestadisticaycartografia/vares/index.htm [Accessed 23 April 2020].

IECA-Instituto de Estadística y Cartografía de Andalucía, 2020. *Padrón municipal de habitantes 1998, 2008, 2018 and 2019* [online]. Available from: https://www.juntadeandalucia.es/institutodeestadisticaycartografia/padron/index.htm [Accessed 7 March 2020].

IECA-Instituto de Estadística y Cartografía de Andalucía, 2020. *Plazas en establecimientos hoteleros por clase y categoría in 2019* [online]. Available from: https://www.juntadeandalucia.es/institutodeestadisticaycartografia/badea/informe/datosaldia?CodOper=b3_1234&idNode=9516 [Accessed 23 April 2020].

IECA-Instituto de Estadística y Cartografía de Andalucía, 2020. *Torremolinos. Inmigraciones por grupos de edad from 2014 to 2018* [online]. Available from: https://www.juntadeandalucia.es/institutodeestadisticaycartografia/iea/resultadosConsulta.jsp?CodOper=104&codConsulta=84145 [Accessed 30 April 2020].

INE-Instituto Nacional de Estadística, 2020. *Viajeros y pernoctaciones por puntos turísticos in 2019* [online]. Available from: https://www.ine.es/jaxiT3/Datos.htm?t=2078#!tabs-tabla [Accessed 23 January 2020].

Lash, S. and Urry, J., eds., 1998. *Economías de signo y espacio: sobre el capitalismo de la posorganización*. Buenos Aires: Amorrortu.

López-Rodríguez, D. and de los Llanos, M., 2019. Evolución reciente del mercado del alquiler de vivienda en España. *Boletín Económico del Banco de España*, 3, 1–18.

Marín Cots, P., Guevara Plaza, N., and Navarro-Jurado, E., 2017. Renovación urbana y masificación turística en la ciudad antigua: pérdida de población y conflictos sociales. *Ciudad y Territorio: Estudios Territoriales*, 193, 453–468.

Montosa Muñoz, J.C., 2012. Población y urbanización en el área metropolitana de Málaga. *Revista de Estudios Regionales*, 93, 143–173.

Mullins, P., 1991. Tourism urbanization. *International Journal of Urban & Regional Research* [online], 15 (3). Available from: doi:10.1111/j.1468-2427.1991.tb00642.x [Accessed 1 September 2019].

Navarro-Jurado, E., 2000. Turismo y turismo cultural en la ciudad de Málaga. Pasado, presente y futuro. *In*: G. Morales Matos, ed. *IV Coloquio de Geografía Urbana proceedings*, 22–24 June 1998, Las Palmas de Gran Canaria. Las Palmas de Gran Canaria: Asociación de Geógrafos Españoles, 163–169.

Observatorio de las Ocupaciones-SEPE, 2019. *Informe del mercado de trabajo en Málaga, 2019. Datos 2018* [online]. Available from: https://www.sepe.es/HomeSepe/que-es-el-sepe/comunicacion-institucional/publicaciones/publicaciones-oficiales/listado-pub-mercado-trabajo/informe-mercadotrabajo-provincial.html [Accessed 21 January 2020].

Observatorio de la Vulnerabilidad Urbana de España, 2011. *Atlas de la vulnerabilidad urbana 2001 y 2011* [online]. Available from: http://atlasvulnerabilidadurbana.fomento.es [Accessed 7 June 2019].

Observatorio Turístico de la Ciudad de Málaga, 2018. *Informe Noviembre 2017–Octubre 2018* [online]. Available from: http://www.malagaturismo.com/es/paginas/informes/362 [Accessed 21 December 2019].

Observatorio Turístico de Málaga-Costa del Sol, 2018. *Principales cifras 2018* [online]. Available from: http://static.malaga.es/malaga/subidas/archivos/0/5/arc_318150.pdf [Accessed 21 December 2019].

OMAU-Observatorio de Medio Ambiente Urbano, 2015. *Territorio y la configuración de la ciudad* [online]. Available from: http://www.omau-malaga.com/agendaurbana/subidas/archivos/arc_219.pdf [Accessed 24 May 2018].

OMAU-Observatorio de Medio Ambiente Urbano, 2018. *Aproximación a intensidades del uso turístico en Málaga* [online] Available from: http://static.omau-malaga.com/omau/subidas/archivos/2/7 /arc_7972.pdf [Accesed 2 June 2019].

Organización Mundial del Turismo, 2008. *Recomendaciones Internacionales para las Estadísticas del Turismo. Asuntos económicos y sociales* [online]. Available from: https ://unstats.un.org/unsd/publication/Seriesm/SeriesM_83rev1s.pdf [Accessed 3 February 2020].

Registro de Turismo de Andalucía, 2018. *Viviendas con fines turísticos* [online]. Available from: https://www.juntadeandalucia.es/organismos/turismoregeneracionjusticiayadministracionlocal/areas/turismo/registro-turismo/paginas/viviendas-turisticas.html [Accessed 23 January 2020].

Rodríguez, I. and Such, M.P., 2014. La política de apoyo a la renovación y reestructuración de destinos turísticos maduros: una evaluación retrospectiva a partir del estudio de casos. *Architecture, City and Environment* [online], 9 (25). Available from: doi:10.5821/ace.9.25.3635 [Accessed 12 December 2019].

Touraine, A., 1991. Face à l'exclusion. *Revue Esprit*, 169, 7–13.

Williams, A. and Hall, C.M., 2000. Tourism and migration: new relationships between production and consumption. *Tourism Geographies* [online], 2 (1). Available from: doi:10.1080/146166800363420 [Accessed 2 March 2020].

6 Tourism in a peripheral territory in the Lisbon Metropolitan Area

The case of Almada

Madalena Corte-Real, Marianna Monte, Maria João Gomes and Luís Manata e Silva

Introduction

This chapter is based on an analysis commissioned by the municipality of Almada, a city in the Lisbon Metropolitan Area, on local and visitor perceptions and descriptions of the territory and how they underpin tourism development strategies. Almada is a former industrial city and marked by a rather disorganised suburbanisation process due to the lack of urban planning. The area under analysis consists of two parishes, Cacilhas and Almada Velha, located on the Tagus River on the bank opposite Lisbon's old city centre.

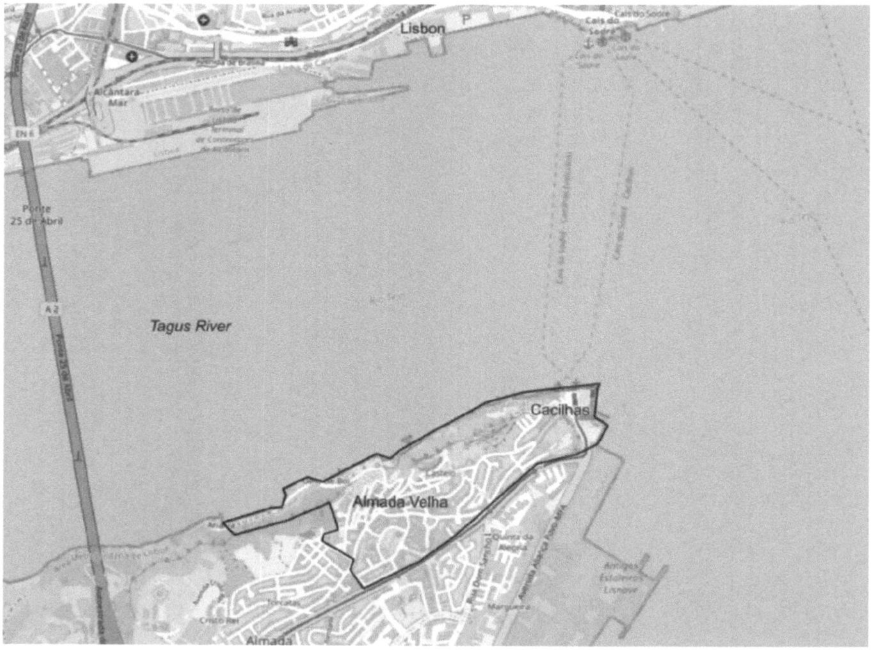

Figure 6.1 Location of the territory under analysis © OpenStreetMap

Since the financial crisis of 2008, Lisbon has seen a significant increase in tourism, strongly promoted by the government and local authorities. This has a huge socio-spatial impact on the older part of the city that many consider excessive due to touristification and gentrification processes. Following the trend of New Urban Tourism and of visiting territories off the beaten track, (small) businesses and the local authority see an opportunity to attract visitors to the old area of Almada.

This chapter, therefore, aims to present the visitors' characteristics, their perception of the territory and to apprehend what attracts them to this area, which is not considered an obvious tourist site. The second goal is to capture the local users' understanding of the territory and their expectations for the impact of tourism. After this introduction, a theoretical framework is presented, followed by a description of the recent evolution of the Lisbon Metropolitan Area, particularly the city of Lisbon, and a characterisation of the area that serves as the case study. The methodology is then explained before setting out the research results. The final section provides a discussion of the outcomes.

Tourism as a development tool

Urban environments are among the most important of all tourist destinations (Edwards *et al*. 2008, Ashworth and Page 2011). In the light of ever more complex, diversified, segmented or even individualised offers in the metropolitan context, ordinary places or socially marginalised areas have also been conquered for tourists who wish to get away from mass tourism (Gravari-Barbas 2017).

In the cities of consumption, the appropriation of territories for recreational purposes has been playing an increasing role (Hannigan 1998, Zukin 1998, Featherstone 1991, Clark 2004, Miles and Miles 2004). In this context, tourism—i.e. consumption and leisure—is considered an important economic booster for the revitalisation of cities (Gravari-Barbas and Guinand 2017). Pollice and Julio (2011) in particular focus on tourism in a weak and marginal post-industrial economy that aims for it to be a strategic sector for development and growth that is fuelled by public and private resources and determines territorial planning.

To make cities more attractive, old port areas become waterfronts of contemplation and recreational enjoyment (Sieber 1993), former industrial facilities are converted into cultural and creative spaces (Mommaas 2004, 2010, Xie 2015), marginalised neighbourhoods in central areas are appropriated and their assets reinvented for consumption purposes (Leite 2002, Hoffman 2003, Zukin 2010, Novy 2018); even subversive urban art like graffiti has been legitimised and commodified (Thörn 2008), now used in placemaking, branding and destination strategies (Evans 2016).

Many tourists today deliberately look for diverse offerings, atmospheres and experiences of cityscapes away from those provided by the well-known touristic sites commodified and promoted for tourist consumption; they seek to discover the city beyond emblematic monuments and experience the 'authentic' city, sharing the everyday fabric with locals (Maitland 2008, 2013, Novy 2019). Robinson (2015) also refers to urban exploration as the attraction of derelict places that are beyond the usual frames of reference for tourists.

Neighbourhoods close to tourist centres, as well as ethnic enclaves and working-class areas with old buildings and obsolete industrial facilities, are part of the universe of peripheral territories that provide new, non-guided forms of exploring cities (Maitland 2013). This includes the Airbnb phenomenon, an online platform where ordinary people rent out their spaces as accommodation for visitors (Guttentag 2013), which illustrates the desire of tourists and, in general, transnational classes, for alternative commercial accommodation (Gravari-Barbas and Guinand 2017).

The tourists are, hence, attracted to, and end up being part of, gentrification processes in these territories. Transformations in retail offers take place with the opening of bars, cultural amenities and shops targeted at new residents and visitors as well as the increase of dwellings transformed into short-term rental at the expense of shops for everyday needs and of affordable permanent housing (Füller and Michel 2014, Mermet 2017, Novy 2018, Cocola-Gant *et al.* 2020).

The options available to travellers are very broad and accessible and their choices can be more informed due to the wide variety of places made popular by media that plays a central role in the spread of symbolic representations and anticipations (Urry 1995, p. 132). Robinson (2015) underlines the importance of photography in visual consumption under the development of urban exploration and tourism. Social media, in particular, is having an increasing impact on the reputation of destinations (Iglesias-Sánchez *et al.* 2019) and the way an area is perceived and constructed (Jansson 2018). These internet-based applications have been widely adopted for researching, planning and sharing stories and experiences allowing users to make and share comments and opinions as well as rate products and services (Thevenot 2007, Xiang and Gretzel 2010, Leung *et al.* 2013, Gretzel 2018). Yoo and Gretzel (2011) and Litvin *et al.* (2008) refer to these sources as electronic word-of-mouth that provide information on a large scale, influencing the consumer decision-making process. They are relevant information sources to apprehend consumer preferences (Dellarocas 2003) and also important forms of interaction in the travel industry to communicate with consumers (Harrigan *et al.* 2017).

In this sense, considering space redefinition and current concerns regarding gentrification processes, different consumption patterns as well as representations outside and inside the territory, a multidimensional approach is used in this case study to understand the meaning of such gentrification. Following Lefebvre's (2000) analysis of the production of space, it considers how the area is perceived, conceived and lived.

The territory in the context of the Lisbon Metropolitan Area[1]

The development of tourism in Lisbon

The city of Almada and the parishes under analysis must be contextualised in the dynamic of Portugal's capital, Lisbon, where tourism and leisure have not only become drivers of urban change, particularly in recent years, but also an important source of revenue (Corte-Real and Monte 2018). In 2018, tourism accounted for 20% of the GDP in the metropolitan area (Deloitte 2019) as it is currently the region with the largest number of tourists at the national level, surpassing the south, known mainly as a sun-and-beach destination (INE 2019). The number of guests grew by 7.6% in a year, reaching 10.4 million guests in 2018. The city of Lisbon, with half a million residents, has a significant weight representing approximately 70% of the total number of guests (Deloitte 2019).

Lisbon and its surroundings, with 2.8 million inhabitants in the metropolitan area, have been marked by decades of urban sprawl with the construction of new housing. While the city lost around a third of its inhabitants between the census of 1981 and 2011, mainly from the historic centre where buildings were in very poor condition, the population of the surrounding municipalities increased by nearly 36%.

Following the 2008 financial crisis, the national government pursued economic recovery, seeking to attract international investment with tax benefits and offering resident permits (Gold Visa Programme) for people outside the EU. On the part of the local authorities, following narratives of urban marketing and neoliberal rationality, the aim was to make Lisbon a more competitive city, calling for visitors and tying up capital flows (Mendes 2017, Barata-Salgueiro *et al.* 2017, Lestegás *et al.* 2018, Sequera and Nofre 2019). Tourism and urban regeneration go hand in hand (Santos 2019, Sequera and Nofre 2019) and by requalifying the waterfront, squares, parks and markets, reducing the circulation of cars in the city, placing elevators in steep areas, increasing pedestrian areas and cycle paths and constructing a new cruise terminal, the intention is to make the city more appealing. The exponential growth of tourism has had a profound impact on the reconfiguration of the urban landscape and on real estate, where many apartments and buildings were directed to short-term accommodation and hotels as well as outlets targeted towards visitors. Old central working-class neighbourhoods are becoming particularly popular among visitors and new residents, contributing to gentrification processes by expelling former residents as a consequence of apartments converted for short-term rental and by reconfiguring local commerce (see Figure 6.2) (Mendes 2017, Corte-Real and Monte 2018, Sequera and Nofre 2019).

The attractiveness of Lisbon has contributed to a significant increase in prices in the real estate market and the demand for accommodation has also driven up prices in neighbouring municipalities, especially those with good accessibility to Lisbon, as is the case of Almada (see Figure 6.3).

Following a great increase in tourism in the historical part of the city, the new strategic plan for tourism presented by the Lisbon Metropolitan Area for 2020–2024 highlights the importance of giving greater prominence to the Tagus River and the better integration of several products (Roland Berger Strategy Consultants 2019). The marketing should focus on a growing promotion of digital channels, addressing the segments of the individual visitor, families and small groups who, under this plan, are best suited to integrating harmoniously with residents, assuming it reduces possible negative impacts.

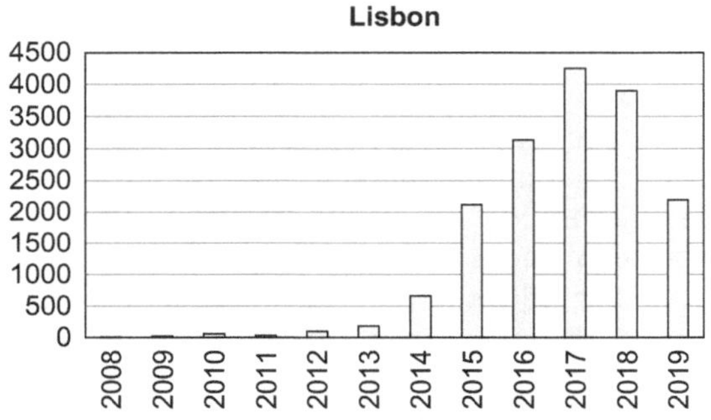

Figure 6.2 Number of new registered short-term rental accommodation (apartments and hostels) in the city of Lisbon (graphic based on data from RNT—Registo Nacional de Turismo [National Tourism Registry])

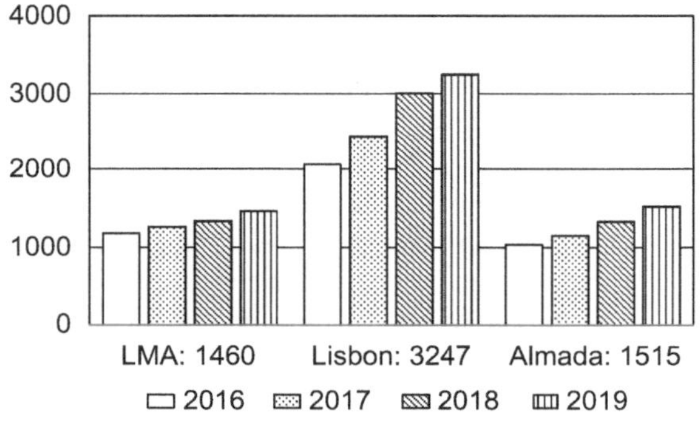

Figure 6.3 Median sales value per m² of family housing (€/m²) in the 4th quarter of 2016–2019 in the Lisbon Metropolitan Area (LMA), Lisbon and Almada (graphic based on data from INE—Instituto Nacional de Estatística)

Characterisation of the territory

The territory under analysis is the oldest part of the city of Almada facing north to the estuary and just a ten-minute ferry journey from the capital's old town (see Figure 6.4. In the parish nearby is the important *Cristo Rei* monument (see Figure 6.6), attracting one million visitors per year (according to the monument's representative in the interview) (Interview 9). The monument is not easily accessible by foot from this area due to its steep topography, but is accessible within a short distance by bus.

The analysed territory has 22,600 inhabitants according to the last census (INE 2012), representing about 13% of the municipality's total population (see Figure 6.5).

Figure 6.4 Ferry from Almada (Cacilhas) to Lisbon. © Authors (2018)

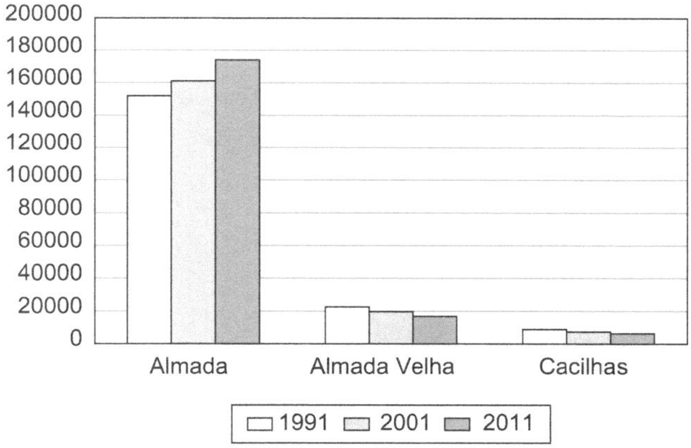

Figure 6.5 Demographic evolution in the last three census in Almada and the two parishes under analysis (Almada Velha and Cacilhas) (graphic based on data from INE)

It has a multimodal transport hub, administrative services of the local government, cultural facilities and it is known for its seafood gastronomy. The area lost some of its vitality after several functions were transferred to newly developed areas, notably the shopping mall, aspects that led to some physical decay and reduction in the number of inhabitants.

The territory had a strong relationship with the river and the sea throughout the 20th century thanks to fishing, maritime trade and the shipping industry, making it a predominantly working-class area. The inauguration of the bridge crossing

Figure 6.6 View of the territory from the ferry boat with the bridge to Almada and the *Cristo Rei* monument in the distance. © Authors (2018)

the Tagus linking Lisbon to Almada in the 1960s fostered a significant population increase in this municipality and a strong urbanisation process.

Nowadays, the abandoned industrial facilities and warehouses, essentially on the riverbank facing Lisbon, are ruins decorated with urban art. This waterfront has two restaurants with terraces and during this study a building was being used by a collective of people working in the cultural sector (see Figure 6.6).

Figure 6.7 Pedestrian street with terraces. © Authors (2019)

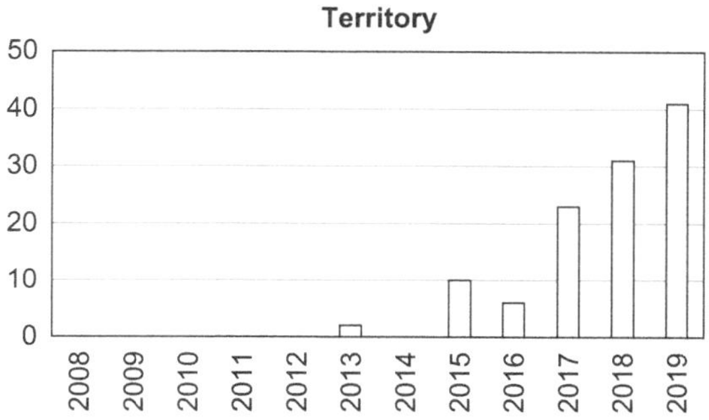

Figure 6.8 Number of new registered short-term rental accommodation (apartments and hostels) in the territory (graphic based on data from RNT).

In the parishes under analysis 113 short-term rental facilities opened starting in 2013 (as can be seen in Figure 6.8), including apartments and hostels, offering accommodation for around 700 people. A main commercial street was pedestrianised and, apart from traditional seafood restaurants, new outlets including a tattoo studio, a vinyl records shop, a bar-bookstore and a market have opened, offering a multicultural environment of gastronomy and entertainment (see Figure 6.7).

The methodology

This study adopts a multi-method approach to understand the perception regarding the territory, namely as a tourist and leisure destination by visitors as well as experts and people who use this area on a regular basis.

Research in national[2] and international[3] newspapers and magazines as well as travel and leisure guides and social media[4] was undertaken to capture the image of the territory projected from the beginning of 2014 until the end of 2018. The online research and content analysis were carried out using the keywords 'Almada Velha' and 'Cacilhas'.

Two focus groups were conducted, one with municipality officials (linked to the departments of economic development, culture, communication, urbanism, tourism, education and environment) and the other with representatives of commerce, associations and the hospitality sector to assess the meaning of the territory, the touristic potential and the impact it can bring. Following this goal, semi-structured in-depth interviews were also conducted with different actors with a very close connection to the territory under analysis, namely from the hotel industry and commerce as well as from cultural activities.

Finally, two street interception surveys were administered, one to 411 users and the other to 287 visitors, both aged over 18 years. We opted for a direct personal approach in which the questionnaire is presented and completed by the interviewer.

As part of the survey of visitors, tourists were considered (who stay more than 24 hours in the visited territory) and day-trippers (who stay less than 24 hours) living outside Almada, namely in other municipalities in the Lisbon Metropolitan Area, those living in other parts of the country and international visitors.

The questionnaires were carried out for ten days in the Easter season (with public and school holidays), on a national (25 April) and an international holiday (1 May). They were conducted in public space, next to commercial areas and public transport terminals as well as parking areas; it was attempted to approach visitors at the end of their stay so that they were able to provide an evaluation. The application to foreigners also required the questionnaire to be translated into several languages, namely English, French and Spanish. Both surveys had as an open and optional question the possibility for respondents to give some suggestions about what could be improved to make Cacilhas and Almada Velha more attractive.

Data analysis

Narratives in the press and social media

The information visitors can obtain on the territory before their visit affects the decision where to go, what to expect and how it can be experienced. In view of the area under analysis, there is not much information available in the municipality or in the tourism offices in Lisbon and none when arriving by ferry or bus at the terminal in Cacilhas.

With regard to mobility, a Portuguese magazine notes that Lisbon and Almada want to strengthen the 'complementarity' of the two municipalities, both to better serve those who live in these cities and for the benefit of the tourist boom in the country. Another article highlights the rising rents in Lisbon and the number of people going to live outside Lisbon, citing, among others, the President of the Lisbon Tenants Association who moved to Almada (Lousada Oliveira 2018).

One magazine has a column in which well-known personalities share their favourite spots. The south bank of the Tagus River seems to be one of the preferences of some protagonists linked to the world of culture and arts. One of them recommends taking the ferry and going to eat tasty delicacies at the different restaurants there that offer the best views of Lisbon and points out that there is no need to take a car to get there (Souto Cunha 2016).

The *Time Out Lisbon* magazine made 70 references to the territory under analysis in approximately two years of online publication (June 2016 to October 2018) where gastronomy is an important topic. Given the physical proximity to Lisbon, the offer of this territory is often associated with the capital in articles that present small collections of topics such as: best tattoo studios (Time Out Lisboa 2018), best happy hours (Dias Real 2020), things to do in Lisbon's museums (Ramos Silva 2018) and programmes for New Year's Eve (Time Out Lisboa 2017). It is also presented as a pleasant journey and an affordable place.

Exploratory research in the international press identified 16 references (in *The Guardian, El País Brasil, Le Monde, O Globo*) to the territory on the topics of tourism, gastronomy and city breaks. According to *The Guardian* (Choat 2014), the best of Lisbon includes a ferry trip that transports the traveller on a quick crossing to a village environment, old buildings, a wide offer of gastronomy, terraces and breath-taking views over the seven hills of Lisbon.

A look at various travel guides reveals that there is almost no information about the territory. In contrast, 59 posts were identified in 55 travel blogs (written by Portuguese, Brazilian, English, French, Spanish, German and Italian bloggers), demonstrating that it is gaining increasing interest. The top three attractions presented in the blog descriptions are again the ferry trip, the views and the gastronomy. It was confirmed that all travellers analysed through their social media come to Almada on their stay in Lisbon. The territory is often perceived as an integral part of Lisbon since in many cities the river is not an administrative separation. For foreign visitors who organise their own itinerary, going to Almada is like an expedition, a curiosity to explore the other bank. It is considered an alternative territory, not very touristic. On one site, it is called 'indie'; it is part of the 'unusual Lisbon' recommendations for places to feel like a local (Borghi 2020), and it also appears on a blog that reveals lost places in Lisbon and, in this context, an old property in ruins by the river. The territory under analysis thus stands out as a non-tourist environment, waiting to be explored. Information gives priority to experiencing the atmosphere rather than visiting museums or monuments.

The ferry trip is always presented as a worthwhile experience. A Spanish blogger says that Almada is a small parish that keeps the secret of having the best views of Lisbon. It offers a complete panoramic view, unlike Lisbon's viewpoints that just have partial views (Montes and Casillas 2015). Another Spanish blog refers to a culinary getaway, where a restaurant is recommended, revealing that the best seafood in the world can be eaten facing Lisbon (KrrteAndo 2016).

The proximity and the speed with which people reach the 'other side' is always highlighted. Apart from an excursion and a car trip around Portugal, these visitors travelled by public transport. Blogs refer to the fascination of walking along the river and the industrial environment with rundown warehouses marked with urban art from which there are dazzling views. Here, the two restaurants with terraces are particularly attractive.

Interest in the landscape and alternative sites can also be observed on Instagram and Twitter. Images reveal that users of these social media enjoy sharing their photographs of the territory. The Tagus and urban views, as well as the street art and the ruins, are the most photographed sites.

Visitors' experience in the territory according to the survey

A significant part of the respondents to the survey (66%) were international tourists from 26 nationalities, mainly European, especially from Spain. Visitors from the Lisbon Metropolitan Area made up 27% of the total and just 7% were from other areas of Portugal. Visitors aged between 18 and 30 made up 37% of the total, 37% were between 31 and 43, with 23% between 44 and 66 and just 3% over 67 years old. Almost all visitors were travelling independently (94%).

Most of the tourists from outside the Lisbon Metropolitan Area (83%) included Lisbon city on their itinerary. Friends and relatives (36%) as well as the internet (34%) were the two most common information sources indicated in the survey. First-time visitors to the territory made up 66 % of all respondents; 18% of the visitors stayed overnight in Almada, and just 11% in the territory. Nearly a third came because of the visit to the *Cristo Rei* monument nearby. Reasons given to explore this area were to "stroll around" and the gastronomy, as well as to get away from mass tourism. Visitors expressed feeling safe and the main criticisms were the poor maintenance of the public space, the run-down houses and the lack of information on what to do in the area.

Internal perception of the territory

Perceived characteristics and potentialities[5]

The focus groups identified the territory as having ageing and Portuguese long-time residents, but they also noted an increasing cosmopolitan atmosphere and a growing number of newcomers. In terms of the tourist potential of this territory, the focus group participants pointed out the panoramic view over Lisbon, the intense cultural activity as well as the sites of cultural and historical interest. The positive points mentioned were the fact that the territory is not overcrowded and that it is safe and genuine.

In the individual interviews it was also stated that there have been some social changes in the population of the territory, which is now more diverse in terms of nationality and age group. The territory is being transformed by both the new businesses opening up and the new residents coming from Lisbon, where they cannot afford rent prices, as well as from abroad, impacting on the real estate market: "I know a couple who were in a rented house, facing the river [...] with a fabulous view and a reasonable rent. The landlord told them 'you have to leave because the building will be completely renovated to be rented to foreigners'" (Interview 12), declared an interviewee.

According to some interviewees, whereas Lisbon is losing its roots, particularly in the historic neighbourhoods, Almada still maintains its essence and the quietness that attracts foreigners to buy and renovate houses to live there. Foreigners "are finally discovering that it is not just Lisbon. They found out that crossing by ferry, in ten minutes, they are here in a calmer area" (Interview 7).

As one interviewee (Interview 8) said, this axis does not offer an obvious consumer product, but it does have potentialities that should be valued like the landscape and the beautiful views to the city of Lisbon and the river; the industrial heritage, including buildings and memories linked to the working-class environment (including social movements, collectives), but also to culture (the area has a strong tradition linked to the theatre, some famous Portuguese rock bands emerged here) and the way in which the occupation of space developed from farms, warehouses and river traffic to an industrial area that is now obsolete.

There is an agreement that the proximity to Lisbon and rapid accessibility by river offers significant growth potential for tourism, in contrast to Lisbon, which is already reaching saturation levels in the circulation of tourists in some of the historic areas.

Overall, it is recognised as a complementary destination to Lisbon based on the creation of offers on well-defined attraction spots. As an interviewee pointed out:

> I think that Almada actually has a lot of potential from the point of view of tourism, but not for mass tourism. It has a lot to do with the landscape, reading the history of the territory in the landscape. It is not something for immediate consumption but requires some reflection, some follow-up.

Expected consequences of tourism

According to the survey of the population that uses this area daily, the majority consider the impacts of tourism to be positive: 80% highlight economic development, 34% refer to the rehabilitation of buildings, 28% expect the upgrading of public space. About a quarter of the respondents (27%) raise concerns about tourism, considering that it will raise real estate prices and prices generally.

The interviewees have also a generally positive view of tourism as an economic driver and promoter of the revitalisation of the area. When asked about the future, these respondents expressed a desire to improve the public space and rehabilitate buildings, avoiding large investors and megalomaniac interventions. "I think the risk is to kill the beauty of it with things that are too big or that cause a mass movement that spoils it" (Interview 2).

Participants in the focus group state that the growth of tourism will bring investment to the territory, contributing to urban regeneration, better organisation and qualification of public space, improved accessibility, the rejuvenation of the population, the emergence of new businesses and the renewal of local commerce. They cautioned that this growth could also lead to a mischaracterisation of the territory and to gentrification processes and, therefore, noted the need for the local municipality to mitigate the negative consequences.

Final remarks

This chapter aimed to present the practice of the territory considering the way it is read both internally and externally. The territory under analysis is two parishes in Almada, a municipality in the Lisbon Metropolitan Area, located in the oldest part, opposite the historical centre of Lisbon.

Being geographically a suburb of Lisbon, this territory had its own dynamics and some autonomy in terms of labour supply, especially linked to the shipbuilding industry, which contributed to development in associative and cultural terms.

These aspects give the area its particularity that is, nevertheless, much seen in the relationship with the capital, given its physical situation, and even as an extension, being strongly influenced by its evolution, currently mainly in terms of demand in real estate and tourism growth.

In Portugal in recent years, tourism has been seen as a panacea for economic leverage, both on the national level as well as in Lisbon. In Almada, marked by abandoned industrial facilities and warehouses as well as some decay in the urban fabric and a depopulation process, tourism seems to be an interesting solution to revitalise the area. Regeneration processes have included the intervention in a commercial artery that was pedestrianised and has numerous traditional restaurants offering seafood. New consumption spaces began to emerge and a market was redefined with an international gastronomic offer. According to the visitors coming to this area, it allows one to be among locals and to eat in the same places as the Portuguese do at affordable prices.

Within the scope of visitors, who travel independently and like to explore urban environments in less touristic areas, word-of-mouth, namely as a broader concept linked to the information provided by other consumers on the internet, is considered a central information source. Walking around the streets and alleys, enjoying urban art and old warehouses, tasting the gastronomy, gazing at beautiful views, crossing the river, getting away from overcrowded tourist areas, being among locals and sitting in terraces and relaxing are aspects that appear to be particularly attractive. There seems to be an interest in territories under a gentrification process, but not yet too commodified. An area relatively well looked after is considered important: issues such as urban hygiene (without litter and graffiti tags), a feeling of safety, and the improvement of public space.

The territory is clearly changing in terms of population dynamics with the presence of new inhabitants alongside the Portuguese long-time residents, first low-income immigrants, now middle-class Portuguese, as well as Brazilians and Europeans.

Regarding how the locals position themselves concerning the development of tourism in the area, it should be noted that generally speaking, the bet is seen as positive, not only by representatives linked to this activity but also by residents, and is recognised as a mechanism for potential economic development, population rejuvenation, building rehabilitation and urban redevelopment. However, there is also a concern regarding possible negative impacts such as the phenomena of gentrification, namely the increase in rent levels, as well as the loss of "authenticity", and in this sense the need for the municipality to mitigate these effects is recognised. In this regard, the importance of preserving what exists and supporting local initiatives was also highlighted.

Notes

1 The most populated area in Portugal, with 18 municipalities that are divided into different parishes.
2 Four national newspapers: *Correio da Manhã*, *Público*, *Diário de Notícias*, *Expresso*; two generalist magazines (*Visão* and *Sábado*) as well as the culture, entertainment and events guide *Time Out Lisbon*.

3 In *The Guardian, El País Brasil, Le Monde, The New York Times, Der Spiegel, Die Zeit, El Mundo, Le Figaro, Corriere della Sera.*
4 For the present articles, mainly blogs will be considered.
5 Based on the focus groups, interviews and the survey of the population who use this area daily.

References

Ashworth, G. and Page, S.J., 2011. Urban tourism research: recent progress and current paradoxes. *Tourism Management*, 32 (1), 1–15.

Barata-Salgueiro, T., Mendes, L., and Guimarães, P., 2017. Tourism and urban changes: lessons from Lisbon. *In*: M. Gravari-Barbas and S. Guinand, eds. *Tourism and gentrification in contemporary metropolises–international perspectives*. Oxon: Routledge, 255–275.

Borghi, V., 2020. Lisbona insolita: 15 luoghi da vedere per sentirsi local. *beBorghi* [online], 9 January. Available from: https://www.beborghi.com/lisbona-insolita-10-luoghi-da-vedere/ [Accessed 15 November 2018].

Choat, I., 2014. 10 of the best restaurants and cafes in Lisbon. *The Guardian* [online], 26 June. Available from: https://www.theguardian.com/travel/2014/jun/26/10-of-the-best-restaurants-cafes-in-lisbon [Accessed 15 November 2018].

Clark, T.N., 2004. Introduction: taking entertainment seriously. *In*: T.N. Clark, ed. *The city as an entertainment machine*. Amsterdam: Elsevier/JAI, 1–13.

Cocola-Gant, A., Gago, A., and Jover, J., 2020. Tourism, gentrification and neighbourhood change: an analytical framework. Reflections from Southern European cities. *In*: J. Oskam, ed. *The overtourism debate. NIMBY, nuisance, commodification*. Bingley: Emerald, 121–135.

Corte-Real, M. and Monte, M., 2018. Gentrificação em Centros Históricos: Impactos do Turismo. *Sociabilidades Urbanas, Dossiê: As novas fronteiras da gentrificação no mundo ibero-americano*, 2 (6), 167–185.

Dellarocas, C.N., 2003. The digitization of word-of-mouth: promise and challenges of online feedback mechanisms. *SSRN Electronic Journal*, 49 (10), 1275–1444.

Deloitte, 2019. *Estudo de Impacte Macroeconómico do Turismo na Cidade e na Região de Lisboa em 2018*. https://www.visitlisboa.com/pt-pt/sobre-o-turismo-de-lisboa/d/598-estudo-de-impacte-do-turismo-2018/showcase

Dias Real, F., 2020. As melhores happy hours em Lisboa. *Time Out* [online], 6 August. Available from: https://www.timeout.pt/lisboa/pt/coisas-para-fazer/as-melhores-happy-hours-em-lisboa [Accessed 15 November 2018].

Edwards, D., Griffin, T., and Hayllar, B., 2008. Urban tourism research: developing an agenda. *Annals of Tourism Research*, 35 (4), 1032–1052.

Evans, G., 2016. Graffiti art and the city: from piece-making to place-making. *In*: J.I. Ross, ed. *Routledge handbook of graffiti and street art*. Abingdon: Routledge, 168–182.

Featherstone, M., 1991. *Consumer culture and postmodernism*. London: SAGE.

Füller, H. and Michel, B., 2014. 'Stop being a tourist!' New dynamics of urban tourism in Berlin-Kreuzberg. *International Journal of Urban and Regional Research*, 38 (4), 1304–1318.

Gravari-Barbas, M., 2017. Tourisme de marges, marges du tourisme. Lieux ordinaires et « no-go zones » à l'épreuve du tourisme. *Bulletin de l'association de géographes français*, 94 (3), 400–418.

Gravari-Barbas, M. and Guinand, S., 2017. Introduction–addressing tourism-gentrification processes in contemporary metropolises. *In*: M. Gravari-Barbas and S. Guinand, eds. *Tourism and gentrification in contemporary metropolises - international perspectives*. Oxon: Routledge, 1–18.

Gretzel, U., 2018. Tourism and social media. *In*: C. Cooper, W. Gartner, N. Scott and S. Volo, eds. *The Sage handbook of tourism management*, Volume 2. Thousand Oaks, CA: SAGE, 415–432.

Guttentag, D., 2013. Airbnb: disruptive innovation and the rise of an informal tourism accommodation sector. *Current Issues in Tourism*, 18 (12), 1192–1217.

Hannigan, J., 1998. *Fantasy city: pleasure and profit in the postmodern metropolis*. London: Routledge.

Harrigan, P., et al., 2017. Customer engagement with tourism social media brands. *Tourism Management*, 59, 597–609.

Hoffman, L.M., 2003. The marketing of diversity in the inner city: tourism and regulation in Harlem. *International Journal of Urban and Regional Research*, 27 (2), 286–299.

Iglesias-Sánchez, P.P., Correia, M.B., and Jambrino-Maldonado, C., 2019. The challenge of linking destination online reputation with competitiveness. *Tourism & Management Studies*, 15 (1), 35–43.

Instituto Nacional de Estatística (INE), 2012. *Censos 2011*. Lisboa: INE.

Instituto Nacional de Estatística (INE), 2019. *Estatísticas do Turismo 2018*. Lisboa: INE.

Jansson, A., 2018. Rethinking post-tourism in the age of social media. *Annals of Tourism Research*, 69, 101–110.

KrrteAndo, 2016. Una escapada culinarian desde Lisboa. *KrrteAndo* [online], 14 November. Available from: https://www.krrteando.com/escapada-culinaria-lisboa/ [Accessed 15 November 2018].

Lefebvre, H., 2000. *La production de l'espace*. 4th ed. Paris: Éditions Anthropos.

Leite, R.P., 2002. Contra-usos e espaço público: Notas sobre a construção social dos lugares na Manguetown. *Revista Brasileira de Ciências Sociais*, 17 (49), 115–134.

Leung, D., et al., 2013. Social media in tourism and hospitality: a literature review. *Journal of Travel & Tourism Marketing*, 30 (1–2), 3–22.

Lestegás, I., Seixas, J., and Lois-González, R., 2018. The global rent gap of Lisbon's historic centre. *International Journal of Sustainable Development and Planning*, 13 (4), 683–694.

Litvin, S.W., Goldsmith, R.E., and Pan, B., 2008. Electronic word-of-mouth in hospitality and tourism management. *Tourism Management*, 29 (3), 458–468.

Lousada Oliveira, O., 2018. Medina quer "mandar" nos transportes da Área Metropolitana de Lisboa. *Visão* [online], 22 April. Available from: https://visao.sapo.pt/iniciativas/2018-04-22-Medina-quer-mandar-nos-transportes-da-Area-Metropolitana-de-Lisboa/ [Accessed 15 November 2018].

Maitland, R., 2008. Conviviality and everyday life: the appeal of new areas of London for visitors. *International Journal of Tourism Research*, 10 (1), 15–25.

Maitland, R., 2013. Backstage behaviour in the global city: tourists and the search for the 'Real London'. *Procedia: Social and Behavioral Sciences*, 105, 12–19.

Mendes, L., 2017. Gentrificação Turística em Lisboa: neoliberalismo, financeirização e urbanismo austeritário em tempos de pós-crise capitalista 2008–2009. *Cadernos Metrópole*, 19 (39), 479–512.

Mermet, A.-C., 2017. Airbnb and tourism gentrification: critical insight from the exploratory analysis of the 'Airbnb syndrome' in Reykjavík. *In*: M. Gravari-Barbas and S. Guinand, eds. *Tourism and gentrification in contemporary metropolises - international perspectives*. Oxon: Routledge, 52–74.

Miles, S. and Miles M., 2004. *Consuming cities*. Hampshire: Palgrave Macmillan.

Mommaas, H., 2004. Cultural clusters and the post-industrial city: towards the remapping of urban cultural policy. *Urban Studies*, 41 (3), 507–532.

Mommaas, H., 2010. Spaces of culture and economy: mapping the cultural creative cluster landscape. *In*: L. Kong and J. O'Connor, eds. *Creative-economies, creative-cities: Asian European perspectives*. London: Springer Dordrecht Heidelberg, 45–59.

Montes, P. and Casillas, E., 2015. Cacilhas, las mejores vistas de Lisboa al otro lado del tajo. *Viaje con Pablo* [online], 12 August. Available from: https://viajeconpablo.com/cacilhas/ [Accessed 15 November 2018].

Novy, J., 2018. 'Destination' Berlin revisited. From (new) tourism towards a Pentagon of mobility and place consumption. *Tourism and Urban Planning in European Cities*, 29 (3), 48–72.

Novy, J., 2019. Urban tourism as a bone of contention: four explanatory hypotheses and a caveat. *International Journal of Tourism Cities*, 5 (1), 63–74.

Pollice, F. and De Iulio, R., 2011. Avaliação Da competitividade turística do território. *Finisterra*, 46 (91), 121–138.

Ramos Silva, M., 2018. 16 coisas que tem de fazer nos museus em Lisboa. *Time Out* [online], 1 August. Available at: https://www.timeout.pt/lisboa/pt/coisas-para-fazer/experiencias-imperdiveis-nos-museus-em-lisboa [Accessed 15 November 2018].

Robinson, P., 2015. Conceptualizing urban exploration as beyond tourism and as anti-tourism. *Advances in Hospitality and Tourism Research (AHTR)*, 3 (2), 141–164.

Roland Berger Strategy Consultants, 2019. *Plano Estratégico para a Região de Lisboa para o horizonte de 2020–2024*. Lisboa.

Santos, J.R., 2019. Public space, tourism and mobility: projects, impacts and tensions in Lisbon's urban regeneration dynamics. *The Journal of Public Space*, 4 (2), 29–56.

Sequera, J. and Nofre, J., 2019. Touristification, transnational gentrification and urban change in Lisbon: the neighbourhood of Alfama. *Urban Studies*, 57 (15), 3169–3189.

Sieber, T., 1993. Public access on the urban waterfront: a question of vision. *In*: R. Rotenberg and G.W. McDonogh, eds. *The cultural meaning of urban space*. Toronto: Abc-clio, 173194.

Souto Cunha, S., 2016. O Gosto dos Outros… Cláudio Garrudo. *Visão* [online], 13 April. Available from: https://visao.sapo.pt/visaose7e/sair/2016-04-13-O-Gosto-dos-Outros.-Claudio-Garrudo [Accessed 15 November 2018].

Thevenot, G., 2007. Blogging as a social media. *Tourism and Hospitality Review*, 7 (3–4), 287–289.

Thörn, C., 2008. Intervention or the need for a new cultural critique. *Monitor,* 5, 48–68.

Time Out Lisboa, ed., 2017. Passagem de ano em Lisboa. *Time Out* [online], 28 December. Available from: https://www.timeout.pt/lisboa/pt/coisas-para-fazer/passagem-de-ano-na-regiao-de-lisboa [Accessed 15 November 2018].

Time Out Lisboa, ed., 2018. Os melhores estúdios de tatuagens em Lisboa. *Time Out* [online], 12 October. Available from: https://www.timeout.pt/lisboa/pt/compras/os-melhores-estudios-de-tatuagens-em-lisboa?page_number=2&zone_id=1428222 [Accessed 15 November 2018].

Urry, J., 1995. *Consuming places*. London: Routledge.

Xiang, Z. and Gretzel, U., 2010. Role of social media in online travel information search. *Tourism Management*, 31 (2), 179–188.

Xie, P.F., 2015. *Industrial heritage tourism*. Bristol: Channel View Publication.

Yoo, K. and Gretzel, U., 2011. Influence of personality on travel-related consumer-generated media creation. *Computers in Human Behavior*, 27 (2), 609–621.
Zukin, S., 1998. *The cultures of cities*. Oxford: Blackwell.
Zukin, S., 2010. *Naked city*. New York: Oxford University Press.

Interviews

Interview 2, 2019. Interview with M. Corte-Real. 7th of February, Lisbon.
Interview 7, 2019. Interview with M.J. Gomes and M. Monte. 11th of February, Almada.
Interview 8, 2019. Interview with M. Corte-Real and M.J. Gomes. 14th of February, Almada.
Interview 9, 2019. Interview with L. Manata e Silva and M. J. Gomes. 25th of February, Almada.
Interview 12, 2019. Interview with L. Manata e Silva and M. Monte. 2nd of February, Almada.

Part II

Protest and frictions

Contesting New Urban Tourism

7 Sustaining a political system

New Urban Tourism in Cuba and related conflicts

Niklas Völkening

New Urban Tourism as a contested policy in Cuba?

To cushion the most severe socioeconomic effects that followed the collapse of the USSR and of the Council for Mutual Economic Assistance (COMECON), the Cuban government promoted international tourism and, among other measures, legalised private accommodation, so-called *casas particulares*, in 1996. The strategy of stimulating tourism in Cuba turned out to be financially successful and simultaneously fostered the emergence of New Urban Tourism, especially in tourism hot spots where many Cubans accommodate international tourists privately. Practices of New Urban Tourism in Cuba are often exciting for tourists and lucrative for hosts, for whom tourism is one of the potentially highest sources of income.

However, Cuba's New Urban Tourism entails causes of conflicts, too. On the one hand, private hosts compete directly with state-run hotels, with an ostensive and sometimes exploitative power imbalance in favour of the state. On the other hand, hosts and other providers of services in New Urban Tourism earn incomes significantly above average, leading to social discontent and apparently contradicting the socialist ideal of an egalitarian society. Furthermore, the rise of New Urban Tourism has pervasive spatial effects regarding the use of space and access to it, which in turn evoke reactions by parts of the local communities.

Based on empirical data collected via semi-structured interviews in Habana Vieja, the historic city centre of Havana (22 interviews), and Trinidad (5 interviews) between March 2017 and April 2019, this chapter reveals the consequences of New Urban Tourism and the concomitant touristification of urban neighbourhoods in Cuba. Especially, spatial changes and altered ways of utilising public urban spaces will be addressed using spatial data collected in comprehensive mappings in both cities. Focal for this chapter is a human geographical analysis of power relations and interests of the actors engaged.

Additionally, the circumstances for Cuba's rise as an international tourism destination since the 1990s will be described, including explanations for the pervasive occurrence of forms of New Urban Tourism within certain urban spaces. Subsequently, the theoretical background of New Urban Tourism is discussed briefly. Afterwards, the local contestations and conflicts accompanying New

Urban Tourism in Cuba will be analysed through Habana Vieja and Trinidad. It turns out that conflicts over New Urban Tourism in Cuba are, surprisingly, motivated by the desire for increasing tourism rather than by the demand to curb it.

Tourism revolution in Cuba to sustain the socialist system

After the victory of Fidel Castro's revolution in 1959, international tourism in Cuba virtually came to a halt for various reasons. Most considerable were the lack of resources to maintain tourism, including the US embargo and travel ban for US citizens (Salinas *et al.* 2017, p. 220), and the framing of tourism as "hedonistic vice" (Sharpley and Knight 2009, p. 242), thus contrasting it with the prevailing socialist ideals.

However, international tourism made a strong comeback after several reforms that tackled the ramifications of the disintegration of the USSR and the COMECON in the early 1990s. As the Cuban economy was highly dependent on subsidised imports from its socialist allies (Zeuske 2016), the abrupt omission of imports threatened the continuity of the socialist-revolutionary system. For instance, Cuban GDP plunged by about 35% between 1989 and 1993 (Salinas *et al.* 2017, p. 221) and the achievements of the revolution, i.e. education, health care and housing, suffered fiercely.

In the wake of this, Fidel Castro declared in 1990 that Cuba had entered a "Special Period in Times of Peace" that was accompanied by fundamental economic and social reforms. These reforms led to enormous growth in international tourism, gaining momentum in the second half of the 1990s (see Figure 7.1). Seminal was the definition of tourism as one of the cornerstones of the new Cuban economy, with a clear objective that also applied to other reforms: to earn hard currency to retain the socialist order (Jatar-Hausmann 1999, p. 49). Already in the early 1990s, tourism had replaced the sugar industry as Cuba's largest source of hard currency income.

The second major building block of post-soviet reforms in Cuba was the legalisation of private enterprises and self-employment (*cuentapropismo*) in 1993. Cubans are currently allowed to work in non-state employment in 201 selected occupations (Mesa-Lago 2018, p. 2). This *cuentapropismo* proved to be advantageous especially in tourism. Additional laws allowed Cubans to run private restaurants (*paladares*) in 1995, to offer private transport (1996) and to rent out private rooms for accommodation (1997) in *casas particulares*.

Monetary measures with implications for tourism caused considerable frictions. In 1994, the *Peso Convertible* (CUC) was introduced as a "tourist currency" parallel to the *Peso Cubano* (CUP) after the US Dollar was temporarily legalised as currency. The dual currency system existed until the turn of the year 2020/21, with 1 CUC, whose value was tied to the US Dollar, worth approximately 24 CUP. While the state paid wages in CUP, payments in tourism were generally made in CUC, leading to enormous inequalities concerning income and purchasing power between state employment and tourism. Although the binary system of CUP and CUC has been abolished, it is unlikely that these income differences will

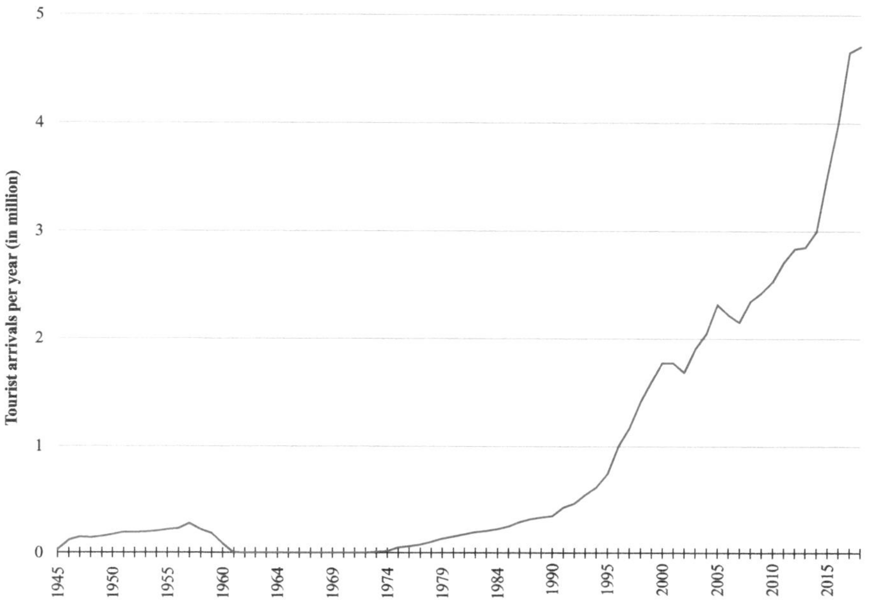

Figure 7.1 Arrivals of international tourists per year (in million) © Niklas Völkening, data from ONEI (2019), Jayawardena (2003) and Garrido (1993)

disappear in the medium term. Tourism earnings, which will presumably continue to be higher, are thus merely denominated in the same currency as the comparatively low incomes from state employment. This leads to paradoxical situations: while doctors, lawyers and professors working for the state struggle to provide for their families, Cubans who are employed in tourism could earn the monthly wage of a state employee in a single day. Since earning hard currency became vital for daily survival, many Cubans try to have a share in tourism, adopting even semi-legal to illegal activities, as it can be difficult to obtain a licence for *cuentapropismo* (Taylor and McGlynn 2009, p. 409, Nau 2016, p. 14).

To sum up the measures taken, the Cuban government introduced market-oriented elements into the national economy like entrepreneurship, international tourism and foreign investments to obtain hard currency for stabilising state and society. In doing so, it used "capitalism to save socialism" (Taylor and McGlynn 2009, p. 412). However, Roland (2010, p. 4) describes tourism as "means to sustain the system". Initially, tourism was meant to be utilised only for a limited period (Salinas *et al.* 2017, p. 222), as deteriorating effects on Cuban society were feared. However, the promotion of tourism soon became a long-term strategy.

Interactions between Cubans and tourists mostly take place within well-defined boundaries. Since tourism revenues are vital for the Cuban state, the level of control the state exercises is extraordinarily high. In many other countries, such strict political regulation of tourism markets is rare and often outsourced to Destination Management Organisations (DMOs) or similar entities (Novy and Colomb 2019, p. 5). In Cuban tourism, the high level of control has persisted throughout the past three decades and the changes of government in the meantime. The "smooth and uneventful" (LeoGrande 2015, p. 378) transition of power from Fidel to Raúl on 31 July 2006, put the "more capital-friendly" (Roland 2010, p. 14) of the Castros in charge and lead to further market-oriented reforms that boosted tourism additionally.

From an economic perspective, these reforms were successful. GDP increased significantly and, according to World Bank data (2018), reached pre-crisis levels in 2005. Until 2018, Cuban GDP grew at an average annual rate of 4.9% (own calculations based on World Bank 2018), although the rate of growth slowed down substantially recently. However, figures on the Cuban economy must be viewed with some caution, as Schmieg (2017, p. 6) estimates that figures published by government authorities overestimate the Cuban GDP more than twofold. The supply situation improved as well and the Cuban leadership seems to be confirmed in its course regarding tourism. Even after the election of Miguel Díaz-Canel as new President of Cuba in April 2018, a shift in the current importance of tourism for the Cuban economy is unlikely—and has not occurred to date. However, the extensive incorporation of tourism in Cuba's national economy was so radical that Jayawardena framed it as a "tourism revolution" (2003, p. 56). This revolution entails socioeconomic and spatial consequences, some of which can be traced back to forms of New Urban Tourism in Cuba.

What is 'new' about New Urban Tourism?

In differentiation from other forms of urban tourism, the 'new' in New Urban Tourism predominantly lies in the "scale, complexity and diversity of consumption experiences which now exist in urban landscapes built specifically for tourism and leisure" (Hall 2017, p. 145). In particular, three dimensions characterise New Urban Tourism: the "extraordinary mundane", specific "encounters and contact zones" and forms of "urban co-production" (Stors *et al.* 2019, p. 8–11) that shape urban space in novel ways and that make it increasingly difficult to distinguish 'tourists' from 'locals' and 'tourism' from 'everyday life' (Maitland 2013, p. 13, Eldridge 2019, p. 423), as tourists blend into the everyday life of cities (Füller and Michel 2014, p. 1306).

Simultaneously, tourism destabilises the mundane and hinders the local societies' ability to reproduce, when tourists and temporary residents gradually supplant it (Cócola Gant 2016, p. 7). This is often attributed to the sharing economy, sometimes framed as "gentrification battlefront" (ibid.) and frequently linked to holiday rentals such as Airbnb. Although critical voices stress the negative consequences of the sharing economy, such as the alteration of socio-economic structures of urban areas and various types of displacement through the conversion of housings into accommodation (ibid.), it need not necessarily be harmful. Thus, Airbnb or *casas particulares* simultaneously benefit the property owner, the tourist and may generally stimulate local economies (Fang *et al.* 2016, p. 264–266).

Nonetheless, New Urban Tourism is a source of conflicts and contestations that seemingly increase in urban settings, which is reflected in the proliferation of vague catchphrases like 'touristification', 'overtourism' and 'tourism-phobia' (Novy and Colomb 2019, p. 1–2). However, buzzwords like the aforementioned disguise the highly political character of tourism and give it the appearance of being all but positive (Novy 2016). In addition to the global intensification of tourism, growing numbers of actors—inhabitants, tourists and investors—claim limited urban space (Koens *et al.* 2018, p. 5), leading to multi-layered conflicts. These contestations are embedded in broader disputes regarding the "right to the city" (Lefebvre 1968), which in essence often involve conflicts between the interests of dwellers on the one hand, and capital seeking profit-maximisation on the other.

Two main reasons for conflicts in the wake of New Urban Tourism are closely connected to this: first, the unequal distribution of profits and costs of tourism, and second, the evident prioritisation of tourism in urban politics paired with perceived neglect of control and regulation (Novy and Colomb 2016). Although conflict and contestation over tourism are not new, the current quality, quantity and placing of resistance against tourism are unprecedented, emerging in reaction to more and seemingly unregulated tourism (Novy 2018).

However, deferrals linked to New Urban Tourism are embedded in particular socioeconomic, political and spatial conditions that constitute the framing context for conflicts and need to be addressed, in order to comprehend the multidimensionality of those contestations. Therefore, policy-oriented (Mordue 2017) and multi-scalar perspectives considering "globalisation, economic restructuring, neoliberalisation and financialisation" (Novy and Colomb 2019, p. 4) are necessary to grasp the entanglements of New Urban Tourism. Nonetheless, tourism is not necessarily the focal point of mobilisation, but may only be one dimension of critique within a broader movement affecting urban development, for instance concerning the distribution of profits generated in public spaces (Novy and Colomb 2019, p. 9).

These remarks on new Urban Tourism were mainly derived from literature considering urban areas of the Global North (Novy and Colomb 2019, p. 5), posing the question of global transferability. Although some studies were published on urban tourism and its effects in the Global South[1] (i.e. Betancur 2014, Janoschka and Sequera 2016, Preston-Whyte and Scott 2017), societies in developing countries hitherto seem to be under-represented in studies on New Urban Tourism. Despite being a global phenomenon, the forms and consequences of New Urban Tourism may vary greatly, depending on the socioeconomic and cultural background of a city and its residents.

Although the importance of digital technology as the "backbone of many New Urban Tourism phenomena" (Stors *et al.* 2019, p. 9) is undisputed, this chapter touches on the topic only lightly. This is because Cuba ranks among the least connected countries in the world in terms of internet access (Grandinetti and Eszenyi

2018). Although a 3G-network was established in December 2018, large parts of the population still lack reliable internet access.

To contribute to New Urban Tourism research, this chapter presents an example from the Global South, exploring the general consequences of New Urban Tourism and touristification within the special context of Cuba. Additionally, conflicts between various actors with differing levels of power will be analysed. Finally, the influence of tourism, especially New Urban Tourism, on urban societies in Cuba and its spatial effects as well as altered ways of usage for public space will be described. As there are only a few research examples on contestations of New Urban Tourism in the Global South, this chapter speaks for multidimensional, comprehensive analyses, which transcend disciplinary boundaries to contribute to a thorough understanding of New Urban Tourism and its effects in the Global South.

Special cases: tourism in Havana and Trinidad

The reasons for New Urban Tourists to visit Cuban cities and to demand respective services are similar to those in other urban destinations. Many tourists desire to experience 'real' Cuba. Especially young Western tourists tend to seek experiences from a local perspective or—even better—being accompanied by locals (Cohen and Cohen 2012). The need for authenticity and originality, for not-yet commodified experiences, is increasing among discerning tourists (Maitland 2013, p. 14), especially as online tourism services, accommodation, and tours facilitate the satisfaction of the thirst for 'authentic' experiences (Pappalepore *et al.* 2014). However, as indicated, this is not fully applicable in the Cuban case.

Various forms of *cuentapropismo* offer opportunities for tourists to get in touch with 'authentic' and 'real' Cubans. Especially *casas particulares*, *paladares*, guided tours and other services offered by *cuentapropistas* are popular among foreign tourists. In recent years, the share of international tourists staying overnight in *casas particulares* remained stable at roughly a quarter (see Figure 7.2). However, to acquire a licence for a private business in tourism and to meet the strict standards often requires significant investments beforehand, limiting these possibilities to Cubans with high income or with relatives abroad sending remittances (Simoni 2017, p. 297), thus excluding many Cubans from private engagement in tourism.

In both exemplary cities chosen for this chapter, forms of New Urban Tourism take up a reasonable proportion of the local tourism market. Habana Vieja and Trinidad also have some further similarities. In both cases, tourism is concentrated in "tourist bubbles" (Judd 1999) that have been valorised for tourist consumption. Habana Vieja and Trinidad are some of the cultural and architectural highlights in Cuba (Látková *et al.* 2017, p. 357), therefore they are among those urban destinations with most foreign visitors. Both were declared UNESCO World Heritage Sites, Habana Vieja in 1982, Trinidad in 1988, further increasing their touristic attractiveness. Furthermore, especially the touristified areas in Habana Vieja and Trinidad feature relatively high and further increasing prices and costs of living,

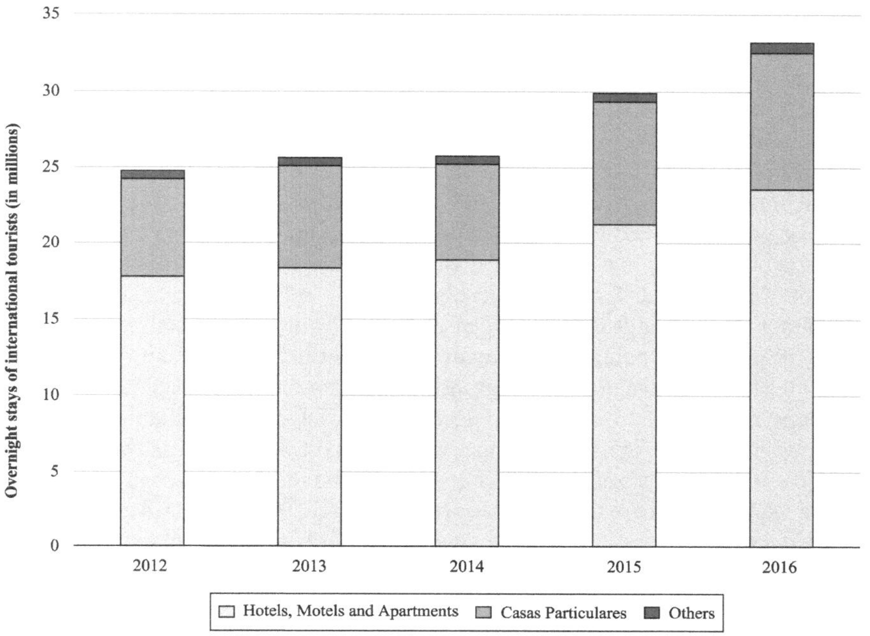

Figure 7.2 Overnight stays of international tourists in Cuba by type of accommodation © Niklas Völkening, data from ONEI (2017)

as several interviewees remarked with concern. In combination with the low average income, this creates spaces that are virtually not financially feasible for most of the local population.

In the section below, Habana Vieja and Trinidad are introduced with a special focus on tourism development, the actors involved and conflicts related to New Urban Tourism. Subsequently, the differences between both urban spaces, which also exist, are discussed.

Habana Vieja—spatial segregation and displacement

Habana Vieja, the historical centre of the Cuban capital, located just 100 miles south of Key West on the Florida Strait, is touristically attractive for various reasons, such as its history, culture and architecture. About 90% of the buildings in the historic centre have high architectural and/or historical value (Peters 2001, p. 5). Nevertheless, during the years following the Cuban Revolution, buildings and infrastructure within Habana Vieja were neglected by the central government (Foster 2003, p. 788). The circumstances of the Special Period further deteriorated building conditions. At the initiative of the Oficina del Historiador de la Ciudad de La Habana (Office of the City Historian of Havana), which is directly subordinated to the state council, programs for renovation and restoration, centrally headed by the Oficina, were introduced in the 1980s. But restoration efforts did not gain momentum until Habana Vieja was designated as a tourism hot spot at the end of the 1990s by order of the Ministry of Tourism. The restorations led to substantial spatial, social and economic transformations. In particular, areas of high potential for touristic commercialisation have shown significant improvement in building and infrastructure conditions since then, while areas of lower touristic interest often remain ruinous (Völkening *et al.* 2019).

Along with improved building conditions, a distinct increase of offerings for tourists is observable. Especially the number of *casas particulares*—as an apparent form of New Urban Tourism—increased sharply in Habana Vieja. The expansion of touristic offerings is at the expense of local dwellers, especially those who earn CUP-income in public service and are therefore excluded from almost any CUC-offerings. The increase of high-price, tourism-oriented restaurants, cafés and shops is paralleled by a drastic decrease of CUP-shops available for local amenities in Habana Vieja (Völkening *et al.* 2019, p. 90). The increase in offerings directed towards tourists is spatially concentrated into the four plazas in north-east Habana Vieja that were extensively restored (Scarpaci 2000, p. 728). Additionally, since real estate commerce has been legal since 2011, social groups with higher incomes have started to spatially segregate from lower-income classes, which are increasingly driven out of Habana Vieja (Völkening *et al.* 2019).

The displacement of dwellers from Habana Vieja in favour of tourist offerings meets the criteria of what Gotham (2005) called "tourism gentrification". Following Marcuse (1985), both "direct displacement" and "displacement pressure" are observable in Habana Vieja. "Exclusionary displacement", though an integral component of Marcuse's triad of displacement, plays a secondary role, as renting living space is rather uncommon in Cuba. While "direct displacement" points out the physical expulsion of dwellers of a certain area (Marcuse 1985, p. 205), "displacement pressure" describes the progressing decrease of living quality for present inhabitants, rendering the area "less and less livable" (Marcuse 1985, p. 207), for instance because social bonds break apart or the possibilities of sustaining a livelihood deteriorate, e.g. due to rising prices of consumer goods or rental fees. Cócola Gant (2016, p. 7) uses "collective displacement" to describe the cumulative or chronological occurrence of these forms of displacement, which progressively favours further touristification at the expense of local communities.

It is important to envision that displacement of local dwellers and the touristification of urban areas are not merely uncontrollable outputs of the interaction between supply and demand, but are subject to deliberate actions and decisions of investors, individual landlords, tourist companies and, ultimately, policy providing the legal framework (Slater 2017). Consequently, focus should be put on actors and their interests, both in theory and in empirical works.

In this context, "displacement by heritage dispossession" (Janoschka and Sequera 2016, p. 1182) is observable in Habana Vieja. At first, public space is cleansed and secured, while informal economic activities are driven out. In a second stage, dispossession and displacement take place, mostly justified by arguments of heritage protection, manifesting the "structural violence of tourism" (Büscher and Fletcher 2017, pp. 657–659) also occurring in Habana Vieja. Nonetheless, a representative survey among the residents of Habana Vieja conducted in 2019 shows that 57% of the population regard tourism as beneficial to their and their families' lives. In contrast, 38% of the population perceive tourism as not beneficial and respectively disadvantageous to their life (Echarri Chávet *et al.* 2019, pp. 12–13).

In conclusion, the restorations accomplished under the guidance of the Oficina were mainly responsible for the rise of Habana Vieja as an international tourism destination. The past three decades have been characterised by a rapid and lasting increase in touristic offerings, both public and private. Forms of New Urban Tourism give Habaneras and Habaneros various possibilities to earn relatively high incomes, simultaneously leading to ongoing displacement of dwellers, the touristification of large parts of Habana Vieja and eventually to conflicts between different groups of actors.

Trinidad—dominance of the private economy and increasing competition

Trinidad, the second example, is a "smaller walkable version of Habana Vieja with key differences" (Scarpaci 2012, p. 82). The town is located on the southern shore of central Cuba, at the foot of the Escambray Mountains. Trinidad is touristically attractive mainly for three reasons. First, it features magnificent colonial architecture, with many buildings in the city centre being in excellent condition, thanks to restorations initiated and instructed by the Oficina del Conservador de Trinidad (Office of the Conservationist of Trinidad). Second, the tropical beaches at Ancón peninsula are located just several kilometres south of Trinidad. Third, Trinidad offers good transport connections and is a springboard for various regional attractions (Scarpaci 2006, pp. 212–213). Consequently, the city is "flooded with international visitors throughout the year" (Tanaka 2013, p. 57).

During the Special Period and the reorientation towards international tourism, the Havana-based Ministry of Tourism initiated touristic development in Trinidad in a top-down attempt, declaring it a "tourist pole" within the centrally planned economy (Gutiérrez Castillo and Gancedo Gaspar 2002). Originally, the Ministry focused on the development of beach tourism on Ancón peninsula. The central plan, which was updated at the beginning of the 2000s and has not yet been realised, earmarked some 4,000 hotel rooms, contradicting the ideas of local authorities and the Oficina del Conservador (Scarpaci 2006, p. 207, 2012, p. 81).

Similar to Habana Vieja, the driving force for restoration—and touristification—in Trinidad was the Oficina del Conservador. The idea of the Conservador comprised the erection of small hotels with few rooms in the historic centre, funded by international investors (Scarpaci 2006, p. 216). To finance the restoration efforts and to sustain the work of the Oficina del Conservador, it raised a tax on the gross revenue of local venues and shops that was reinvested in the restoration of further buildings (Scarpaci 2006, p. 214).

However, the gentle restoration of the city centre was not necessarily focused on tourism development, but primarily on heritage conservation (Scarpaci 2012, p. 81) In further development, emphasis was put on the restoration of the residential areas of the borough of Tres Cruces (ibid.). The cultural events organised by the Oficina del Conservador, which are solely for residents, serve the same purpose, offering an "important space of expression for locals in a town dominated by tourism" (Tanaka 2013, p. 58). Although the restoration of ruined buildings was geared towards tourism in Trinidad, too, the revenues of the Oficina del Conservador were also used to restore family housings and infrastructure not directly related to tourism.

Basically, the Oficina del Conservador was able to assert its position on the development of tourism in Trinidad, as for many years there were no major hotels within the historic centre (Scarpaci 2006, p. 209). Even today, with some hotels having opened in central Trinidad, local *casas particulares* offer six times more rooms than the hotels on site. Consequently, a relatively large proportion of the local population can participate in and profit financially from burgeoning tourism.

Such conservation politics incorporating social awareness certainly are rare. While dwellers in many city centres across Latin America have to succumb to urban development predominated by the interests of international capital (Betancur 2014), Trinidad is a great exception (Scarpaci 2006, p. 27).

The Oficina del Conservador plays a crucial role in the restoration and the increase of attractiveness for tourism in Trinidad. The main beneficiary of this is the local population engaged in New Urban Tourism, while the state and people employed in public non-touristic services benefited significantly less. However, displacement of dwellers are far less common than in Habana Vieja, although recent years have revealed increasing competition by state-run venues, in particular to the disadvantage of *cuentapropistas*.

State versus privately run accommodation in Cuba

In both Habana Vieja and Trinidad, the restoration of cultural heritage, which is the foundation for touristic commodification, was not centrally governed and scheduled (Scarpaci 2012, p. 82), but rather emerged incidentally as a result of the commitments of individual actors. The possibility for both spaces to turn into hot spots of tourism was additionally accomplished by legislative changes and ensuing reactions by local authorities, residents and international investors.

Therefore, the cause and the effect of New Urban Touristification should not be confused. Instead of a few pioneering tourists 'discovering' places unspoiled by tourism that thenceforward turned into professionalising destinations, the spread of touristic offerings is rather the tangible manifestation of subjacent processes, which depend on political and economic dynamics not necessarily linked to tourism (Koens *et al.* 2018, p. 8). In Cuba, those are mainly political decisions and the application of (political) power causing socioeconomic deferrals and shifts in spatial appropriation. Different than in many other cities, Cuban municipal authorities play just a minor role in facilitating and directing tourism. Instead, the centralised Cuban government and its ministries ultimately control tourism development via central plans that other governmental bodies—such as the Oficinas in Havana and Trinidad—must subordinate to.

In Habana Vieja and Trinidad, New Urban Tourism gives rise to various conflicts and contestations. One focal source for conflict is the "restratification" of society (Nau 2016, p. 13) along an economic axis and the widening income gap between people who earned CUP and those who earned CUC (LeoGrande 2015, p. 395). These differences lead to the emergence of a "class of new rich" (Simoni 2017, p. 305) at the hot spots of international tourism, with only a few residents profiting from tourism. And while incomes rose significantly for some Cubans, costs of living increased strongly, too (Hoffmann 2015, p. 5). This poses problems to maintain the provision of basic supplies, especially for marginalised groups. Much stronger than in many cities of the Global North, New Urban Tourism contributes to increasing disparities among the population, putting dwellers, who profit from tourism, under pressure from above and below.

Pressure from above is applied by the state through increasing competition by its venues—for instance concerning accommodation and gastronomy through which the state seeks to safeguard the highest possible revenue. Although its intentions are presumably appreciable, such as cross-financing the welfare system, the repercussions for *cuentapropistas* are grave. While in Habana Vieja the touristic demand seems to grow in step with supply, the number of *casas particulares* and *paladares* in Trinidad increases faster than the demand of international tourists. As state-run hotels and restaurants push progressively into local markets, competition becomes fierce. The *cuentapropistas* of Trinidad are relatively powerless compared to state-backed hotel chains. Prospects for effective protest and contestation are limited, and open protest is still hard to carry out.

Pressure from below is applied by those groups that are excluded from tourism and that ogle the high incomes generated in tourism. This occasionally evokes signs and utterances of envy and concerns about the egalitarian Cuban society—partly to be heard as whispers on the street, partly it is broadcast on state television. But more often, Cubans acknowledge the necessity of revenues generated in tourism to sustain the welfare system in Cuba. Nonetheless, the seminal factor for preventing discontent and protest against tourism and accompanying inequalities is the hope many Cubans put into tourism, awaiting their own participation in tourism. The opportunity to obtain financial benefits through tourism may enlarge the willingness to tolerate or even promote it (Scheyvens 2007).

This offers an interesting difference in attitude towards New Urban Tourism in many other cases. In most European cities, dwellers are among the advocates of increased regulation, the limitation of short-term rentals and other forms of New Urban Tourism (Novy and Colomb 2019, p. 8). However, in Cuba, large parts of the population argue for an increase of licences for forms of New Urban Tourism, as they might profit from it. On the other hand, the state seeks to increase control over private engagements in tourism by decreasing the awarding licences for *casas particulares* and restricting employment opportunities in tourism.

While some mobilisations against New Urban Tourism intertwine their critique of tourism with an anti-capitalist stance (Novy and Colomb 2019, p. 8), this actually expectable connection is rarely formulated in Cuba. Instead, although the capitalist character of tourism is generally acknowledged, it is often tolerated—if not appreciated—as a necessity to support the socialist system.

The particular case of Cuban New Urban Tourism

As shown by the examples of Habana Vieja and Trinidad, New Urban Tourism also leads to discontent and conflicts in Cuba, especially in tourism hot spots which are affected by multi-scalar conflicts which have altered utilisation of public space. Conflicts often arise between *cuentapropistas* and the state as a result of a competitive situation and between Cubans who are financially better off through tourism and their fellow citizens who remain in relative poverty.

In both cases presented, the dominant structure for tourism development is a top-down hierarchy, controlled by the central Cuban government. However, local authorities, like city historians or conservationists, may participate in shaping urban space and the possibilities of appropriation for the local population, though with limited influence. Eventually, dwellers have few possibilities to assert their interests against governmental bodies—on local as well as regional or national levels. From this perceived power imbalance, several sources for conflict and contestation arise, mostly concerning access to tourism and the monetary resources it promises.

Urban tourism is the driver for economic development in many Cuban cities. Tourism in general is vital to sustain the social system, while partially it contributes to the erosion of society. To a certain extent, this is specific to Cuba, struggling to combine the imperatives of socialist society and capitalist tourism. For Cubans, engagements in New Urban Tourism offer opportunities of income far above the average. Despite criticism towards it, *cuentapropismo* in tourism seems to be worthwhile for many Cubans. While this might be beneficial for individuals, it is potentially harmful for society and social order. Simultaneously, New Urban Tourism fuels competition between private providers of tourism services in Cuba and the state, mainly in an economic arena that might also spill over into social or political issues.

In conclusion, conflict and contestation around New Urban Tourism in Cuba targets the transforming forces of tourism, as well as the distribution of revenue generated in tourism. As the analysis of Habana Vieja and Trinidad indicates, societies in the Global South are confronted with tourism-related problems partially similar to those of societies in the Global North. However, in other regards, they face challenges that are perceived as more seminal than questions of authenticity, for example if economic opportunities outlast social discontent.

Note

1 The term "Global South" attempts to productively incorporate historical experiences of colonisation, imperialism and socioeconomic dependencies and inequalities into the notion (Dados and Connell 2012, p. 13).

References

Betancur, J.J., 2014. Gentrification in Latin America. Overview and critical analysis. *Urban Studies Research* [online]. Available from: http://downloads.hindawi.com/archive/2014/986961.pdf [Accessed 4 December 2020].

Büscher, B. and Fletcher, R., 2017. Destructive creation. Capital accumulation and the structural violence of tourism. *Journal of Sustainable Tourism*, 25 (5), 651–667.

Castro Ruz F., 1990. *Discurso en el Acto Central por el XXX Aniversario de lo Comites de Defensa de la Revolucion* [online]. 28 September, Havana. Available from: http://www.cuba.cu/gobierno/discursos/1990/esp/f280990e.html [Accessed 11 December 2020].

Cócola Gant, A., 2016. Holiday rentals: the new gentrification battlefront. *Sociological Research Online*, 21 (3), 112–120.

Cohen, E. and Cohen, S.A., 2012. Current sociological theories and issues in tourism. *Annals of Tourism Research*, 39 (4), 2177–2202.

Dados, N. and Connell, R., 2012. The global south. *Contexts*, 11 (1), 12–13.

Echarri Chávez, M., Korstanje, M., and Robert Beatón, M.O., 2019. Visión Comunitaria del Turismo. Consideraciones desde la Práctica en el Centro Histórico de La Habana, Cuba. *Rosa dos Ventos. Turismo e Hospitalidade*, 11 (1), 2–22.

Eldridge, A., 2019. Strangers in the night. Nightlife studies and new urban tourism. *Journal of Policy Research in Tourism, Leisure and Events*, 11 (3), 422–435.

Fang, B., Qiang, Y., and Law, R., 2016. Effect of sharing economy on tourism industry employment. *Annals of Tourism Research*, 57, 264–267.

Foster, S., 2003. From Harlem to Havana. Sustainable urban development. *Tulane Environmental Law Journal*, 16, 783–805.

Füller, H. and Michel, B., 2014. 'Stop being a tourist!' New dynamics of urban tourism in Berlin-Kreuzberg. *International Journal of Urban and Regional Research*, 38 (4), 1304–1318.

Garrido, E.V., 1993. *Cuba y el Turismo*. La Habana: Editorial de Ciencias Sociales.

Gotham, K.F., 2005. Tourism gentrification: the case of New Orleans' Vieux Carre (French Quarter). *Urban Studies*, 42 (7), 1099–1121.

Grandinetti, J. and Eszenyi, M.E., 2018. La Revolución Digital: mobile media use in contemporary Cuba. *Information, Communication & Society*, 21 (6), 866–881.

Gutiérrez Castillo, O. and Gancedo Gaspar, N., 2002. Una década de desarrollo del turismo en Cuba (1990–2000). *Economía y Desarrollo*, 131 (2), 72–73.

Hall, C.M., 2017. Tourism urbanisation and global environmental change. *In*: E. Gössling and C.M. Hall, eds. *Tourism and global environmental change. Ecological, social, economic and political interrelationships*. London: Routledge, 142–155.

Hoffmann, B., 2015. Kuba–USA: Wandel durch Annäherung. *Focus Lateinamerika*, 2, 1–8.

Janoschka, M. and Sequera, J., 2016. Gentrification in Latin America. Addressing the politics and geographies of displacement. *Urban Geography*, 37 (8), 1175–1194.

Jatar-Hausmann, A.J., 1999. *The cuban way: capitalism, communism, and confrontation*. West Hartford, CT: Kumarian Press.

Jayawardena, C., 2003. Revolution to revolution. Why is tourism booming in Cuba? *International Journal of Contemporary Hospitality Management*, 15 (1), 52–58.

Judd, D.R., 1999. Constructing the tourist bubble. *In*: S.S. Fainstein and D.R. Judd, eds. *The tourist city*. New Haven, CT: Yale University Press, 35–53.

Koens, K., Postma, A., and Papp, B., 2018. Is overtourism overused? Understanding the impact of tourism in a city context. *Sustainability*, 10 (12), 4383.

Látková, P., et al., 2017. Tour guides' roles and their perceptions of tourism development in Cuba. *Tourism Planning & Development*, 15, 347–363.

Lefebvre, H., 1968. *Le droit à la ville* [*The right to the city*]. Paris: Anthopos.

LeoGrande, W.M., 2015. Cuba's perilous political transition to the post-castro era. *Journal of Latin American Studies*, 47 (2), 377–405.

Maitland, R., 2013. Backstage behaviour in the global city. Tourists and the search for the 'Real London'. *Procedia: Social and Behavioral Sciences*, 105, 12–19.

Marcuse, P., 1985. Gentrification, abandonment, and displacement. Connections, causes, and policy responses in New York City. *Journal of Urban and Contemporary Law*, 28, 195–240.

Mesa-Lago, C., 2018. *Voices of change in Cuba from the nonstate sector*. Pittsburgh, PA: University of Pittsburgh Press.

Mordue, T., 2017. New urban tourism and new urban citizenship. Researching the creation and management of postmodern urban public space. *International Journal of Tourism Cities*, 3 (4), 399–405.

Nau, S., 2016. Kubas Gesellschaft im Wandel der Zeit. *Geographische Rundschau*, 10, 12–18.

Novy, J., 2016. The selling (out) of Berlin and the de- and re-politicization of urban tourism in Europe's "Capital of Cool". *In*: C. Colomb and J. Novy, eds. *Protest and resistance in the tourist city*. London: Routledge, 52–72.

Novy, J., 2018. Urban tourism as a bone of contention. Four explanatory hypotheses and a caveat. *International Journal of Tourism Cities*, 5 (1), 63–74.

Novy, J. and Colomb, C., 2016. Urban tourism and its discontents. An introduction. *In*: C. Colomb and J. Novy, eds. *Protest and resistance in the tourist city*. London: Routledge, 1–30.

Novy, J. and Colomb, C., 2019. Urban tourism as a source of contention and social mobilisation. A critical review. *Tourism Planning & Development*, 16 (4), 358–375.

ONEI Oficina Nacional de Estadística e Información, 2017. *Capítulo 15: Turismo. Anuario Estadístico de Cuba 2016*. Havana.

ONEI Oficina Nacional de Estadística e Información, 2019. *Capítulo 15: Turismo. Anuario Estadístico de Cuba 2018*. Havana.

Pappalepore, I., Maitland, R., and Smith, A., 2014. Prosuming creative urban areas. Evidence from East London. *Annals of Tourism Research*, 44, 227–240.

Peters, P., 2001. *Rescuing old Havana*. Arlington: Lexington Institute.

Preston-Whyte, R. and Scott, D., 2017. Urban tourism in Durban. *In*: C.M. Rogerson and G. Visser, eds. *Urban tourism in the developing world. The South African experience*. London: Routledge, 245–264.

Roland, L.K., 2010. Tourism and the commodification of Cubanidad. *Tourist Studies*, 10 (1), 3–18.

Salinas, E., Mundet L., and Salinas, E., 2017. Historical evolution and spatial development of tourism in Cuba, 1919–2017. What is next? *Tourism Planning & Development*, 15 (3), 216–238.

Scarpaci, J.L., 2000. Reshaping Habana Vieja: revitalization, historic preservation, and restructuring in the socialist city. *Urban Geography*, 21 (8), 724–744.

Scarpaci, J.L., 2006. *Plazas and Barrios. Heritage tourism and globalization in the Latin American centro historico*. Tucson, AZ: University of Arizona Press.

Scarpaci, J.L., 2012. Urban heritage, representation and planning. Comparative approaches in Habana Vieja and Trinidad, Cuba. *In*: D. Fairchild Ruggles, ed. *On location. Heritage cities and sites*. New York: Springer, 61–88.

Scheyvens, R., 2007. Exploring the tourism-poverty nexus. *Current Issues in Tourism*, 10 (2–3), 231–254.

Schmieg, E., 2017. *Kuba »aktualisiert« sein Wirtschaftsmodell. Perspektiven für die Zusammenarbeit mit der EU. SWP-Studie 2/2017*. Berlin: Stiftung Wissenschaft und Politik.

Sharpley, R. and Knight, M., 2009. Tourism and the state in Cuba. From the past to the future. *International Journal of Tourism Research*, 11 (3), 241–254.

Simoni, V., 2017. Business, hospitality, and change in Cuba's private tourism sector. A view from casas particulares in Viñales. *Tourism Planning & Development*, 15 (3), 293–312.

Slater, T., 2017. Planetary rent gaps. *Antipode*, 49 (S1), 114–137.

Stors, N., et al., 2019. Tourism and everyday life in the contemporary city. An introduction. *In*: T. Frisch et al., eds. *Tourism and everyday life in the contemporary city*. Oxon: Routledge, 1–23.

Tanaka, M., 2013. Dressed up and sipping rum. Local activities within the touristic space of Trinidad, Cuba. *In*: D. Stevenson and A. Matthews, eds. *Culture and the city. Creativity, tourism, leisure*. London: Routledge, 64–76.

Taylor, H.L. and McGlynn, L., 2009. International tourism in Cuba. Can capitalism be used to save socialism? *Futures*, 41 (6), 405–413.

Völkening, N., Benz, A., and Schmidt, M., 2019. International tourism and urban transformation in old Havana. *Erdkunde*, 73 (2), 83–96.

World Bank, 2018. *World development indicators* [online]. Washington, DC. Available from: http://data.worldbank.org [Accessed 16 November 2020].

Zeuske, M., 2016. *Kleine Geschichte Kubas*. 4th ed. München: C.H. Beck.

8 Embattled consumptionscape of tourism

Networked urban contention against inbound tourist shoppers in Hong Kong

Tin-yuet Ting and Wei-Fen Chen

Introduction

This chapter examines the recent development of shopping tourism driven by travellers from emerging economies, especially China, and the critical responses it elicits from urban dwellers.

Since the 2000s, Hong Kong has been one of the main travel destinations accommodating the huge growth of mainland Chinese shopping tourists. Hong Kong has only 7.4 million residents but welcomes nearly 50 million tourists from mainland China each year (So 2017). Moreover, compared with tourists from other countries, Chinese tourists are more likely to prioritise shopping as their main motivation for travelling (Mangin 2015). Showing a preference for access to specific consumer goods, the massive influx of tourist shoppers has inevitably changed the urban landscape and livelihoods in Hong Kong. Over the years, local residents' concerns about these drastic changes have led to episodes of networked urban contention.

In particular, residents of the northern and western districts of Hong Kong, the border areas adjacent to mainland China, have organised direct interventions on the internet, often using Facebook pages and WhatsApp groups. These intervention activities have included confronting shopping tourists, storming stores in major tourist nodes to force them to close and calling for local residents to engage in political consumption, for example, boycotting tourism-oriented businesses and promoting traditional local stores with the use of social media and mobile technology. Departing from conventional protests, which have generally been led by elite actors and guided by political ideologies, these citizen actions represent an overlooked phenomenon in which urban dwellers directly and spontaneously challenge the latest form of leisure tourism based on everyday considerations.

Using the case of Hong Kong, this chapter illustrates how the shopping sprees of inbound tourists have shaped an 'enclave' consumptionscape, which is disconnected from the local lifestyle to serve the needs of these tourists and the response of digitally savvy urban dwellers to this trend. Specifically, we conceptualise a tourist consumptionscape as a series of urban configurations that reflect government policies and their corresponding arrangements to create a consumer-friendly landscape and redefine the socio-material relationships of the city for the development of shopping tourism. This development echoes the neoliberal initiative of "spatial fix" (Harvey 2001, p. 308), which seeks to transform the current urban landscape into an abstract space of capital accumulation. In the processes of urban reconstruction and dispossession, the emergence of the tourist consumptionscape separates urban dwellers from their local environment and lifestyle. This leads to conflict and confrontation between local residents and travellers over urban resources, ranging from commodities to infrastructure and space.

We argue that the proliferation of the tourist consumptionscape serves as the source and site of networked urban contention. This preposition builds upon extant research suggesting a link between everyday life and political initiative in the city. In Hong Kong, networked urban contention has been fuelled by the everyday experience of encountering inbound tourists, exacerbated by a shortage of commodities, a segregated marketplace and a disruption of the social order caused by shopping tourism in the local consumptionscape (Chen and Ting 2019). Often linked with a reactionary desire to rehabilitate the glory of the past and populist anti-traveller discourse against visitors from mainland China, these contentious activities are closely associated with the struggle for "collective consumption" (Castells 1977, p. 445), as shown in the following analysis, to reclaim access to the urban consumptionscape, restore the local urban livelihood system and foster consumer citizenship in the city. They emerge as bottom-up attempts to balance "neoliberal governmentality" (Lorenzini 2018, p. 155), which delegates responsibility for socio-spatial governance to local communities based on predominant market mechanisms.

In what follows, we begin by conceptualising the effect of shopping tourism on urban landscape and livelihoods and illustrate how this effect has led to urban discontent in the border towns of Hong Kong. Then, we discuss the research context and methods used in the study. Next, we identify three main sets of contentious activities organised with the use of social media and mobile technology at the grassroots level and assess their potential and limitations in tackling issues in the local consumptionscape shaped by an excessive amount of inbound tourists. Overall, this chapter contributes to the study of New Urban Tourism by revealing the growing tension between local communities and inbound travellers through the lens of cross-border consumption and mundane networked politics.

The rise of tourism-oriented consumptionscapes in global cities

Shopping tourism and its urban discontent

Scholarly debate on the role of leisure tourism in urban development is often linked with discussions of the political shift from a state economy to a market economy at the transnational level (Bianchi 2018). From Fordism to post-Fordism, the fluidity of the new world economic order has encouraged local governments to compete for the influx of travellers and capital, as the speed and mobility of goods, information, money and people increasingly determine the balance of power in the global capitalist system. Guided by these neoliberal initiatives and agendas, national governments have attempted to integrate local communities into global markets through various processes, such as urban reconstruction and dispossession (Harvey 2005).

Therefore, the promotion of tourism has become a fundamental component of urban development, as it seeks to create "experiences of place" (Hazbun 2004, p. 321) for tourists and to generate economic rewards for local businesses to thrive and for transnational corporations to invest in the local markets of host societies. In this trend, shopping tourism has led to the socio-spatial (re)configuration of large cities. First, local governments have facilitated the construction of luxury malls and transportation systems to connect key shopping destinations (Josiam *et al.* 2005). For example, the surge in mainland Chinese shopping tourists in Hong Kong over the last two decades is one of the by-products of development projects facilitating cross-border transportation (Leung 2019). Second, the development of shopping tourism is often accompanied by top-down discourses that (re)define normative ideas about the roles of local communities by prioritising the new consumption-oriented economy (Chen and Ting 2019). For instance, Hong Kong is often dubbed a "shopper's paradise" to attract inbound tourists (Hong Kong Tourism Board 2020). Shopping and discounts have become major themes of communication with travellers, emphasising consumption as a must-have experience during a trip to Hong Kong (Tang 2020). This may explain the urban transformations in terms of physical environments and city cultures, turning local communities into global markets (Hall 2006).

Whilst the development of shopping tourism can create market niches for local communities, economic interests may not translate into the well-being of local residents, as income and interest may only benefit specific retail giants or monopolies with limited connections with local communities (Rainer 2016). Indeed, the proliferation of shopping tourism often hampers local dwellers' access to urban resources and other aspects of an urban lifestyle, and thus goes hand in hand with local dwellers' confusion and hostility (Chen and Ting 2019). For instance, tourists' excessive purchases of basic necessities, such as vaccines and other health products, can lead to a shortage of supplies. As the regular shopping outlets of local residents become tourists' shopping destinations, locals have to compete with inbound buyers to secure enough resources for their daily use.

Moreover, shopping tourism can undermine local livelihoods and lifestyles through redevelopment projects that revitalise and commodify neighbourhoods in standardised tourist enclaves. Facilitated by neoliberal policies and agendas, global brands have established flagships and mega retail stores in major cities to target tourist shoppers from emerging economies selling luxuries such as leather goods, jewellery, electronics and cosmetics (Hales 2018). Therefore, urban planning accommodating shopping tourism can lead to another form of urban contention, as it segregates the marketplace by dividing it into prime locations for tourists and peripheral locations for locals (Chen and Ting 2019). As rents in business areas increase due to global brands opening flagships and mega retail stores in city centres, small traditional shops that target local consumers, such as local diners, clinics and grocery stores, are pushed to peripheral locations or have to close due to declining profits (Concepcion 2013). This is particularly striking in the case of Hong Kong, where competition for prime locations due to neoliberal policies has led to an explosion in land prices in city centres, restricting local residents' access to land, resources and decent living conditions (Wassener and Hui 2013. As a result, urban discontent can arise especially among local residents, who suffer first-hand from the undesirable costs of shopping tourism.

These critical studies on tourism development and its effects on the urban landscape and livelihoods highlight how urban restructuring for shopping tourism can become a source of discontent and generate recognisable forms of conflict over time. It is our contention that these urban conflicts are keenly felt and pragmatically implemented by lay citizens in relation to their daily activities. Paying more attention to the daily struggles of locals and their daily encounters in the city can thus offer fruitful insights into the emerging urban conflicts due to the rise of shopping tourism.

Mundane networked politics in the urban space

Increasingly, local residents without explicit political affiliations have increasingly coordinated and self-mobilised to address local agendas and issues in the case of Hong Kong. To deepen the discussion, we draw on studies of social movements and new media to better understand how urban dwellers, especially those who operate outside social movement organisations and political interest groups, engage in anti-shopping tourism through a bottom-up approach. Examining how urban contention emerges in the city is a central focus in the field of critical urban studies and social movement studies. In particular, Castells' (1977, p. 445) influential work theorises "collective consumption" among locals as the predominant mode of urban contention under global capitalism. For him, collective actions in the city are organised around the demands of locals for collective consumption, which refers not only to their basic needs for commodities, but also to their need for other socialised goods, such as access to schools, hospitals and public transport, to reproduce their labour power. The link between the city and the grassroots is thus established to understand the wave of new urban social movements in the context of urban daily life.

Despite differences in perspectives, urban sociology studies support this argument, suggesting that it reveals the subterranean ways in which political potential is realised by the ordinary actions that people take in their everyday city lives (de Certeau 1984, Lefebvre 1991). Recent studies particularly highlight the importance of investigating the practical engagement of lay citizens in urban development projects in the formation of contemporary political life (Anand 2011, Knox 2017). Turning our attention to the importance of daily activities in the city, these studies depart from disembodied statecraft and focus on political agency and citizen action that revolve around shared urban experiences at the grassroots. They inform our understanding of how the urban encroachment of tourist consumptionscapes constitutes urban contention, by recasting local residents' collective actions as mundane responses to fundamental issues.

Particularly in today's networked environments, the common use of mobile social media can create new opportunities for lay citizens to engage in contentious politics on a daily basis traversing political and everyday arenas (Ting 2019). Whilst the pervasiveness of Information and Communication Technology (ICT) can facilitate the mobilisation and coordination of networked and ad hoc forms of urban protests (Ting 2020), it can also shape the contours of collective actions by enabling new modes of engagement (Earl and Kimport 2011) and by personalising contentious politics (Bennett and Segerberg 2013). Integrating these research insights, we seek to uncover the political texture of mundane activities in the networked urban settings of Hong Kong. Specifically, we analyse three sets of contentious activities among digitally savvy urban dwellers, each showcasing a specific link between shopping tourism and urban contention and the potential and limitations of such mundane networked politics.

Research context and methods

China has become the world's fastest growing tourism market. In the last 15 years, Hong Kong, located in southeastern China and as a Special Administrative Region (SAR) of the People's Republic of China, has received more mainland Chinese tourist shoppers than other global cities. Whilst Hong Kong consists of several islands and peninsulas, the northern and western parts of its territory are adjacent to mainland China. In terms of transportation, mainland Chinese tourists often visit Hong Kong by air, ferry, car and train. Due to the Individual Visit Scheme launched in 2003, which simplifies the logistical arrangements for mainland Chinese tourists seeking to visit Hong Kong individually rather than as part of group tours, the number of mainland Chinese tourists visiting Hong Kong has increased considerably over the last couple of decades, rising from 4.3 million in 2004 to 47 million in 2014 (So 2017). On average, each tourist spends more than HK$7,000 (approximately US$900) during an overnight visit and HK$2,300 (approximately US$300) during a same-day visit (HKSAR Census and Statistics Department 2018).

Whilst an increase in shopping tourism can create market niches for host societies, tourist shoppers and excessive business activity can also pose a challenge to sustainable urban development. In Hong Kong, most shopping malls have been accommodating large chain stores or high-end boutiques (Wassener and Hui 2013). Particularly the northern and western districts of Hong Kong, border areas adjacent to mainland China, have experienced gentrification; commercial properties have changed the urban landscape to meet the needs of tourists rather than those of local residents. Over time, marketplaces in these areas have been transformed into stores selling goods demanded by tourist shoppers, such as pharmacies selling Western medicine and groceries including baby formula. The emergence of this tourist consumptionscape has therefore encroached on the livelihoods of local residents by limiting their collective consumption.

In recent years, a series of contentious activities launched by local residents have emerged at the grassroots. They sought to subvert the predominance of the urban consumptionscape and to effect changes at the mundane level, through the use of social media and mobile technology. Unlike conventional social movements, which have sought to remedy the fallacies of deliberative democracy and to make systematic changes at the societal level, the contours of networked urban contention examined in this chapter emerged in response to the uneven effect of shopping tourism on the border towns of Hong Kong. In these cases, citizen activists are not pursuing well-articulated altruistic political ideals or structural transformations (Ting 2017), but are seeking to restore their urban livelihoods of the early 2000s, before the undesirable changes brought about by policy-led leisure tourism and regional integration.

This study uses data obtained from netnography and qualitative analysis of news coverage to examine the interactions between shopping tourism and networked urban contention in Hong Kong's border towns. As an interpretive research method, netnography denotes unobtrusive, observational analysis of contents and discourses on online platforms (Langer and Beckman 2005). It is considered to be an effective method for understanding cultural practices and for examining the point of view of participants in online spaces (Kozinets 2012). Netnographic data were collected from the social media pages of seven community and activist groups between 2012 and early 2020. Revealing ingroup communication and activity archives, these data enabled ethnographic field observations and were used to investigate the contours of citizens' contentious activities and their interpretations of the changing urban consumptionscape. The qualitative analysis of the news coverage of 12 local and international media outlets helped track media attention to and public perceptions of shopping tourism and corresponding networked activism over time. It also generated empirical material for data triangulation with those collected using netnography. This study thus adopted a holistic approach by using materials collected from the two data sets to supplement each other.

Contentious activities assembled in networked urban settings

Reclaiming access to the urban consumptionscape

The most conspicuous examples of networked urban contention against tourist shoppers from mainland China are 'reclaim actions', through which local residents mobilise and coordinate via social media and online forums to disturb or expel inbound shopping tourists. Whilst local residents in Hong Kong tend to speak Cantonese, mainland Chinese shoppers often use their mother tongue, Mandarin, also known as Putonghua. The language used thus marks a clear division between tourists and locals. During such reclaim actions, activists have hurled abuse at Putonghua-speaking shoppers, targeting those carrying large suitcases, a sign of a shopping spree. These activities, along with crowding out the malls and commercial space in their neighbourhoods to interfere with business, are highly controversial. Rather than expressing discontent or petitioning government actions, they are intended by urban dwellers as direct interventions to reclaim access to the urban consumptionscape.

Since 2012, multiple reclaim actions organised by local communities and activist groups through Facebook pages and WhatsApp networks have drawn public attention. For instance, citizen activists have often self-organised on the internet to march in groups along Canton Road, a high street with numerous luxury shops and a popular destination for mainland tourists. During some of these actions, citizen activists marched along the street with signs reading "Go back to China". They shouted at random Chinese shopping tourists, accusing them of overwhelming local communities and hogging the city's already scarce resources (FlorCruz 2014). In recent years, these reclaim actions have spread from the northern districts, such as Sheung Shui and Shatin, to the western districts, such as Tuen Mun and Tung Chung, where new facilities targeting tourist shoppers have been built. In Tuen Mun, for instance, responding to online calls to action, a large number of local residents have in the past few years confronted shopping tourists at the cross-border bus terminus, where mainland tourists often meet before returning to mainland China. They urged the tourists to buy basic necessities in their hometowns, accusing them of damaging already overcrowded cities and supporting skyrocketing rents that leave little room for local small businesses to survive (Zhao 2019).

As suggested by the name given to these the events, local residents legitimise their actions as 'reclaiming' their urban neighbourhoods, which were once reserved for local dwellers. By confronting individual tourist shoppers and disrupting the operations of tourism-oriented businesses, such as pharmacies and luxury stores, these reclaim actions enable local residents to fight for their urban space and resources encroached on by the expansion of the tourist consumptionscape. According to a local resident in a press interview, the aim of their reclaim actions is to restore the community's commercial landscape, scale of rents and public hygiene conditions (Su 2019). Similar ideas can be seen in the slogans chanted or written on placards, such as "Save our Sheung Shui! Protect our home!" and "Tuen Mun residents take charge of your own district".

However, some reclaim actions, notably harassment and mobbing, have been carried out without public agreement. During these actions, some citizen activists have used xenophobic slogans, such as "Hong Kong people do not welcome Chinese people!". Radical locals have also negatively described shopping tourists, calling them 'locusts' (FlorCruz 2014), a derogatory term that has spread in recent years to depict mainland Chinese tourists as ignorant, devastating, vulgar, selfish, voracious and grasping. This division between ingroup and outgroup explains some occasional violent confrontations. In addition, people have posted comments containing hate speech on the social media pages of activist groups in local communities that read "Locusts, go back to your mainland!", among others. As a result, some citizens and the local government have condemned these events as barbaric and uncivilised activities that humiliated mainland visitors and tarnished the city's image (FlorCruz 2014).

Restoring the local urban livelihood system

Recognising that the emergence of the tourist consumptionscape involves larger socio-spatial arrangements, local residents have started to grasp that more can be achieved by intervening in the local urban systems, such as public transportation and tourism-related regulations, that have led to unsustainable shopping tourism. Accusing local authorities of turning a blind eye to the influx of shopping tourists (Zhao 2019), in recent years more and more activist groups have extended their networked activism from shopping malls to the daily operations of public services and law enforcement.

In the four towns of the northern districts adjacent to mainland China, namely Sheung Shui, Fanling, Lo Wu and Lok Ma Chau, local residents have frequently self-organised on the internet and gathered at Mass Transit Railway (MTR) stations. They have claimed to 'help' MTR staff enforce compliance with regulations often taken lightly by passengers through actions such as reporting travellers' oversized luggage. However, they have often also taken it upon themselves to confront shopping tourists and check their luggage in front of the stations, sometimes leading to scuffles between tourists and residents (Ng and Nip 2012). During these events, local residents have sought to strictly enforce the weight limits themselves, performing tasks that should have been taken care of by local authorities. In addition, to compel local authorities, subversive social media memes have been created and spread on the internet, featuring passengers carrying a large quantity of goods often overlooked in regulation enforcement. Others have used online satire and mocked MTR staff and police officers for allegedly ignoring signs of parallel trading, violating their own policies and Hong Kong laws, respectively.

On another front, direct interventions in local urban systems have been organised among residents in western districts of Hong Kong, such as Tung Chung. Since the Hong Kong–Zhuhai-Macao Bridge opened in 2018, more than 86,000 passenger trips have been made to Tung Chung across the bridge, making Tung Chung a new target for shopping tourism (Cheng 2018). As the influx of visitors has sparked complaints from residents, local communities have launched similar types of networked collective actions on social media platforms to crack down on illegal tour groups. Legally, tour guides from mainland China must be accompanied by a local liaison in Hong Kong, but some guides ignore these regulations and lead tour groups alone. Therefore, online community groups have called on local residents to gather in tourist shopping areas to identify unlicensed travel agents (Zhang 2018). During weekends, local residents have mobilised on the internet to approach escorted tour groups in Tung Chung, displaying large banners that read "Report illegal tour guide" outside shopping malls. They have asked mainland tour guides to show their licences, checked whether these groups were accompanied by local agents and urged tourists seated outside shops to leave. By drawing public attention through media exposure, these activists have compelled the police to investigate and take action against some of the suspected illegal tour guides.

These networked collective actions of direct intervention in local urban systems are the result of the disappointment of citizens with the local government's inability to address the strain placed by the influx of mainland tourists on the urban landscape and livelihoods. Recognising ineffective governance in their communities, urban dwellers have played an increasingly proactive role in regulating the city's daily operations and correcting its social order in the face of an excessive number of inbound visitors.

Fostering consumer citizenship in border towns

Whilst local residents have taken part in networked collective actions facilitated by ICT, they have also actively engaged in connective actions in which dispersed individuals address common problems through digital media (Bennett and Segerberg 2013). In the border towns of Hong Kong, digitally savvy citizens have used mobile social media to enable acts of political consumption and strengthen solidarity within and between their neighbourhoods. Common to these networked practices are attempts to connect and make sense of isolated individual efforts to extend and sustain networked urban contention against shopping tourism in people's daily lives.

Increasingly, urban dwellers in the northern and western districts have self-organised on social media platforms to support traditional local stores and boycott tourism-oriented shops in local communities. Congregating on the Facebook pages and online forums of local communities, they have sought to promote a local-friendly economy by regularly disseminating information on local businesses refusing to cater to inbound tourists or affected by soaring rents and the segregated marketplace. They have sometimes created lists and maps on social media to identify and locate these local businesses, calling others to support their stores and avoid shopping in tourism-oriented stores. Others have tested and verified these local businesses on behalf of communities and groups. Often posted with pictures or videos, they have shared their personal consumer reviews via online forums and Facebook pages, praising shopkeepers who prioritise Cantonese-speaking customers over tourists, to provide proof and justify the localness of these stores.

In addition to advocating for shopping in local-oriented shops and boycotting tourism-oriented ones, other forms of adversarial social media campaigns have been launched to punish tourism-oriented businesses, accusing them of making profits at the expense of the lives of local residents. For instance, an online community group from Hung Hom recently mobilised local residents via Facebook and WhatsApp to file complaints with the Food and Environmental Hygiene Department against restaurants catering to package tours from mainland China. They successfully led government officers to conduct several checks in these restaurants, interrupting their activities (Siu 2019). Rather than gathering and acting together at the same time and in the same place in networked collective actions, these digitally mediated practices of connective actions make it possible to carry out individualised acts of political consumption on a day-to-day basis, even when local residents are simply shopping and passing by on their own.

Not all forms of networked activism have directly led to changes in the local consumptionscape. Some have sought to create a more robust network among local residents by connecting individual anti-shopping tourism efforts. Digital storytelling and curation using mobile social media, in particular, tend to promote ingroup solidarity and mutual recognition through the circulation of shared experiences and the witnessing of common livelihoods (Ting 2015). Prominent examples include local residents distributing stories of their personal struggles in border towns, such as being involved in disputes with shopping tourists and expressing anger and resentment at the loss of the local culture and lifestyle due to the changing urban landscape, and recording and denouncing excessive shopping and 'uncivilised' behaviour among travellers using captioned pictures and real-time streaming with mobile devices. Disseminated through loosely connected and digitally mediated networks, these networked practices have not only demonstrated citizens' attempts to subvert the imbalance in the local consumptionscape, but also helped to create transient "hybrid public spaces" (Castells 2012, p. 221) for peer sharing and learning and for building consent among consumer citizens in the processes of networked urban contention against shopping tourism.

Conclusions

Using the case of Hong Kong, this chapter examines local residents' contentious activities revolving around inbound shopping tourism and discusses their interventions in networked urban settings. Shifting scholarly attention from hospitality to hostility, we illustrate the formation of mundane networked politics to restore local market orders and urban livelihoods in the border towns of Hong Kong. In line with previous studies on the 'dark side' of urban tourism, we show that the emergence of the consumptionscape in prime locations in Hong Kong is closely linked with shopping tourism, designed specifically to meet the needs of tourist shoppers.

This chapter makes two major contributions to the study of New Urban Tourism. First, departing from a neoliberal account that approaches tourism mainly as a product of homogenous public policies around the world, which leaves little room for discussing political agency and citizen action in the emergence of tourism-oriented consumptionscapes, we highlight how the well-being of local residents and their daily experiences have increasingly become the source and site of networked urban contention in response to the undesirable outcomes of neoliberal urban development projects. Through the lens of local residents' struggles, we address how the interests generated by shopping tourism are unevenly distributed and can deepen social inequalities and conflicts in host societies at the grassroots level. We also show how networked urban contention reinforces the division between ingroup locals and outgroup travellers, reflected in daily consumption and spatial conflicts.

Second, whilst consumer activism is often seen as coherent and progressive in terms of political ideals, this chapter illustrates how self-mobilised consumer activism can have a 'humble' start, which can be disconnected from altruistic considerations of the common good and developed from coping strategies to solve the mundane problems brought about by excessive shopping tourism. Whilst activism is often considered as prioritising the common good and asking individuals to endure for collective goals, such as recycling for a cleaner planet and limiting product choices to vegan cosmetics to support animal rights, the contentious activities discussed in this chapter suggest solution-based contours born from people's discontent and inconvenience at the everyday level. The corresponding appeals and actions have been developed largely based on citizens' mundane considerations to restore their urban landscape and lifestyle. Therefore, our analysis suggests a different way to consider how local consumer activism can develop vis-à-vis shopping tourism.

References

Anand, N., 2011. Pressure: the politechnics of water supply in Mumbai. *Cultural Anthropology*, 26 (4), 542–564.

Bennett, W.L. and Segerberg, A., 2013. *The logic of connective action: digital media and the personalization of contentious politics*. Cambridge, MA: Cambridge University Press.

Bianchi, R., 2018. The political economy of tourism development: a critical review. *Annals of Tourism Research*, 70, 88–102.

Castells, M., 1977. *The urban question: a Marxist approach*. London: Edward Arnold.

Castells, M., 2012. *Networks of outrage and hope: social movements in the Internet age*. Chichester: Wiley.

Chen, W.-F. and Ting, T.-Y., 2019. Shopping tourism and conflicts: neoliberal consumptionscape in host societies. *Tourism, Culture & Communication*, 19 (2), 155–160.

Cheng, K., 2018. Activists 'reclaim' Tung Chung amid influx of mainland tourists via mega bridge [online]. *Hong Kong Free Press*, 12 November. Available from: https://www.hongkongfp.com/2018/11/12/activists-reclaim-tung-chung-amid-influx-mainland-tourists-via-mega-bridge [Accessed 24 May 2020].

Concepcion, J., 2013. Soaring rents in prime locations push out local shops and restaurants [online]. *South China Morning Post*, 20 February. Available from: https://www.scmp.com/property/hong-kong-china/article/1154023/soaring-rents-prime-locations-push-out-local-shops-and [Accessed 24 May 2020].

De Certeau, M., 1984. *The practice of everyday life*. Berkeley, CA: University of California Press.

Earl, J. and Kimport, K., 2011. *Digitally enabled social change: activism in the internet age*. Cambridge, MA: MIT Press.

FlorCruz, M., 2014. 'Anti-Locust' protests in Hong Kong call for restrictions on Chinese mainland tourists [online]. *International Business Times*, 9 February. Available from: https://www.ibtimes.com/anti-locust-protests-hong-kong-call-restrictions-chinese-mainland-tourists-1556457 [Accessed 24 May 2020].

Hales, C., 2018. Australian luxury retail boom fuelled by Chinese shoppers and micro-influencers [online]. *South China Morning Post*, 10 January. Available from: https://www.scmp.com/lifestyle/fashion-luxury/article/2127434/australian-luxury-retail-boom-fuelled-chinese-shoppers-and [Accessed 24 May 2020].

Hall, C.M., 2006. Urban entrepreneurship, corporate interests and sports mega-events: the thin policies of competitiveness within the hard outcomes of neoliberalism. *The Sociological Review*, 54 (2), 59–70.

Harvey, D., 2001. *Spaces of capital: towards a critical geography*. Edinburgh: Edinburgh University Press.

Harvey, D., 2005. *The new imperialism*. Oxford: Oxford University Press.

Hazbun, W., 2004. Globalisation, reterritorialisation and the political economy of tourism development in the Middle East. *Geopolitics*, 9 (2), 310–341.

HKSAR Census and Statistics Department, 2018. Table E553*: Per capita spending and length of stay of visitors by country/region of residence* [online]. Available from: https://www.censtatd.gov.hk/hkstat/sub/sp130.jsp?productCode=D5600553 [Accessed 24 May 2020].

Hong Kong Tourism Board, 2020. *Shop* [online]. Available from: http://www.discoverhongkong.com/eng/shop/index.jsp [Accessed 24 May 2020].

Josiam, B.M., Kinley, T.R., and Kim, Y.K., 2005. Involvement and the tourist shopper: using the involvement construct to segment the American tourist shopper at the mall. *Journal of Vacation Marketing*, 11 (2), 135–154.

Knox, H., 2017. Affective infrastructures and the political imagination. *Public Culture*, 29 (2), 363–384.

Kozinets, R.V., 2012. *Netnography: doing ethnographic research online*. Los Angeles, CA: SAGE.

Langer, R. and Beckman, S.C., 2005. Sensitive research topics: netnography revisited. *Qualitative Market Research*, 8 (2), 189–203.

Lefebvre, H., 1991. *The production of space*. Oxford: Blackwell.

Leung, K., 2019. New cross-border transport links bring big rise in mainland Chinese visitors to Hong Kong over Lunar New Year holiday but many returned home the same day [online]. *South China Morning Post*, 7 February. Available from: https://www.scmp.com/news/hong-kong/hong-kong-economy/article/2185316/new-cross-border-transport-links-bring-big-rise [Accessed 24 May 2020].

Lorenzini, D., 2018. Governmentality, subjectivity, and the neoliberal form of life. *Journal for Cultural Research*, 22 (2), 154–166.

Mangin, V., 2015. The world's most influential shoppers [online]. *BBC Capital*, 25 August. Available from: https://www.bbc.com/worklife/article/20150824-the-worlds-most-influential-shoppers [Accessed 24 May 2020].

Ng, J. and Nip, A., 2012. Protesters rally in Sheung Shui against cross-border parallel traders [online]. *South China Morning Post*, 16 September. Available from: https://www.scmp.com/news/hong-kong/article/1037962/protesters-rally-sheung-shui-against-cross-border-parallel-traders [Accessed 24 May 2020].

Rainer, G., 2016. Constructing globalized spaces of tourism and leisure: political ecologies of the Salta Wine Route (NW-Argentina). *Journal of Rural Studies*, 43, 104–117.

Siu, P., 2019. Hong Kong residents fed up with influx of mainland Chinese tourists call on government to stop them visiting more than once a month [online]. *South China Morning Post*, 13 January. Available from: https://www.scmp.com/news/hong-kong/hong-kong-economy/article/2181895/hong-kong-residents-fed-influx-chinese-tourists [Accessed 24 May 2020].

So, A.Y., 2017. Anti-China sentiment in post-handover Hong Kong [online]. *Asian Dialogue*, 2 June. Available from: http://theasiadialogue.com/2017/06/02/anti-china-sentiment-in-post-handover-hong-kong [Accessed 24 May 2020].

Su, X., 2019. Protest planned for Hong Kong's Sheung Shui against mainland Chinese parallel traders puts pharmacies on alert [online]. *South China Morning Post*, 12 July. Available from: https://www.scmp.com/news/hong-kong/politics/article/3018266/pharmacies-frequented-mainland-chinese-visitors-alert [Accessed 24 May 2020].

Tang, C., 2020. An ultimate guide to Hong Kong shopping: what to buy and where to shop [online]. *China Highlights*. Available from: https://www.chinahighlights.com/hong-kong/shopping.htm [Accessed 25 May 2020].

Ting, T.-Y., 2015. Digital narrating for contentious politics: social media content curation at movement protests [online]. *M/C Journal: Journal of Media and Culture*, 18 (4). Available from: http://journal.media-culture.org.au/index.php/mcjournal/article/view/995 [Accessed 25 May 2020].

Ting, T.-Y., 2017. Struggling for tomorrow: the future orientations of youth activism in a democratic crisis. *Contemporary Social Science*, 12 (3–4), 242–257.

Ting, T.-Y., 2019. Everyday networked activism in Hong Kong's Umbrella Movement: expanding on contemporary practice theory to understand activist digital media usages. *International Journal of Communications*, 13, 3250–3269.

Ting, T.-Y., 2020. From 'be water' to 'be fire': nascent smart mob and networked protests in Hong Kong. *Social Movement Studies*, 19 (3), 362–368.

Wassener, B. and Hui, M., 2013. Soaring rents in Hong Kong push out mom and pop stores [online]. *The New York Times*, 4 July. Available from: https://cn.nytimes.com/china/20130704/c04hongkong/en-us [Accessed 24 May 2020].

Zhang, K., 2018. Two arrested as Tung Chung protesters clash over influx of visitors from mainland China to Hong Kong [online]. *South China Morning Post*, 11 November. Available from: https://www.scmp.com/news/hong-kong/society/article/2172695/two-arrested-tung-chung-protesters-clash-over-influx-visitors [Accessed 24 May 2020].

Zhao, S., 2019. We don't want your money: Chinese shoppers told to spend cash elsewhere by frustrated Hongkongers who just want some peace and quiet [online]. *South China Morning Post*, 3 February. Available from: https://www.scmp.com/news/hong-kong/hong-kong-economy/article/2184891/we-dont-want-your-money-chinese-shoppers-told-spend [Accessed 24 May 2020].

9 Between political protest and tourism gentrification

Impacts of New Urban Tourism in Hamburg's Schanzenviertel

Anja Saretzki and Karlheinz Wöhler

Introduction

The Schanzenviertel is one of Hamburg's smallest and youngest districts[1] and one of the city's most touristified quarters, where tourism-induced socio-spatial transformation can be observed best. The former mixed residential and industrial area has been developed into the centre of the alternative scene in Hamburg since the 1980s. The so-called Rote Flora, an autonomous cultural centre that has been squatted, is at the heart of the Schanzenviertel (see Figure 9.1). It originated from the protest against gentrification and evolved into a symbol of far-left opposition. At the same time, it is a tourist sight which is advertised as "the cultural centre" of "Hamburg's hippest neighbourhood" (Hamburg Tourismus n.d., hamburg.de n.d., transl. A.S.) and a favoured destination of New Urban Tourists.

Urban tourism is the fastest growing segment in the international tourism market with a global market share of almost 30% in 2019. Europe received about 60% of all international trips (= 114 million international city trips in 2017, Messe Berlin and IPK International 2019, 2020). Urban tourism is also booming in Germany. Hamburg—one of Germany's top three destinations—serves as an example: while overnight stays in Germany increased by 29% between 2008 and 2018, the rate of increase in Hamburg was 88% for the same period (Hamburg Tourism 2019). The increase not only in tourist numbers but also changing forms and modes of urban tourism, such as the sharing economy, tourism off the beaten track or party-tourism, have given rise to feelings of "overtourism" and "tourismphobia" (Milano *et al.* 2019). A recent UNWTO study on residents' perceptions of city tourism found that the positive outcomes of tourism, for instance, generation of wealth and income, creation of intercultural exchanges, new offers and more jobs, barely outweigh the negative tourism impacts, such as overcrowding and the rising costs of housing, goods and services (World Tourism Organization and IPSOS 2019). A survey in Hamburg's districts of St. Pauli and Schanzenviertel shows similar results: too many tourists, Disneyfication, gentrification and the production of waste were regarded as the most negative effects (Stimmen von St. Pauli 2014). By contrast, a representative survey on behalf of Hamburg Tourism found that the residents of Hamburg's central districts (including the Schanzenviertel) rate

Figure 9.1 The Rote Flora © Anja Saretzki

tourism chiefly as a positive economic factor (77%) and only 14% felt bothered by tourists (Hamburg Tourismus 2017).

The pressure of tourism has developed in Hamburg's cityscape to a different degree. The Schanzenviertel is among the parts of the city most affected due to its attractiveness for New Urban Tourism. This attractiveness is fed by the special 'Schanzen-feeling'.[2] The Schanzenviertel is not only a former working-class and post-industrial district at the inner city's fringe that offers well-preserved Wilhelminian-style perimeter block housing but is also known as a refractory neighbourhood where alternative political drafts are articulated, manifest themselves and achieve relevance in society as a whole (Tietgen 2014). The Schanzenviertel perceived itself as a distinctive district and hotbed of alternative and countercultural concepts of life. Hence, it is a "great place for consuming authenticity" (Zukin 2010, p. 16) and it fits perfectly into the promises of New Urban Tourism.

It is hypothesised that the Rote Flora as an original protest against the neoliberal logic of exploitation is a substantial driver of touristification and tourism gentrification due to the emergence of New Urban Tourism. The transformation of the Schanzenviertel into a gentryscape and tourismscape serves as an example of how the concurrence of different stakeholders, practices and processes results in touristification and how the stakeholders deal with potential frictions and conflicts within the rise of New Urban Tourism. We analyse the discourses of gentrification and touristification to illustrate the unwanted interaction of protest against urban upgrading with tourism and leisure activities and to provide insights into the challenges of future urban tourism development.

Challenges of New Urban Tourism

New Urban Tourism is a designation for a relatively recent phenomenon in urban tourist studies. This kind of urban tourism focuses on tourists' immersion in the urban everyday life. Tourists experience a city off the beaten track which is beyond special tourism precincts but also beyond typical tourist behaviour patterns. Novy (2018) points out that especially former working-class and post-industrial districts at the inner city's fringe—such as the Schanzenviertel—are subject to increasing numbers of tourists. These kinds of urban areas promise the generation of alternative, more profound and putatively authentic experiences which the New Urban Tourist is looking for (Condevaux *et al.* 2016, Maitland 2010, Stors *et al.* 2019). Füller and Michel (2014, p. 1306) emphasise that "it is a very specific kind of urbanity or urban experience" the New Urban Tourist seeks.

The emergence of New Urban Tourism is based on an increasingly saturated market for conventional forms of tourism resulting from increased travel know-how and the postmodern desire for individuality, autonomy and flexibility in consumption and mobility. Additionally, political-economic developments, such as airline deregulation and the expansion and diversification of the urban accommodation sector, along with technological evolution and innovations, such as internet-based travel services, online platforms and the sharing economy, allow for travelling more independently and flexibly (Bock 2015). New Urban Tourism accentuates the tourist's role as a prosumer who produces the city while consuming it: strolling around the Schanzenviertel means also becoming part of the production of urban space. As a form of creative tourism, New Urban Tourism is bound up with the concept of "living like a local" (Richards 2013) and tourists look for a creative interchange not just with the place and its history but also with the local everyday life, as Maitland (2010) argues. This practice leads to increasing contacts between tourists and residents and influences residents' perceptions of tourism (Dirksmeier and Helbrecht 2015).

At the same time, the touristification of everyday life (Wöhler 2011) effects residents' "as if tourists" attitudes (Lloyd and Clark 2001): similar to the New Urban Tourists, they consume the "new urban culture devoted to aesthetic pursuits" (Judd 2003, p. 32), they use the same urban infrastructure and follow the same activities. In doing so, the boundaries between tourists (as out-of-town visitors), visitors from other districts (so-called internal tourists) and residents, as well as distinctions between tourism and everyday life, blur (Ashworth and Page 2011, Maitland 2010). The indistinguishability between tourism and non-tourism activities and the touristification of everyday life encourages urban change and contributes to the mutation of primary residential neighbourhoods to entertainment and leisure districts—a development the Schanzenviertel has witnessed in the last decade.

The Schanzenviertel as a gentryscape and a tourismscape

The Schanzenviertel developed from a middle-class suburb to an industrial area and a working-class neighbourhood in the middle of the 19th century. The establishment of Hamburg's central slaughterhouse at the end of the 19th century and the factories of Steinway & Sons (pianos) and Montblanc-Simplo (writing utensils) caused the influx of working-class families and established a district where people lived, worked and found leisure amenities. A concert hall named Flora was the main attraction, surrounded by a variety of pubs (Siebecke 2012, Vogelpohl 2012). After World War II, the Schanzenviertel experienced a decline: deindustrialisation gave rise to an impoverishment of the neighbourhood. Most factories were abandoned and buildings fell into neglect. However, the neighbourhood's central location, low rents and spacious vacant edifices made it attractive for students, artists and creative workers. The 1990s new economy boom with several successful pioneering companies especially transformed the quarter, followed by advertising agencies, music businesses, restaurants, bars and small fashionable shops (Oßenbrügge *et al.* 2009). The Schanzenviertel experienced a significant change that formed its character today as a nightlife district and New Urban Tourism destination.

There is nearly unanimous agreement in Hamburg that the Schanzenviertel is considered to be a gentrified but also a touristified neighbourhood by now. An analysis of the transformation process[3] has to work with both concepts of gentrification and touristification. From a scholarly point of view, both concepts lack a clear definition (for gentrification see e.g. Maloutas 2012, for touristification see Cócola Gant 2018, Novy 2018). However, there is a consensus that both processes are linked to one another and mutually reinforce each other (Cócola Gant 2018, Gravari-Barbas and Guinand 2017, Novy 2018). Gotham (2005, 2018) coined the phrase "tourism gentrification" to draw attention to tourism as a driver of gentrification. He claims that the consumer taste for gentrified spaces is created and marketed by powerful organised political and economic interests. This supply-side argument neglects consumer sovereignty as opposed to demand-side arguments and their focus on people as gentrifiers (Zukin 2010). Cócola Gant (2018), on the other hand, argues that tourism also occurs in places that have not been planned as tourist spaces but overlaps with gentrified areas. Hence, Sequera and Nofre (2018, p. 851) refer to "urban touristification, resulting from either (i) the transformation of a working-class neighbourhood into a site for tourist consumption, entertainment and leisure, or (ii) the transformation of an already gentrified neighbourhood into a place for tourist consumption, entertainment and leisure". Their reasoning de-links the direct association between touristification and gentrification because touristification lacks the element of class hierarchy: it causes, *inter alia*, cross-class displacement instead of working-class displacement and a worsening of community liveability instead of class wars.

However, displacement is a significant part of the process for every approach. Cócola Gant (2018) suggests three interrelated forms of displacement as a result of tourism: residential, commercial and place-based displacement. The latter means that residents experience a domination of space by visitors and feel a sense of dispossession from the places they inhabit. They sustain a reduction in their quality of life or, as Sequera and Nofre (2018) put it, a worsening of community liveability. This is especially true regarding the touristification of the urban night (Nofre *et al.* 2018, Rouleau 2017): residents suffer from rubbish and shards of broken glass on the pavements, high noise levels and the rowdy behaviour of partygoers.

The discussion of different approaches showed the variety of interrelations and problems regarding tourism in urban neighbourhoods. Concerning the Schanzenviertel, gentrification and touristification seem to be linked to and mutually reinforce each other. It was not planned as a tourist space, even though it had a certain past regarding entertainment facilities (Tietgen 2014). However, the Schanzenviertel is seen as a distinct neighbourhood in Hamburg and its distinctiveness makes it a focus of discourses on tourism gentrification and protest.

Discourses of gentrification and touristification

Discourses of gentrification and touristification are historically and spatially embedded. Hence, discourses that form the Schanzenviertel have to be understood as spatio-temporally tied constructs (Bauriedl 2008). As "practices that systematically form the objects of which they speak" (Foucault 2002, p. 54) they centre the Schanzenviertel as a place where those social relationships unfold that make it a gentryscape and tourismscape. From a material point of view, Schanzenviertel discourses are seen as a spatial practice that affects social and physical space, but the spatial dimension also has an impact on discursive formations (Bauriedl 2008). In order to grasp the material productivity of the Schanzenviertel discourses and their influencing factors, the analytical focus is on four partial discourses.[4] The discursive objects and concepts, subject positions, or rather enunciative modalities, strategies and the local mapping of partial discourses in the physical space are outlined below.

Data was gathered during field visits in 2019 and at the beginning of 2020. On-site observations and conversations with government officials and local residents provided access for the gathering of media resources and planning documents, brochures and additional materials that make up the majority of data analysed. Neighbourhood-level, citywide, regional and national periodicals were analysed as well as web sites from urban administration, citizens' initiatives, tourism marketing, tour operators, city guides, city magazines, Hamburg blogs and travel blogs. Data came from a great variety of sources with quite opposing viewpoints: not only publications from local authorities but also radical left-wing magazines, such as *Zeck*, were analysed.

Rote Flora discourse

The Rote Flora is an autonomous cultural centre that is associated with Hamburg's radical left-wing scene. The building—the remnant of a former revue theatre and concert hall named Flora, built in 1888—has been squatted in since 1989 (Decker *et al.* 2018). Not designated as a historic monument, the building was mostly pulled down in 1988. The main entrance was the only part to be incorporated in the New Flora, a musical theatre for nearly 2,000 visitors. The new theatre was planned for the musical *Phantom of the Opera* and was part of the intention to establish Hamburg as a "musical capital" (Hamburg Tourismus 2010). A campaign promoted under the slogan "Schmeißen wir das Phantom aus dem Viertel" (Let's throw the Phantom out of the district) was organised to prevent the project. The group in charge constituted the core of the later Flora squatters, the so-called FloristInnen.[5] They warned residents of a tourist flow that broached the district's appropriation by tourists and external visitors (Blechschmidt 1998). This was the first-time occurrence of touristification as a 'Schanzenviertel problem'. The Flora theatre, as a potential visitor honeypot, became the object of a discourse on the future of the Schanzenviertel.

Massive protests had an impact. In the autumn of 1988, the investor had already shelved his plan and built the new theatre in a different location in the neighbouring district. The Flora ruin was vacant at first until the FloristInnen squatted in the building (promoted by a chain of events, see Blechschmidt 1998), now called the Rote Flora. While action groups formed by residents against the musical project disbanded, the radical left-wing scene settled in the district. Today, the Rote Flora is a self-proclaimed political and cultural district centre with a radical left-wing bias and its own local newspaper, the *Zeck* (Anon 2004). The Flora ruin was reconstructed by the FloristInnen and offers space for political group meetings, an archive of social movement documents, practice rooms for local musicians, a motorcycle repair shop, a bar and a venue for concerts and parties. The Flora's local political claim has not only been the organisation of a base for the radical left-wing scene in Hamburg but also the district's preservation (Hoffmann 2011, p. 84, 266f.). The FloristInnen tried to resist the growing gentrification and touristification in the Schanzenviertel. The Flora discourse uses the preservation strategy intensely: maintaining the district's socio-structural composition and defending the Schanzenviertel against outside influences. The concept of struggle played a pivotal role: the struggle against the musical project with the premonition of "invading bulks of tourists" (Anon 1988, transl. A.S.) and a resulting transformation of the Schanzenviertel at first and, in later years, the struggle against gentrification and touristification, especially against internal tourism (Blechschmidt 1998, Rote Flora Plenum 2011). It has always been a struggle on a discursive level that was fought out via pamphlets, contributions in alternative journals, such as the *Zeck*, or interviews in mainstream media, as well as on a non-discursive level via demonstrations or militant actions, such as assaults on building lots. The actions, campaigns and political positions of the FloristInnen experienced great resonance in the media, for example during the 2017 G20 summit. Mainstream media particularly caused a reproduction of the Flora discourse over time and made it adaptable for the rest of Hamburg. The local press became a discourse amplifier because it provided a forum not merely for the FloristInnen but also politicians, who made use of the press to distinguish themselves by calling for vacating the Rote Flora or expressing sympathy for the Flora's positions. The FloristInnen acted as advocates for the residents' interests and showed themselves to be the Schanzenviertel's legitimate representatives. Other groups adopted their arguments, even if their militant actions were treated critically (Hoffmann 2011). Hence, the Rote Flora became the benchmark of a discourse which proved to be the Schanzenviertel's hegemonic discourse.

However, in the long run, the Schanzenviertel's gentrification and touristification was unstoppable. After preventing the New Flora as a visitor magnet, the counter-project of the gentrification-critical Rote Flora turned out to be a tourist honeypot and accelerated the process of revaluation, as some commentators noted (Anon 2004, Hoffmann 2011). State-led gentrification as "gentrification from above" was replaced by a kind of "gentrification from below" (Birke 2014, p. 91). The Rote Flora's subculture as a countermodel for the commodified culture of the New Flora became apparently a more sustainable location factor and fostered the development of the Schanzenviertel as a trendy neighbourhood that is attractive not only for the creative class as a home and a workplace but also for all kinds of New Urban Tourists as a sight and a nightlife district. A combination of the authenticity of Wilhelminian-style buildings, left-wing residents with an alternative lifestyle and the "hip grunge ethos" (Mele 2000), represented by the Rote Flora, makes the Schanzenviertel ideally fitted to the urban gentrification processes depicted, for instance, in the research of Mele (2000) or Zukin (1991) in New York and, more recently, the work of Baudry (2017) for Rome, Novy (2009, 2018) for Berlin or Pappalepore *et al.* (2014) for London.

The FloristInnen's supposed hegemony in the Rote Flora discourse proved to be delusive in the long run. The individual incorporated cultural capital of the FloristInnen and other alternative residents, who are not just creative but also rebellious and interesting, generates "real cultural capital" (Zukin 1990). The protest against gentrification and touristification turns out to be the Schanzenviertel's cultural capital that entails its symbolic upgrading and hence to gentrification and touristification. A discourse shift takes place: the concept of subculture could no longer be linked to the fight against gentrification and touristification. Instead, subculture was considered to be a location factor. On the one hand, the Rote Flora today is the epicentre of the radical left-wing scene in Hamburg. But simultaneously, the iconic building serves as a tourist sight, a venue for concerts and parties and the Schanzenviertel's emblematic centre. The Rote Flora helps to foster the location of trendy bars and coffee shops. The grungy look and a feeling, which is especially informed by the Rote Flora, make all the difference for the Schanzenviertel and establish the perfect destination for New Urban Tourists. Meanwhile, the Rote Flora has gained the status of a brand due to media response within the scope of the Flora discourse. The Schanzenviertel is informed by the Rote Flora and, in the eyes of many residents, "her persistence within the district's dynamics is immensely important" (Anon 2000a, p. 9, transl. A.S.). Residents have, however, no homogeneous attitude to the Rote Flora. This became obvious in the aftermath of the 2017 G20 summit. Advocates consider the Flora to be part of the neighbourhood's DNA, whilst others call for consequences and a distancing from violence, even though a closure or demolition was no option (Arnsperger 2017).

The FloristInnen realised their role in the process of revaluation and commodification of the Schanzenviertel and are aware of their "ambivalence about changing between disturbing factor and location factor" (Rote Flora Plenum 2011, para. 17, transl. A.S., see also Anon 2000b, 2004, Blechschmidt 2018, gruppe demontage 2001). On the one hand, they assume their role ironically when they call the Rote Flora "an authentic backdrop for a fancy audience with purchasing power" (Anon 2001, p. 9), which makes for a very marketable "Bronx feeling for high-income earners" (ibid., all transl. A.S.). On the other hand, they criticise the neoliberal exploitation of the Schanzenviertel and emphasise Flora's role in raising the awareness of processes of gentrification and touristification in the district. However, the attempt to regain sovereignty over the Rote Flora discourse was not successful, as subsequent strands of the discourse will show.

Discourse of inclusion and exclusion

The successful prevention of the musical project had a double effect on the Schanzenviertel's residents: they learned how to organise and articulate their own interests and a 'we-feeling' developed and strengthened local patriotism. Both effects became enmeshed in the discourse of inclusion and exclusion: the question was asked who belongs and who has a right to the Schanzenviertel. The inclusion/exclusion discourse, thereby, affiliated to the general "right to the city" discourse in Hamburg. Lefebvre's demand for the right to the city (1996) aims at privileging the city's use value over its exchange value and guaranteeing use value for the city's residents. For today's metropolises, even the question "who qualifies as a *local*" (Novy 2018, p. 436) seems crucial, considering the interwovenness of current patterns of (tourism) mobility and place consumption (ibid.). Furthermore, the hierarchy of rights is questionable: who has primary rights and what kind of rights (Saretzki 2020). As a residential neighbourhood, a commercial zone and a leisure space, the Schanzenviertel satisfies diverse requirements (Vogelpohl 2012). Tourists and partygoers are objects of the inclusion/exclusion discourse for most residents. The development of a 'we-feeling' is contrasted with reference to 'the Others': "non-locals, who dominate the neighbourhood" or "suburbanites flocking over the weekend" (ARGE Kirchhoff/Jacobs 2011, p. 49, transl. A.S.). Local action groups, such as the Schanzenbeirat or the AnwohnerIni Schanzenviertel refer to the Schanzenviertel as "our neighbourhood", the Schanzenpark as "our park" (Gaukeley 2006, p. 4) and "Reclaim the Schanzenfest" (a neighbourhood festivity, Lopez 2006, p. 7) to emphasise their particular claims on the whole Schanzenviertel. They projected their discontent with local urban politics onto the figure of the tourist, who was held responsible for the district's negative developments.

The inclusion/exclusion discourse provoked the district's self-concept of a hospitable, cosmopolitan and tolerant neighbourhood (Standpunkt Schanze 2017, p. 2), and, due to its emphasis on particular interests, it uncovered the "bogus identity of a self-consistent district" (gruppe demontage 1997, p. 9, transl. A.S.). The FloristInnen agreed on the problem of touristification and the 'We'/'the Others' contrast when they condemned the tourism commodification of local life and called for a "resident only" infrastructure (Anon 1988, p. 9), but they also reflected the neighbourhood's contradictions and their own position in this process (Birke 2014, Blechschmidt 1998, Gruppe Blauer Montag 1997). It becomes apparent that interests and positions of residents and FloristInnen diverged over time but do not became independent, as the next discourse will show.

Discourse of threat and displacement

The discourse of threat and displacement proves to be a typical gentrification discourse. It ties in with the inclusion/exclusion discourse and the Rote Flora discourse as the starting point of gentrification and touristification. However, every discourse operates with different objects and concepts. The displacement discourse is chiefly about public space. There is undoubtedly residential and commercial displacement. The transformation of residential space into tourist dwellings was realised as a problem,[6] but public protest was directed primarily at the prevention of hotel projects (e.g. the Mövenpick Hotel im Wasserturm or the Pyjama Park Hotel).[7] Commercial displacement applies to the substitution of cheaper stores for everyday commodities by upscale boutiques and predominantly to the substitution of retail stores by gastronomic amenities (Bezirksamt Altona 2012, p. 25). However, the discourse focus in the Schanzenviertel is on place-based displacement.

The Schanzenviertel was declared an urban redevelopment area between 1999 and 2012. Under the heading of the so-called "careful urban renewal", an inclusive planning approach (Pfotenhauer 2000), the *steg* (Stadterneuerungs- und Stadtentwicklungsgesellschaft Hamburg)[8] was commissioned for the co-ordination of the programme. The opening of the so-called Piazza (Figure 9.2) was a cornerstone in the redevelopment process: a complete renovation of the street in front of the Rote Flora with a significant broadening of sidewalks that gives restaurants and bars the possibility of spilling out extensively into the public space for their commercial purposes. The Piazza forms part of Schulterblatt street, the main commercial street in the Schanzenviertel, and boosted the neighbourhood's 'gastronomisation'. Up to now, nearly a hundred gastronomic amenities have been located in a small triangle of streets in the heart of the district (ARGE Kirchhoff/Jacobs 2011, p. 25). Additionally, there are several kiosks in the area that sell alcohol far into the night and foster the development of the so-called '*cornern*', a kind of hanging around at street corners while drinking and talking.[9] This behaviour turned out to be a significant factor of the place-based displacement in the Schanzenviertel (Twickel 2017). Most residents saw the Piazza as a trigger for the massive touristification of the neighbourhood and complain of increasing noise intensity, litter problems in the streets and backyards and the reckless behaviour of tourists and partygoers (ARGE Kirchhoff/Jacobs 2011, p. 48). They raised their concerns in interviews with social scientists, consultants,

Figure 9.2 The Piazza © Anja Saretzki

urban renewal agents and the media, in meetings of local action groups and public panel discussions calling for, *inter alia*, a limitation of open-air gastronomy and behaviour instructions for tourists (Standpunkt Schanze 2019). The residents dominated the displacement discourse even though the Bezirksamt Altona and the *steg* as her agent tried to regain control over the discourse. The *steg* used their newsletter, the *Schanze Quartiersnachrichten*, to make the renewal a success and to downplay the problems of touristification. Contrasting the concepts of 'threat' and 'displacement' with 'careful urban renewal' was attempted. Bezirksamt Altona and *steg* claimed to enhance the neighbourhood while preserving its social mixture and identity at the same time (Bezirksamt Altona 2012). However, the district's identity was not only the identity of a lively neighbourhood but, with the Rote Flora and its history of protest, of a rebellious one.

Marketing discourse

The marketing discourse raised the struggle for the Schanzenviertel from a local to a global level. The district became a distinct feature in Hamburg's global competitiveness as a tourist destination. The Hamburg Tourismus GmbH (henceforth HHT), Hamburg's destination management organisation, aims at fostering tourism growth, improving the city's quality of life and supporting Hamburg's development as a tolerant and liveable city (Hamburg Tourismus 2010, p. 11). The HHT sees the city as a "place to be" that can "promise a unique flair" and "authenticity based on history and tradition" (ibid., p. 120). The Schanzenviertel corresponds to this image, offering urban renewal plus rebellious history and Wilhelminian-style façades plus edginess. It reveals tourism's potential to become unlocked. Hence, the marketing discourse picked up the Flora discourse as a location discourse and included the counter-culture identity cultivated in the displacement discourse. Once again, the Rote Flora became the object of a discourse: instead of being a contested place, the Rote Flora transmuted into the metaphor for an exciting destination. The concept of subculture was no longer linked to the concept of struggle but to radiating liveliness and attractiveness. The HHT pursues a growth strategy and, therefore, needs a unique selling proposition to hold its own in global competition. Militant actions, the claim for exclusive rights and the protest against displacement turned out to be the neighbourhood's exploitable DNA: "resistance becomes not only incorporated into the process of gentrification, but is *distinctly commodified*" as Naegler (2012, p. 13) argued. The Rote Flora discourse is affiliated with the marketing discourse in seeing the Flora as a location factor. The HHT's press officer described the Schanzenviertel as a decisive part of the city with the Rote Flora as an "eye-catcher" (Arnsperger 2017). However, he underlines the fact that the HHT does not use the Rote Flora for advertising the city (Heinemann 2017). The website of the HHT and that of the Bezirksamt Altona show the Rote Flora as the focal point of the presentation of the district. The Flora is visually present and descriptions of the district use terms such as "independent" (regarding stores), a "special atmosphere", "unique", "trendy", "hip", 'creative' and 'iconic'. Hence, an attitude to life is marketed that needs the Rote Flora as a gimmick. This is also true for the representation of the Schanzenviertel in travel blogs or on social media websites: on Instagram, it belongs to the most tagged hotspots in Hamburg.[10] For the HHT, the marketing of the Schanzenviertel is still in its infancy and needs to be unlocked.

Conclusion

New Urban Tourism is—as are most forms of tourism—a longing for authentic tourist experiences, but it is focused on non-tourist urban environments. The encounter of tourists and residents of the affected neighbourhood (and probably other city users) often causes problems when residents feel displaced in several ways. Protest and resistance are accompanied by increasing touristification in a densely populated area such as the Hamburger Schanzenviertel. Hence, there is a special situation in the district regarding the Rote Flora: resistance against touristification merges with political protest and a critique of neoliberal capitalism. The analysis of different discourses around the touristification of the Schanzenviertel showed not only the formation of a discourse coalition but also paradox alliances. The findings of the study suggest that the Schanzenviertel's culture of resistance is harnessed to draw attention to Hamburg as an attractive tourist destination. As a refractory space, the Schanzenviertel became an authentic place and an attractive destination. The Schanzenviertel's alternative lifestyle, with the protest against gentrification and touristification, effectuates an 'Othering' and re-exoticises banal urbanity and commercial culture and generates real cultural capital in Zukin's sense (Zukin 1990). Thus, the incorporation of resistance may result in the commodification of resistance. Urban renewal strategies profit from such developments, as the Piazza example has shown. The Schanzenviertel's distinctiveness represents an authentic urban environment and a perfect destination for New Urban Tourists. However, it is contested regarding the various and conflicting ways of use.

Protest and resistance to urban tourism can take various forms (Colomb and Novy 2017), entail various consequences (from incorporation to commodification of resistance, Naegler 2012) and challenge the city in various ways (from resistance to tourism to resistance that uses tourism, Gravari-Barbas and Jacquot 2017). However, it is hard to imagine that protest and resistance may stop tourism in cities.

Notes

1 The districts Sternschanze (the administrative designation of the district) and HafenCity were newly formed in 2008 as a result of Hamburg's zoning reform. The Sternschanze district covers 0.5 km^2 and is densely populated
2 The short form "Schanze" (entrenchment) is used colloquially. The term is derived from a former fortification.
3 A detailed analysis of the gentrification process on the whole is beyond the scope of this contribution.
4 The methodological approach of the study is based on Diaz-Bone (2006) and Keller (2011).
5 Used here as a synonym for "different", in the course of the discourse evolving groups of squatters and activists.
6 According to the Bezirksamt Altona (the district authority responsible), the misappropriation of residential space via Airbnb or other holiday rental platforms has been a significant problem in the Schanze's tight housing market since 2013. For counteractive measures, see https://www.hamburg.de/wohnraumschutz/ [Accessed 30 April 2020].
7 For further information, see the website http://www.schanzenturm.de/ [Accessed 30 April 2020].
8 The *steg* was founded by the city of Hamburg in 1989 but has been a private company since 2003.
9 *Cornern* is a German neologism that is derived from the English word corner. While the corresponding behaviour is a widespread nightlife practice, the term *cornern* is especially common in Hamburg.
10 https://www.mopo.de/hamburg/vorzeigbare-hansestadt-das-sind-die-top-10-der-hamburger-instagram-motive-28555322 [Accessed 26 April 2020].

References

Anon., 1988. Flora—was sonst. *Neues Schanzenleben*, 8, 2–9.
Anon., 2000a. Erster Mai–mitten drin statt nur dabei. *Schanze Quartiersmanagement*, July, 2, 9.

Anon., 2000b. [K]eine Flora mit Verträgen? Interview mit vier AktivistInnen aus der Roten Flora. *Zeck*, 94, 5–9.

Anon. 2001. Eurhythmie? Auf die Füße treten! Kretschmer, die Stadt, der Standort und Gentrification. [Press release] 19 June 2001. *Zeck*, 100, 8–9.

Anon. 2004. 15 Jahre besetztes autonomes Zentrum–15 Jahre Rote Flora. *Zeck*, 123, 7–9.

ARGE Kirchhoff/Jacobs, 2011. *Soziale Er-hal-tungs-ver-ord-nung für das Gebiet Sternschanze. Gutachten zur Überprüfung der Anwendungsgrundlagen einer Sozialen Er-hal-tungs-ver-ord-nung gemäß BauGB § 172 Abs.* 1 Satz 1 Nr. 2. Hamburg.

Arnsperger, M., 2017. So denken Nachbarn im Schanzenviertel jetzt über die Rote Flora. *Focus Online* [online]. Available from: https://m.focus.de/politik/deutschland/hamburg-so-denken-nachbarn-im-schanzenviertel-jetzt-ueber-rote-flora_id_7349762.html [Accessed 26 April 2020].

Ashworth, G. and Page, S.J., 2011. Urban tourism research: recent progress and current paradoxes. *Tourism Management*, 32 (1), 1–15.

Baudry, S.L., 2017. Rome: a cultural capital with a poor working-class heritage. *In*: M. Gravari-Barbas and S. Guinand, eds. *Tourism and gentrification in contemporary metropolises. International perspectives*. London: Routledge, 134–152.

Bauriedl, S., 2008. Räume lesen lernen: Methoden zur Raumanalyse in der Diskursforschung. *Historical Social Research*, 33 (1), 278–312.

Bezirksamt Altona, 2012. *Begründung zum Bebauungsplan Sternschanze 6*. Hamburg: Bezirksamt Hamburg-Altona.

Birke, P., 2014. Autonome Sehenswürdigkeit. Die Rote Flora und die Hamburger Stadtentwicklung seit den späten 1980er Jahren. *Sozial. Geschichte Online* [online], 13, 80–104. Available from: https://nbn-resolving.org/urn:nbn:de:hbz:464-20140530-141748-4 [Accessed 25 April 2020].

Blechschmidt, A., 1998. "Gleichgewicht des Schreckens". Autonomer Kampf gegen Umstrukturierung im Hamburger Schanzenviertel. In: StadtRat, eds. *Umkämpfte Räume*. Hamburg: Verlag Libertäre Assoziation/Verlag der Buchläden Schwarze Risse–Rote Strasse, 83–101.

Blechschmidt, A., 2018. Niemals aufgeben–die Geschichte der Roten Flora seit 1989. *In*: E. Decker, H. Schultze and G. Zint, eds. *Die Geschichte der Flora am Schulterblatt*, 2nd ed. München: Dölling & Galitz, 47–49.

Bock, K., 2015. The changing nature of city tourism and its possible implications for the future of cities. *European Journal of Futures Research* [online], 3 (20). Available from: doi:10.1007/s40309-015-0078-5 [Accessed 18 April 2020].

Cócola Gant, A., 2018. Tourism gentrification. *In*: L. Lees and M. Phillips, eds. *Handbook of gentrification studies*. Cheltenham: Edward Elgar, 281–293.

Colomb, C. and Novy, J., 2017. *Protest and resistance in the tourist city*. London: Routledge.

Condevaux, A., Djament-Tran, G., and Gravari-Barbas, M., 2016. Before and after tourism(s). The trajectories of tourist destinations and the role of actors involved in "Off-The-Beaten-Track" tourism: a literature review. *Via@Tourism Review* [online], 9. Available from: https://journals.openedition.org/viatourism/413 [Accessed 14 April 2020].

Decker, E., Schultze, H., and Zint, G., 2018. *Die Geschichte der Flora am Schulterblatt*. 2nd ed. München: Dölling & Galitz.

Diaz-Bone, R., 2006. Zur Methodologisierung der Foucaultschen Diskursanalyse. *Historical Social Research*, 31 (2), 243–274.

Dirksmeier, P. and Helbrecht, I., 2015. Resident perceptions of new urban tourism: a neglected geography of prejudice. *Geography Compass*, 9 (5), 276–285.
Foucault, M., 2002. *Archaeology of knowledge*. London: Routledge.
Füller, H. and Michel, B., 2014. "Stop Being a Tourist!" New dynamics of urban tourism in Berlin-Kreuzberg. *International Journal of Urban and Regional Research*, 38 (4), 1304–1318.
Gaukeley, G., 2006. Schrumpfender Sternschanzenpark: Wie der Umbau das *Nah-er-ho-lungs-ge-biet* des Viertels frisst. *schanze | 20357*, Sep/Oct, 2, 4–5.
Gotham, K.F., 2005. Tourism gentrification: the case of New Orleans' Vieux Carre (French Quarter). *Urban Studies*, 42 (7), 1099–1121.
Gotham, K.F., 2018. Assessing and advancing research on tourism gentrification. *Via@ Tourism Review* [online], 13. Available from: https://journals.openedition.org/viatourism/2169 [Accessed 25 March 2020].
Gravari-Barbas, M. and Guinand, S., 2017. Introduction: addressing tourism-gentrification processes in contemporary metropolises. *In*: M. Gravari-Barbas and S. Guinand, eds. *Tourism and gentrification in contemporary metropolises. International perspectives*. London: Routledge, 1–21.
Gravari-Barbas, M. and Jacquot, S., 2017. No conflict? Discourses and management of tourism-related tensions in Paris. *In*: C. Colomb and J. Novy, eds. *Protest and resistance in the tourist city*. London: Routledge, 31–51.
Gruppe Blauer Montag, 1997. Thesen zur Situation im Schanzenviertel. *Zeck*, 65, 6–7.
Gruppe demontage, 1997. Jenseits von Eden. Zur Diskussion um die Situation im Schanzenviertel. *Zeck*, 65, 8–9.
Gruppe demontage, 2001. Städtische Modernisierung und Restlinke. Zur Auseinandersetzung um die Rote Flora in Hamburg. *Zeck*, 100, 12–18.
Hamburg.de., n.d. *Schanzenviertel: Beliebtes Szeneviertel* [online]. Available from: https://www.hamburg.de/schanzenviertel/ [Accessed 31 August 2020].
Hamburg Tourismus, 2010. *Marketingplan 2010–2015*. Hamburg: Hamburg Tourismus.
Hamburg Tourismus, 2017. *Zustimmung der Hamburger Einwohner zum Tourismus. Repräsentative Einwohnerbefragung 2017*. Hamburg: Hamburg Tourismus.
Hamburg Tourismus, 2019. *Zahlen, Fakten, trends 2018*. Hamburg: Hamburg Tourismus.
Hamburg Tourismus, n.d. *Rote Flora und Schulterblatt. Pulsierendes Viertel* [online]. Available from https://www.hamburg-tourism.de/shoppen-geniessen/szene-nachtleben/rote-flora/ [Accessed 31 August 2020].
Heinemann, C., 2017. "Wir machen mit der Roten Flora keine Werbung". Interview mit Sascha Albertsen von Hamburg Tourismus. *Hamburger Abendblatt*, 16 September [online]. Available from: https://www.abendblatt.de/hamburg/article211942501/Wir-machen-mit-der-Roten-Flora-keine-Werbung.html [Accessed 26 April 2020].
Hoffmann, K.D., 2011. *"Rote Flora". Ziele, Mittel und Wirkungen eines linksautonomen Zentrums in Hamburg*. Baden-Baden: Nomos.
Judd, D.R., 2003. Visitors and the spatial ecology of the city. *In*: L.M. Hoffman, S.S. Fainstein, and D.R. Judd, eds. *Cities and visitors: regulating people, markets, and city space*. Malden, MA: Blackwell, 23–38.
Keller, R., 2011. *Diskursforschung. Eine Einführung für SozialwissenschaftlerInnen*. 4th ed. Wiesbaden: VS Verlag.
Lefebvre, H., 1996. *Writings on cities* (translated and introduced by E. Kofman and E. Lebas). Oxford: Blackwell.
Lloyd, R. and Clark, T.N., 2001. The city as an entertainment machine. *In*: K.F. Gotham, ed. *Critical perspectives on urban redevelopment*. New York: Emerald, 357–378.

Lopez, F., 2006. Reclaim the Schanzenfest. *schanze | 20357*, Sep/Oct, 2, 7.

Maitland, R., 2010. Everyday life as a creative experience in cities. *International Journal of Culture, Tourism and Hospitality Research*, 4 (3), 176–185.

Maloutas, T., 2012. Contextual diversity in gentrification research. *Critical Sociology*, 38 (1), 33–48.

Mele, C., 2000. *Selling the lower east side. Culture, real estate, and resistance in New York City*. Minneapolis, MN: University of Minnesota Press.

Messe Berlin & IPK International, 2019. *ITB world travel trends 2018/2019: what are the trends to look out for?* Berlin: Messe Berlin.

Messe Berlin & IPK International, 2020. *ITB world travel trends 2020: latest insights & outlook*. Berlin: Messe Berlin.

Milano, C., Novelli, M., and Cheer, J.M., 2019. Overtourism and tourismphobia: a journey through four decades of tourism development, planning and local concerns. *Tourism Planning & Development*, 16 (4), 353–357.

Minutes of the 49th meeting of the Stadtteilbeirat Sternschanze. 25 October 2017 [online]. Standpunkt Schanze. Available from: http://www.standpunktschanze.de/stadtteilbeirat-sternschanze-protokoll-der-49-sitzung-am-25-10-2017/ [Accessed 30 April 2020].

Minutes of the 64th meeting of the Stadtteilbeirat Sternschanze, 24 April 2019 [online]. Standpunkt Schanze. Available from: http://www.standpunktschanze.de/stadtteilbeirat-sternschanze-protokoll-der-64-sitzung-vom-24-04-2019/ [Accessed 30 April 2020].

Naegler, L., 2012. *Gentrification and resistance. Cultural criminology, control, and the commodification of urban protest in Hamburg*. Berlin: LIT.

Nofre, J., et al., 2018. Tourism, nightlife and planning: challenges and opportunities for community liveability in La Barceloneta. *Tourism Geographies*, 20 (3), 377–396.

Novy, J., 2009. Kreuzberg's multi- and inter-cultural realities: are they assets? *In*: V. Aytar and J. Rath, eds. *Selling ethnic neighborhoods: the rise of neighborhoods as places of leisure and consumption*. New York: Routledge, 68–84.

Novy, J., 2018. "Destination" Berlin revisited. From (new) tourism towards a pentagon of mobility and place consumption. *Tourism Geographies*, 20 (3), 418–442.

Oßenbrügge, J., Pohl, T., and Vogelpohl, A., 2009. Entgrenzte Zeitregime und *wirt-schafts-räum-li-che* Konzentrationen. Der Kreativsektor des Hamburger *Schan-zen-vier-tels* in zeitgeographischer Perspektive. *Zeitschrift für Wirt-schafts-geo-gra-phie*, 53 (4), 249–263.

Pappalepore, I., Maitland, R., and Smith, A., 2014. Prosuming creative urban areas. Evidence from East London. *Annals of Tourism Research*, 44, 227–240.

Pfotenhauer, E., 2000. Stadterneuerung–Sanierung. *In:* Häußermann, H., ed. *Großstadt. Soziologische Stichworte*. 2nd ed. Wiesbaden: VS Verlag für Sozialwissenschaften, 247–257.

Richards, G., 2013. Creating relational tourism through exchange. *In*: *ATLAS Annual Conference*, November 2013, Malta.

Rote Flora Plenum, 2011. Flora bleibt unverträglich: Zu den kommenden Kämpfen um die Rote Flora [online]. *Florableibt*. Available from: http://florableibt.blogsport.de/images/floradruckversion.pdf [Accessed 19 April 2020].

Rouleau, J., 2017. Every (nocturnal) tourist leaves a trace: urban tourism, nighttime landscape, and public places in Ciutat Vella, Barcelona. *Imaginations*, 7 (2), 58–71. Available from: doi:10.17742/IMAGE.VOS.7-2.3 [Accessed 3 March 2019].

Saretzki, A., 2020. Haben Touristen ein Recht auf Stadt? *In*: J. Reif and B. Eisenstein, eds. *Tourismus und Gesellschaft. Kontakte–Konflikte–Konzepte*. Berlin: ESV, 111–128.

Sequera, J. and Nofre, J., 2018. Shaken, not stirred. New debates on touristification and the limits of gentrification. *City*, 22 (5–6), 843–855.

Siebecke, G., 2012. *Die Schanze: Galão-Strich oder Widerstandskiez? Streifzug durch ein klammheimliches Klavierviertel*. Hamburg: VSA.

Stimmen von St. Pauli, 2014. Die Ergebnisse der Bewohner*innenbefragung in St. Pauli. *Umfrage von Juni bis September 2014* [online]. Available from: http://stimmen-von-st-pauli.net/wp-content/uploads/2014/09/stimmen_von_stpauli_umfrage-ergebnisse_20-09-2014.pdf [Accessed 2 April 2020].

Stors, N., et al., 2019. Tourism and everyday life in the contemporary city. An introduction. *In*: T. Frisch, et al. eds. *Tourism and everyday life in the contemporary city*. London: Routledge, 1–23.

Tietgen, J., 2014. *St. Pauli & Schanzenbuch*. Hamburg: Junius.

Twickel, C., 2017. Um die Ecke gebracht. *Zeit Online*, 19 October [online]. Available from: https://www.zeit.de/2017/43/cornern-st-pauli-sternschanze-hamburg [Accessed 30 April 2020].

Vogelpohl, A., 2012. *Urbanes Alltagsleben. Zum Paradox von Differenzierung und Homogenisierung in Stadtquartieren*. Wiesbaden: Springer VS.

Wöhler, K., 2011. *Touristifizierung von Räumen. Kulturwissenschaftliche und soziologische Studien zur Konstruktion von Räumen*. Wiesbaden: VS Verlag.

World Tourism Organization and IPSOS, 2019. *Global survey on the perception of residents towards city tourism: impact and measures*. Madrid: UNWTO.

Zukin, S., 1990. Socio-spatial prototypes of a new organization of consumption: the role of real cultural capital. *Sociology*, 24 (1), 37–56.

Zukin, S., 1991. *Landscapes of power. From Detroit to Disney World*. Berkeley, CA: Univ. of California Press.

Zukin, S., 2010. *Naked city. The death and life of authentic places*. New York, NY: Oxford University Press.

10 The empty boxes of Venice

Overtourism—conflicts, politicisation and activism

Marta Torres Ruiz

The perception of 'overtourism' (UNWTO *et al.* 2018) combined with urban transformations caused by New Urban Tourism (Fainstein *et al.* 2003, Maitland 2007, Novy 2010) has recently led to protests and activism in cities around the world. These politicisation processes of city tourism have gained relevance in tourism and urban studies. The introduction of non-deterministic theory, such as regulation theory (Fainstein *et al.* 2003), has given priority to conflict and agency in the research and analysis of tourism. Conflicts between visitors and residents have been discussed as a form of class struggle (Maitland 2007). However, the question of how these conflicts are constituted is still open.

This research tries to bridge this gap by focusing on civic groups who complain about overtourism in their place of residence. It covers a specific case in a European city and a prominent reference in the discussion of overtourism: the vacant houses of Venice and their discoursivisation as empty boxes. The Venetian housing shortage is a source of conflict where different interests of tourists and citizens overlap. It is expected that it will help to find answers at the interface of overtourism and New Urban Tourism. The aim of the following case study is to trace lines of conflict provided by empirical material, such as arguments and narratives of interviewed residents, landlords and activists of Venice. The chapter argues that besides structural class antagonisms and antagonisms between tourists and residents, there is at least one other constituent relationship, namely that between residents and city administration.

Conflicts, politicisation and activism

New Urban Tourism has not gone unchallenged and tourism-related conflicts have arisen in different cities. While having a similar starting point, namely experiences of conflict with tourism, the topics that have been addressed in scientific studies present manifold approaches just as the involved actors battle a wide spectrum of adverse effects. In the following, the approach to these conflicts is chosen via the terms and research fields of overtourism, New Urban Tourism and tourism related activism.

Overtourism, conflicts and negativity

In the recent years overtourism has been gaining attention in the tourism industry and in tourism study (Dodds 2019, Milano *et al.* 2019a, Pechlaner et al. 2019, McKinsey &

Company and World Travel & Tourism Council 2017, OECD 2020). Simultaneously, the term 'overtourism' has been used as an expression for protests of a large number of civic groups and their frustration with the growing tourism industry within cities (Milano *et al.* 2019b). Therefore, research of tourism has partly shifted to conflict-related studies dealing with the global politicisation of tourism in urban spaces.

The rapidly growing attention to this term is followed by the formal implementation of overtourism through a definition by the World Tourism Organisation (UNWTO): "the impact of tourism on a destination, or parts thereof, that excessively influences perceived quality of life of citizens and/or quality of visitor experiences in a negative way" (UNWTO *et al.* 2018, p. 4). It is almost impossible to sum up overtourism in a simpler formula, as it is discoursified by so many media, disciplines and actors. Almost all scientific literature on the topic begins by identifying massive growth in tourist numbers. As a result, the perceived impact on destinations is assumed to be environmental, infrastructural, economic and social— such as loss of quality of life for residents, loss of quality of experience for visitors, straining tourist carrying capacities (UNWTO *et al.* 2018), gentrification processes, loss of cohesion, overcrowding (Milano *et al.* 2019a), anti-tourist sentiments (Pechlaner *et al.* 2019), xenophobia and many more. The methods for dealing with the subject are as comprehensive as the solutions offered. For example, a study by the European Parliament (Peeters *et al.* 2018), lists 121 responses and measures against overtourism. It is a topic usually associated with European cities but also has been mentioned in relation to all kinds of destinations (Milano *et al.* 2019a). Since overtourism is clearly neither related to an actor, a motivation, an actual or perceived effect, nor a solution or a place, UNWTO's open definition leaves only one common denominator: negativity.

The assessment of overtourism as a problematic phenomenon is nothing new in tourism studies. Criticism of tourism has often been treated through abstract terms such as loss of authenticity, place and culture. As far as urban areas and the city are concerned, there is historical evidence of disgust with the crowds of people, as visible in a quote of an Englishman in 1865: "The cities of Italy are now deluged with droves of the creatures" (Boorstin 1967, p. 85). More than 100 years later Jost Krippendorf predicted the future increase of city tourism (Krippendorf 1975). He argued that after the total occupation of the landscape with second homes, there would be nothing left for tourists but the city. This would lead to a "lethal imbalance" (ibid., p. 62). His publication documents a more than critical pathologisation of tourism that was abundant in those years (Young 1973, Turner and Ash 1975).

Despite the long history of discussion, there are two things that may have changed in the course of the last two decades. First, tourism studies seem to have a growing interest in civil activists, their demands and arising social conflicts as part of politicisation processes in cities (Colomb and Novy 2017). Second, civic groups all over the world are organising in transnational networks based on the common confrontation with overtourism, like SETnet (Sud d'Europa contra la Turistització—South Europe against Touristification) and the Civil Society Network of Mediterranean Historical Cities. In the same way, research has partly shifted to a transnational level (Hayes and Zaban 2020) referring to the complexity of post-industrial tourism as a global phenomenon of a neoliberal economy (Bianchi 2009). Allegations by residents or civic groups about gentrification, privatisation of public properties, deregulation at administrative and legal level and the simultaneous promotion of tourism, are subsumed under this term and internationally discussed. From this point of view, overtourism seems rather to represent a broad set of global political demands and struggles than a purely local economic, social or environmental problem. Still the line of argumentation in overtourism is drawn as per definition through the negative impact of tourism in relation to its destination. The division between visitors and residents is maintained.

New Urban Tourism, conflict and agency

In New Urban Tourism the explanation of conflict has shifted from the division between visitors and residents to argumentation of class struggle (Maitland 2007). Conflict and agency have been introduced into a theoretical framework and have gained priority in analysis (Fainstein *et al.* 2003).

Maitland (2007) suggests that conflicts resulting from tourism-related transformations in cities are an outcome of class opposition, rather than the pure distinction between hosts and guests. He therefore refers to the "cosmopolitan consumer class" (Fainstein *et al.* 2003) in overlap to the "creative class" (Florida 2002) that build a "dominant cosmopolitan creative class" in conflicted relation to the "poorer local population" (Maitland 2007, p. 85).

Fainstein *et al.* (2003) introduce regulation theory (Aglietta 1979) into the field of New Urban Tourism—a non-deterministic theory, that postulates economic, social and political transformation as radically constituted by their relation to each other (Marchart 2013). They therefore use regulation theory as a form of structural analysis that "builds in agency [...] while accepting that a given regime of accumulation structures systems" (Fainstein *et al.* 2003, p. 241). In these terms, agency as a form of regulation and non-determinism—or in other terms contingency, conflict and compromise (Marchart 2013)—are key aspects of an approach to the analysis of urban transformation processes. The primacy of conflict and its outcome as partial compromises extends in regulation theory not only to the political institutions, but also to the whole of the social spectrum (Marchart 2013).

In this way, civil protest and activism are opened for analysis in the field of New Urban Tourism, which can be extended to overtourism. At the interface of both research fields, conflict and opposition are common denominators, emphasising the relationship between residents and visitors (overtourism) and socio-economic struggles (New Urban Tourism). The lines of conflict that can be found in the discourse and activism of civic groups are the subject of this chapter.

Civic groups, politicisation and activism

Civic groups involved in anti-tourism or the impairment of tourism are not a new phenomenon either. Their emergence has been addressed since the 1950s (Tourist go home 1959). However, there are signs of a growing politicisation of the issue of city tourism, through protests and the formation of resistance by civic groups on a global scale (Colomb and Novy 2017). "Politization from below" (ibid.) or protest politics directed against neoliberal shifts in urban politics and the lack of democratic participation in urban transformations are central aspects of their criticism, which has led to activism and diversified forms of group organisations.

The perception of these shifts has justified the formation of groups of activists in European cities, such as ABDT (Barcelona), UdrugaGrad (Dubrovnik) and Gruppo25aprile (Venice) in the last decade. Their formation was triggered by decisions by the city administration to rededicate public property for tourism purposes or to enlarge tourism infrastructures. For example, the group Srđ je Grad was formed 2013 in Dubrovnik during a referendum against the lease of public land on Mount Srđ for a private golf resort project. Yet, the positions of these heterogenic groups in terms of discoursivisation of tourism are diverse and subject to change. In the current discourse there are demands for more sustainable forms of tourism or, in a more radical way, an immediate decrease of tourism. 'The living city' is opposed to the tourism industries, such as in the external representation of Comitato No grandi Navi which 'combats' (see Figure 10.1) the growth of ports in Venice.

Most of the members of different groups interviewed in several European cities state that they see themselves as civic, as something that has to do with the people.

Figure 10.1 External representation of Comitato No grandi Navi in Venice (Translation of wording: "Support the battle for the living city. Support the committee. Each donation can make a difference"). © Marta Torres Ruiz

Yet their position within or outside of politics differs. For example, two of the groups interviewed ran on civic lists (*lista civica*)—an Italian form of candidature independent of political parties—and gained seats for the Venice City Council in 2020. Both explain that the step towards participation into what they call "real politics" has not always been a goal. A group interviewed in Barcelona says that they would see themselves as political agents and as a representation of "the people" and that they mediate between "them" and the city government. They therefore are part of political decision-making in an advisory role. In an effort to widen the participation of residents towards social groups with fewer economic resources and, as one member states, "with almost no voice", this group is developing less formal formats of meeting. The formats of meeting and action are diverse, starting from regular assemblies, to specific campaigns, demonstrations or eviction prevention and more directed events for their neighbourhood, like open-air cinemas, street picnics, etc. All of the groups met conduct studies and carry out workshops on their concerns and draw on open data from their governments. The form of organisation and funding differs from non-formal to formalised associations. The organisations found are mainly based on horizontal hierarchies.

Even though the different groups within a city do not necessarily agree on all the objectives, they tend to collaborate. In Venice for example there are several groups related to tourism, but with different focuses: housing (ASC Venezia), ecology (Venezia Pulita), ports (Comitato No grandi Navi) or others with a broader understanding of the topic (Gruppo25aprile, OPA). Comitato No grandi Navi and Gruppo25aprile supported each other at a big demonstration with around 8,000 to 10,000 participants in 2019 (Cricket 2019). In Barcelona, there are several groups that relate to a specific quarter (Barris), and one group, ABDT, acting as their all-connecting representation. The European networks set up by these groups are also developing in parallel. SETnet and the Civil Society Network of Mediterranean Historical Cities overlap in cities but collaborate with different groups within these cities.

The vacant houses of Venice

The city of Venice is an important reference point for the discussion on overtourism. The striking comparison between rapidly growing tourist numbers and shrinking population of the city is a recurring theme in this respect. International tourist arrivals in Venice have increased by around 1.2 million between 2011 and 2019 (Statista 2020). The municipality hast lost around 10,000 inhabitants between the years 2009 and 2019 (Città di Venezia 2019). The tight housing rental market is seen as a major driver for residents to leave the city who can no longer afford to rent on the island. The following case study focuses on the different conflicts, ranging from the struggles between tourist and residents while finding a place to rent, to social-economic opposition and regulation or deregulation policies.

The rapid decline in tourism due to COVID-19 (Gössling *et al.* 2020) has surprisingly not resulted in a decline in discontent with tourism in the researched activist groups in Venice. The claims and demands of several members of different interrelated groups and initiatives I interviewed in September 2020, are still far-reaching: housing, cruise ships, ecology of the lagoon, precarious economy, loss of neighbourhoods, lack of vision for the city, noise, pollution and corruption. On the recurring topic of housing, two persons independently used the same intriguing metaphor referring to the city and its public housing properties: the 'empty box' or 'empty boxes'. They were used in connection to two issues of contention: privatisation of public housing and the lack of regulation for short-term housing rentals.

Empty boxes and occupiers: privatisation of public housing

Usually, when it comes to overtourism and housing, we can find claims about Airbnb and the conversion of long-term rentals for residents into short-term rentals for visitors (Nieuwland and van Melik 2020). In a city like Venice with a decreasing population and increasing numbers of tourists, not only private housing but also public housing becomes part of the conflict. While walking through Venice with one group, one member points to a large old vacant building: "It was public housing, but the administration wouldn't do anything about it" (group member M.). The lack of affordable and public housing is seen as a factor leading to a radical depopulation of the city or at least to a major imbalance. A member of a civic group explains that, due to this process, Venetians like him are treated as an endangered species:

> There is a digital (population) counter in Venice, which tells us when we will have reached zero inhabitants, like pandas. I think this is the wrong message. [...] But there is an enormous contrast between the shrinking population and the number of vacant apartments. This shows an imbalance, which is upside down in Venice.
>
> (Group member N.)

The relation between large numbers of vacant housing and the shrinking population (what N. calls "contrast") falls into what Gentili and Hoekstra (2019) call the "Italian paradox" or "houses without people and people without houses". The housing shortage stays in paradoxical relation to the high vacancy of apartments on the island, because high vacancy not necessarily means high supply.

In 2017 the Municipality of Venice has had a total of 5,184 publicly owned homes administrated by the public economic entity ATER (Azienda territoriale per l'edilizia residenziale di Venezia). Of them, 791 were vacant, because they were in need of major maintenance work. Between 2011 and 2018 ATER had planned the restoration of 190 vacant homes, but so far only completed around 10% of the projects. In almost the same period between 2009 and 2018, ATER sold 364 apartments. For the years 2019 and 2020 ATER has planned the restoration of 203 vacant residences (all numbers OCIO 2020a). The lack of success in maintenance work and reassignment of vacant social housing is explained by the public administration, in this case ATER, with lack of funding (La Voce di Venezia 2019). For this reason, a new regional law (L.R. 39/2017) was adopted in 2017 that allows the sale of public property on the basis of strict regulations. ATER declares that, following the law, they plan to sell apartments with a 20% discount below the market value to the tenants if they are occupied, and to also sell some vacant apartments to third parties not to make profit but to get the very much needed funds for the refurbishment works (La Voce di Venezia 2019).

For the civic groups the combination of privatisation of social housing together with increasing numbers of visitors and the decreasing number of residents becomes not only an argument of discontent with tourism, but also leads to allegations and accusations against the city administration:

> "The politicians see the city as an 'empty box'. They want to empty the city and refill it with tourists [...]. The apartments are deliberately left empty by the administration and then sold cheaply to friends of the government or to large hotels. Many are in public ownership".
>
> (Group members A. and M.)

The so-called '*svendita*' (sell-out) of municipal housing to private individuals, together with the seemingly low success of ATER in the refurbishment of vacant public housing, has become a source of distrust among residents. The possible transformation of previously privatised apartments or other public properties into Airbnb flats or hotels has led to the presumption that the city wants to replace residents with tourists. The 'empty box' derogatorily expresses the lack of participation and inclusion of residents in these conversion processes.

This frustration has in turn led to activism, such as illegal occupations. ASC Venezia currently occupies a total number of around 70 public apartments in the island of Giudecca (see Figure 10.2) and in the area of Cannaregio (on the main island).

The group refers to itself as "occupiers" (*occupanti*), unlike squatters, with whom they do not fully identify. They explicitly only occupy or 'self-assign' (*auto-assegnazione*) public apartments—and in contrast to squatters never private properties—that have been vacant for several years and that have not been assigned to applicants by the public administration for the above-mentioned reasons. The refurbishment of these houses is carried out by the members of the group together with certified craftsmen, as is part of their political message. To this end, they are also working with experts and architects that took part in the Venice Architectural Biennale, experimenting with new building materials. The group visually documents the renovations to prove that they are possible even with minimal resources. The occupiers deliberately emphasise transparency. A symbolic monthly rent is paid to the public administration to show the occupancy of the apartments:

> "We only occupy public apartments that are vacant and not given away. We do not occupy private apartments, we do not take anything from anybody, the owners or anybody who is entitled to a public apartment, that is important

to us. We respect the right to property. The public housing that is vacant has been paid for with our grandparents' tax money, so we consider it a public good. They often stand empty for years. We work transparently, we do not hide anything [...]. The administration says that there is no money to refurbish these flats in order to rent them out as social housing. So, we do it ourselves, also to show that it is possible. We get people who are certified and do the necessary renovation work".

(Group member N.)

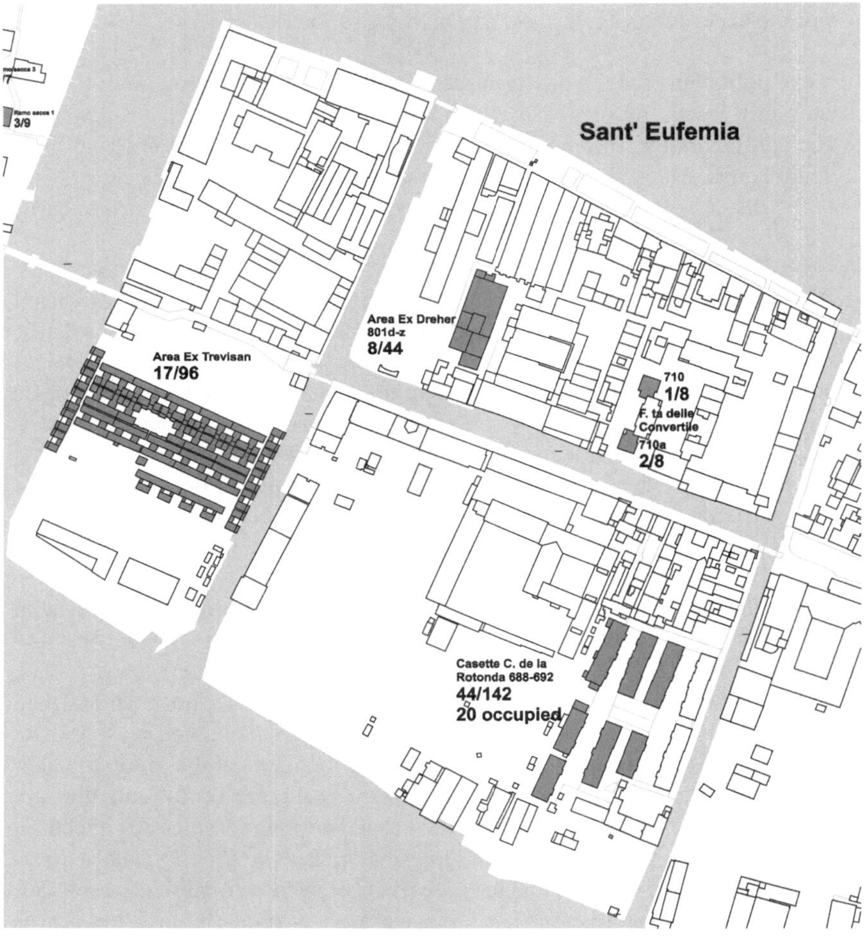

Figure 10.2 Extract of a map of vacant public housing in Sant' Eufemia, Giudecca (Venice). Numbers: vacant public flats/total public flats (here marked in grey). Location and numbers are based on a study (unpublished) for *Progetto Giudecca Green – Per il recupero del verde e della residenza pubblica in isola,* R.Bolani, 2019. © Marta Torres Ruiz

As N. says, in the beginning they were a precarious group, mostly students. But especially since the economic crisis of 2008, they have noticed that more and more people are joining their group, because they cannot afford the rents in the city. On how to solve this conflict and on the role of the police during evictions he concludes:

> "The problem is the empty flat, you can't solve the problem with police, you have to fill it. As long as 'the box' is empty, it will go on".
>
> (Group member N.)

The use of the 'empty box' in connection with the emphasis on self-identification as occupiers and the concept of self-allocation of social housing describes a process of self-determination on the part of the activists. During the interviews, public housing vacancy has been problematised on different levels, such as shortage of housing, lack of public economic resources or misuse due to privatisation and conversion into touristic accommodations. The concept of vacancy and the 'empty box' for public housing has been used repeatedly in the context of an ongoing struggle for the role and participation of residents in urban policy.

Airbnb and evictions: lack of regulation for short-term rents

In addition to the privatisation of social housing, the interviewed groups also criticise the renting of private housing via Airbnb or other platforms. They say the owners would not want to rent to residents, because of their more secure legal standing and the difficulties of formal eviction processes:

> "The advertisements are not directed to *residenti*, they are difficult to evict, and landlords prefer someone without this status. Therefore, they are rented to tourists. Now, for example, during Corona, where there are hardly any tourists, everyone is desperately looking for students, they are treated like angels. They come for nine months and do not have the full rights of residency".
>
> (Group member N.)

From a regulatory point of view, the short-term rental of flats or whole houses in Italy falls under a partially different legal definition than the long-term rental. Yet it does not qualify for the same regulations as hotel rooms, even if the space is actually leased as tourist accommodation. For the owners, it is possible to rent out a limited number of flats for short periods of 30 consecutive days, although in this case special taxation applies (Smorto 2016). The lack of stricter regulation and of a consistent public surveillance of this market has been criticised by residents. OCIO, a civil monitoring group on housing in Venice, proclaims the possibility and necessity to regulate the tourist leases (Medium 2020). In its studies the group shows evidence for a significant increase in the total numbers of short-term leases in Venice, between the years 2016 (around 34,000 beds) and 2018 (around 52,000 beds) (OCIO 2020b). Even if other research shows different total numbers of apartments for rent on Airbnb (Corona 2019), there is unity about there being an increase in the numbers of this lease-form.

Due to the high level of legal protection, renting long-term has become a challenged concept in Venice. The Fair Rent Act established in 1978 in Italy restricted the rights of landlords and at the same time highly protected residents from contract cancelations and evictions. This sort of regulation has been claimed being an obstacle for the rental housing market. The rental law was therefore adapted and partly liberalised in 1998 with the aim of attracting owners to rent out their apartments on a long-term basis (Gentili and Hoekstra 2019). However, this is not yet seen as a sufficient protection for some owners. An Airbnb landlord with two apartments in Venice asks:

> "What is better? A vacant flat or one with Airbnb? I think Airbnb is the better solution. You insure the property and it is in operation. To be in operation, it means that somebody must take care of it, maintaining it continuously in good health. Some of my friends told me to rent to Venetians, but as *residenti* they have such strong rights. They could also squat the flat and live there for free, not paying the rent for up to 1.5 years".
>
> (Landlord L.)

Asked about these problems, two members of a civil network who are searching for a living space since their current apartment has been inhabitable during the '*acqua alta*' of 2019, say:

> "There have certainly been individual cases of people who were unable to pay their rent due to the crisis in 2008, but this is not the normal case. E. herself is a *residenti*, and we have always paid the rent [...]. But it is possible that some of them have destroyed the apartments. When we moved into our apartment, it was completely abandoned".
>
> (Network members D. and E.)

Besides the fear of occupations of private property by residents, their statements show another preoccupation: the maintenance and insurance of the buildings. As the interviewed landlord says, Airbnb includes insurance for the property. On the website of Airbnb, the two introductory arguments for using the platform as a host are "security" and "full control" (Airbnb 2020). The insurance, "full control" and ability to set up own rules for the lease, address the concerns that have led my interview partner to use the platform.

The presumed exaggerated concern for the protection of tenants has become an argument against long-term rentals for owners in Venice. At the same time, the lack of regulation for short-term rentals has become a point of complaint for tenants, whose contracts have been cancelled with the intention of converting the flats into touristic accommodations. Activist forms of eviction prevention have therefore been carried out by the civic group Gruppo25aprile.org in 2019 (Gasparini 2019). The group prevented the eviction of an older woman in 2019 with a signatures collection campaign. In the case of vacant private housing, the issues of maintenance, security and control of property, as well as regulation and deregulation in tenancy law, play a greater role. The state and the city authorities are challenged in their decisions on the protection of long-term tenants as well as on the regulation of short-term tourist rentals.

Conclusion: the empty box and the public

Overtourism (UNWTO *et al.* 2018) has gained attention in recent years as a label, that bundles a wide range of demands and latent conflicts. At the same time civil movements and resistance have gathered momentum as expressions of global politicisation processes in cities dealing with rapidly growing tourism. In the intersection of overtourism and New Urban Tourism, conflicts and activism have become a common field of research. In that regard, research on overtourism generally focuses on the indicators of negativity and forms of neglect based on the division between visitor and resident in a broad spatial and disciplinary context (Dodds 2019, Milano *et al.* 2019a, Pechlaner *et al.* 2019, McKinsey & Company and World Travel & Tourism Council 2017, OECD 2020). New Urban Tourism specifically relates to the social, economic and spatial outcomes resulting from the overlap of different users in cities and has thus introduced theories of political economy to analysis (Fainstein *et al.* 2003, Maitland 2007). The emergence of activism, civic groups and global networks in the discussion, classifies tourism as a stage for political struggles on a transnational level. The research on civic groups presented in this chapter shows evidence of a diversified form of civic engagement and participation in political representation, that resorts to terms of political theory (such as to the people). The form, hierarchies and representatives of these groups yet present a heterogeneous scene lacking a clear socio-economic classification.

The lines of conflict around vacant housing in Venice can be read from different perspectives. In the case of social housing, the occupiers who cannot afford the rents in the city and their resistance to third parties buying out social housing, indicate a socio-economic opposition. In the case of private housing, the diverging interests of long-term residents (affordable housing) and private owners (protection of rental income and assets) can be observed through their different positioning to short-term rentals (Airbnb). The lack of regulation in this field is framed as a battle between residents and visitors. But one more point of contest was found: the one between city users and city government or, more generally, between citizens and the state. The different cases can be re-read through the lens of agency (Hall 1983): the residents self-organise, occupy apartments and defy eviction to ensure their housing in the city. The landlords use Airbnb to secure their rental income and shield their property from a supposed overreach by regulation. The city partly privatised public housing in a proclaimed effort to manage and refurbish other parts of existing housing in the city. For these cases, the 'empty box' has been used by the interview partners as an expression for the perceived lack of residents' representation, needs and concerns in city politics—or, in a broader extent, for the hollowing out of the public. Criticism against privatisations of public housing and the lack of regulation of tourist leases has been expressed through this concept, and was clearly directed against the city administration. Activism and civic self-determination, such as the occupation or "self-assignment" of public housing was thus legitimised—as a group member has said, "as long as the box is empty the problems will continue". In the case of vacant housing in Venice, several lines of conflict occur simultaneously. This results in a constitution of conflicts, which is not only structured by opposition of two predetermined parties, but as a multiplicity of ongoing antagonistic struggles over power distribution and participation.

Civic engagement in relation to urban tourism and its impacts has, as illustrated, led to a preoccupation of tourism studies with possible responses, which in turn raise the question of how these conflicts are constituted. The introduction of political economy theories in New Urban Tourism (Maitland 2007, Fainstein *et al.* 2003) provides methodological approaches for these studies. Fainstein *et al.* (2003) argue towards balancing the role of government in tourism politics and on its relevance in different scales of regulation and deregulation on a municipal, national or global level. With this understanding, the results of this investigation show that regulation can be extended to other forms of engagement, such as non-institutionalised activism. Agency can therefore be seen as a partial way out of conflict (contingency, conflict and compromise) through state and city regulation or deregulation, the recourse to third-party agents but also through self-organised institutions like civic groups. Regulation theory (Aglietta 1979) can assist in this respect by shifting to a non-deterministic understanding that argues towards agency (Marchart 2013). Marchart (2013) notes that regulation theory can be regarded as compatible with the post-structural deconstruction of Marxism, such as Laclau and Mouffe's theory of discourse and hegemony (Laclau and Mouffe 1991). Following this path, the increasing interest of tourism studies in contestation would thus also provide an opportunity for the introduction of post-structural conflict theories of political theory and political philosophy. Against this background, structuralist theories on the constitution of conflicts (Maitland 2007) could be questioned and extended to the study of hegemonic discursive practices.

References

Aglietta, M., 1979. *A theory of capitalist regulation. The US experience*. London: Verso.

Airbnb, 2020. [Online]. Available from: http://web.archive.org/web/20201018082417/https://www.http://airbnb.com/host/homes [Accessed 20 October 2020].

Bianchi, R.V., 2009. The 'Critical Turn' in tourism studies: a radical critique. *Tourism Geographies* [online], 11 (4), 484–504. Available from: https://www.tandfonline.com/doi/pdf/10.1080/14616680903262653 [Accessed 20 October 2020].

Boorstin, D.J., 1967. *The image*. New York: Atheneum Paperbacks.

Città di Venezia, 2019. *Movimento demografico: excel-datasheets 2009 and 2019* [online]. Available from: https://www.comune.venezia.it/it/content/movimento-demografico [Accessed 20 October 2020].

Colomb, C. and Novy, J., eds., 2017. *Protest and resistance in the tourist city*. London: Routledge.

Corona, A., 2019. Inside Airbnb Infokit: Venice. *Adding data to the debate* [online], 12 August. Available from: http://insideairbnb.com/venice/report_en.html [Accessed 20 October 2020].

Cricket, D., 2019. "No to big ships" demonstration in Venice. *Youthreporter* [online], 16 June. Available from: https://www.youthreporter.eu/de/beitrag/no-to-big-ships-demonstration-in-venice.15523/ [Accessed 20 October 2020].

Dodds, R., 2019. *Overtourism: issues, realities and solutions*. Berlin: Walter de Gruyter GmbH.

Fainstein, S.S., Hoffman, L.M., and Judd, D.R., 2003. Making theoretical sense of tourism. *In*: S.S. Fainstein, et al., eds. *Cities and visitors: regulating people, markets, and city space*. Oxford: Blackwell, 239–253.

Florida, R.L., 2002. *The rise of the creative class: and how it's transforming work, leisure, community and everyday life*. New York: Basic Books.

Gasparini, A., 2019. A Castello vince la solidarietà, sospeso lo "sfratto" della 94enne, 2019. *Venezia Today* [online], 14 May. Available from: https://www.veneziatoday.it/attualita/sfratto-sospeso-anziana-castello-gruppo-25-aprile.html [Accessed 20 October 2020].

Gentili, M. and Hoekstra, J., 2019. Houses without people and people without houses: a cultural and institutional exploration of an Italian paradox. *Housing Studies*, 34 (3), 425–447.

Gössling, S., Scott, D. and Hall, C.M., 2020. Pandemics, tourism and global change: a rapid assessment of COVID-19. *Journal of Sustainable Tourism*, 29 (1), 1–20.

Hall, S., 1983. The problem of ideology: Marxism without guarantees. *In*: B. Matthews, ed. *Marx 100 years on*. London: Lawrence & Wishart, 57–86.

Hayes, M. and Zaban, H., 2020. Transnational gentrification: the crossroads of transnational mobility and urban research. *Urban Studies*, 57 (15), 3009–3024.

Krippendorf, J., 1975. *Die Landschaftsfresser: Tourismus und Erholungslandschaft–Verderben oder Segen?* Bern: Hallwag.

La Voce di Venezia, 2019. Ater Venezia presenta piano di vendita delle case. *La Voce di Venezia* [online], 30 October. Available from: https://www.lavocedivenezia.it/ater-venezia-piano-vendita-case-2019/ [Accessed 20 October 2020].

Laclau, E. and Mouffe, C., 1991. *Hegemonie und radikale Demokratie. Zur Dekonstruktion des Marxismus*. Wien: Passagen.

Maitland, R., 2007. Tourists, the creative class and distinctive areas in major cities. The roles of visitors and residents in developing new tourism areas. *In*: G. Richards and J. Wilson, eds. *Tourism, creativity and development*. London: Routledge (Contemporary geographies of leisure, tourism, and mobility), 73–86.

Marchart, O., 2013. *Die Prekarisierungsgesellschaft: Prekäre Proteste. Politik und Ökonomie im Zeichen der Prekarisierung*. Bielefeld: Transcript Verlag.

McKinsey & Company and World Travel & Tourism Council, 2017. *Coping with success. Managing overcrowding in tourism destinations* [online]. Available from: https://www.mckinsey.com/industries/travel-transport-and-logistics/our-insights/coping-with-success-managing-overcrowding-in-tourism-destinations [Accessed 20 October 2020].

Medium, 2020. Regolamentare le locazioni turistiche è possibile, oltre che necessario. *Medium* [online], 5 August. Available from: https://medium.com/ocio-venezia/regolamentare-le-locazioni-turistiche-%C3%A8-possibile-oltre-che-necessario-b5742d2dded2 [Accessed 20 October 2020].

Milano, C., Cheer, J.M. and Novelli, M., 2019a. *Overtourism, excesses, discontents and measures in travel and tourism*. Wallington: CABI.

Milano, C., Cheer, J.M., and Novelli, M., 2019b. Overtourism and degrowth: a social movements perspective. *Journal of Sustainable Tourism*, 27 (12), 1857–1875.

Nieuwland, S. and van Melik, R., 2020. Regulating Airbnb: how cities deal with perceived negative externalities of short-term rentals. *Current Issues in Tourism*, 23 (7), 811–825.

Novy, J., 2010. What's new about new urban tourism? And what do recent changes in travel imply for the 'tourist city' Berlin? *In*: Richter, J., ed. *The tourist city Berlin: tourism and architecture*. Berlin: Braun, 190–199.

OCIO, 2020a. *Analisi Manutenzioni–ATER* [online]. Available from: http://ocio-venezia.it/files/2020-02-06_analisi-manutenzioni-ater.pdf [Accessed 20 October 2020].

OCIO, 2020b. *Gli squilibri del turismo Veneziano* [online]. Available from: https://ocio-venezia.it/pagine/affittanze-dati/ [Accessed 20 October 2020].

OECD, 2020. Organisation for economic cooperation and development: tourism trends and policies 2020. *OECDiLibrary* [online], 4 March. Available from: https://www.oecd-ilibrary.org/industry-and-services/oecd-tourism-trends-and-policies_20767773 [Accessed 20 October 2020].

Pechlaner, H., Innerhofer, E., and Erschbamer, G., 2019. *Overtourism: tourism management and solutions*. New York: Routledge.

Peeters, P., et al., 2018. *Research for TRAN Committee–overtourism: impact and possible policy responses*. Brussels: European Parliament, Policy Department for Structural and Cohesion Policies.

Smorto, G., 2016. *Impulse paper no. 02: on the business authorisation/licensing requirements imposed on peer-providers and platforms in the accommodation/tourism sector in Paris, Rome, Milan and London* [online]. Available from: https://bit.ly/2WwQSqa [Accessed 20 October 2020].

Statista, 2020. Number of international tourist arrivals in the Italian city of Venice from 2011 to 2019. *Statista* [online], 8 June. Available from: https://www.statista.com/statistics/732572/international-tourist-arrivals-in-venice-italy/ [Accessed 20 October 2020].

Tourist Go Home, 1959. Short Film. Directed by S. Jackson and R. Weyman (Reg.), W. Hewitson and P. Jones (Prod.). Canada: National Film Board of Canada (Montreal). Available from: https://www.nfb.ca/film/tourist_go_home/ [Accessed 20 October 2020].

Turner, L. and Ash, J., 1975. *The golden hordes: international tourism and the pleasure periphery*. London: Constable.

World Tourism Organisation (UNWTO), et al., 2018. *'Overtourism'?–understanding and managing urban tourism growth beyond perceptions* [online]. Madrid: UNWTO. Available from: https://www.e-unwto.org/doi/book/10.18111/9789284419999 [Accessed 20 October 2020].

Young, G., 1973. *Tourism blessing or blight*. Middlesex: Penguin Books.

11 Powerful ways of (not) knowing New Urban Tourism conflicts

Thin problematisation as limitation for tourism governance in Berlin

Christoph Sommer

Challenging the ways of knowing conflict-laden tourism

Until recently tourism seemed to be an ever-growing phenomenon. The effects of a tourism overload sparked a so-called 'overtourism' debate. Even if today, a COVID-19 induced 'undertourism' captures lots of attention, both debates express a dimension of conflict and prompt to reconsider urban tourism governance. Indeed, research dealing with "the debate on overtourism" (Oskam 2020, p. 3) includes harsh and reasonable governance criticism. According to Dodds and Butler (2019, p. 10) one "key enabler of overtourism is the attitude, or mindset, of companies, governments, [and] marketing organisations [...] that favour growth above all else". From a political-economic perspective, the governance of conflict-laden urban tourism is often discussed as a manifestation of an entrepreneurial or post-political urban governance (e.g. Novy and Colomb 2019). The chapter at hand offers a different perspective. Instead of questioning if and why current modes of tourism governance are determined by a growth-focused mindset (which could be discussed as expression of neoliberalism), it is detailed *how* destination governance actors *frame* frictions associated with New Urban Tourism.

Assuming that the construction of governance problems is linked to the availability of solutions, it is relevant to better understand how New Urban Tourism problems are known or even denied by governance actors (e.g. destination management organisations (DMOs), administrative bodies responsible for tourism, mayors). Not saying that better knowledge does guarantee better governance, the distinct governance habits of knowing conflict-laden New Urban Tourism are presumed to limit or extend governance options. Therefore, the chapter asks: How exactly do the governance actors of destinations render tourism as a governance issue doable? More specifically, how far do distinct governance habits of knowing conflict-laden New Urban Tourism (powerfully) predetermine governance options?

Using Berlin as a case, generic governance techniques of problematising conflict-laden tourism are empirically reconstructed. Berlin represents an informative case as New Urban Tourism conflicts have been intensively debated in Germany's capital for the last decade. Additionally, Berlin is a city-state in the German federal system. This makes it interesting to ask how actors involved in Berlin's destination governance (i.e. the public–private DMO visitBerlin, see below) use their comparatively extensive institutional power to frame New Urban Tourism conflicts. Building on document analysis, the text empirically shows that the power of (not) knowing urban tourism conflicts turns up in what I call—and conceptually develop in this paper—'thin problematisations' mobilised by governance actors involved in tourism issues. As a conclusion, I address the question of how to challenge such simplifications academically and in practice-related ways.

Thinking governance problems and solutions relationally

The key conceptual assumption guiding my case study is that knowing how to reduce the complexity of New Urban Tourism conflicts represents an essential and necessary practice accomplished by destination governance actors. In contrast to "good governance" research taking tourism problems as given in advance, this piece builds on stances pointing to the construction and contestation of problems as powerful aspects of urban governance. Basically, it is assumed that problems and adequate solutions co-emerge in a mutually constitutive manner.

Regarding the governance of urban tourism, Novy and Colomb (2017, p. 6) pointedly stress that "the process of defining, conceptualising and measuring tourism [...] itself" needs to be regarded as "deeply political". They argue that it is, in particular, the tourism industries' view on tourism (in terms of hotel guest statistics etc.) which limits the understanding of tourism as it neglects, for example, other temporary visitors like day-trippers or tourists visiting friends or relatives. On a more general level, Hall (2011) argues in a similar way. According to him (2011, p. 437) the "shift in approach in the tourism policy literature [...] from the notion of government to that of governance" implies analytically taking into account even "the definition of policy problems" (ibid.). Referring to Bachrach and Baratz (1962), Hall (2009) also reminds us taking into account a second face of power in tourism governance settings: to consider the possibility that certain potentially controversial issues might not even enter the tourism governance arena as they are actively forced back before they get discussed at all.

Hence, analysing the governance of New Urban Tourism conflicts starts with an explicit analytical focus on the construction and contestation of problems and solutions. The notion of governance is used primarily as a lens to analyse "the actors and interests that make urban policy decisions, set and control agendas, define problems and propose solutions" (McCann 2017, p. 314). The case study at hand focuses on actors involved in destination governance at federal state level. As a city-state in the German federal system, Berlin has a two-tier government system (a citywide administration and 12 local boroughs). Actors involved in destination governance at federal state level first and foremost involve the DMO visitBerlin, the Senate Department for Economics, the Senate Chancellery (= the office of the governing mayor) and the House of Representatives. Within Berlin's tourism governance setting, the DMO visitBerlin (organised as public–private partnership) is often assumed to be *the* authority responsible for issues relating to tourism marketing and development (Novy 2017). At the same time, the Senate

Department for Economics is a powerful player, since it is as the city's administrative body responsible for tourism, authorised to issue directives towards visitBerlin.

As mentioned above, it is presumed that ways of knowing *doable* problems and *viable* solutions in respect to conflict-laden urban tourism play a key role in evidence-based destination governance processes. As a crucial aspect of tourism-related policymaking, processes of measuring, interpreting or also denying tourism conflicts are embedded in institutional arrangements (polity) and processes of decision-making (politics) alike. These far-flung knowledge processes are assumed to be necessarily selective and reductionistic as the messiness of the world out there needs to be somehow organised (AutorInnenkollektiv 2010). However, rendering New Urban Tourism conflicts doable is not limited to one single (e.g. statistical) logic of knowing. As Valverde (2011, p. 277) convincingly shows, such necessarily reductionistic knowledge production in urban governance involves "epistemologically hybrid" approaches. The "pragmatic approach" of urban governance builds on "old and new gazes, premodern and modern knowledge formats, in a nonzero-sum manner and in unpredictable and shifting combinations" (ibid., p. 281). Considering these conceptual thoughts, the chapter aims to empirically pin down how exactly New Urban Tourism conflicts in Berlin are powerfully constructed by the destination governance actors involved.

Evidence from Berlin: rendering New Urban Tourism conflicts doable

Berlin represents an informative case to study destination governance habits of (not) knowing tourism frictions. Conflicts of—in particular—New Urban Tourism have been intensively debated in Berlin since 2010. The touristic search for urban experiences off the beaten track is a vital feature of the city's tourism. In various inner-city neighbourhoods, a broad range of disputes have arisen (see e.g. Novy 2017, Sommer and Kip 2019, Müller *et al.* 2019), mainly concerning (retail) gentrification, noise and the expansion of tourist accommodations. Nevertheless, governance processes of knowing (defining, measuring, denying etc.) conflict-laden urban tourism have remained under-researched (with a few exceptions, e.g. Sommer and Helbrecht 2017, Füller *et al.* 2018), as has the relational governmental *construction* of doable tourism problems and practicable solutions. The piece at hand aims to address this gap in research.

Regarding methods, this chapter mainly builds on an analysis of policy documents, assessing 34 policy documents (2011–2016) of the DMO visitBerlin, of the Senate Department for Economics and of the Senate Chancellery as well as media statements of representatives of the latter. This allowed capturing the institutionally objectified knowledge produced by Berlin's major institutions responsible for tourism governance. In the following, three empirically reconstructable governance techniques of problematising conflict-laden urban tourism are discussed: interpreting conflicts in the media (see "Interpreting conflicts in the media"), knowing the problem perception of residents statistically (see "Knowing problem perceptions statistically") and regulating problem-conclusion-combinations conceptually (see "Regulating problem-conclusion-combinations on paper"). They represent crucial ways of (not) knowing conflict-laden tourism.

Interpreting conflicts in the media

Framing urban tourism conflicts in the context of Berlin's success as a destination is a well-established destination governance approach to tame tourism problems. This framing builds on the city's evolution towards a "tolerant metropolis" (visitBerlin 2014, p. 3, transl. by the author) with overnight stays "having increased fourfold since 1990" (visitBerlin 2016, p. 7, transl. by the author). The development is qualified as a "catch up process" (visitBerlin 2014, p. 3), which Berlin has undergone since 1990, when the socialist German Democratic Republic merged with the Federal Republic of Germany and Berlin ceased to be a divided city. This narrative is mainly reproduced through media statements by destination governance actors. On closer examination the following argumentative patterns structure the contextualisation of tourism conflicts within the successful urban development trajectory.

First, the success of Berlin as city/destination is framed as a comeback story. According to this narrative, Berlin is allegedly in the process of "swinging back in the normal state of a world city" (e.g. visitBerlin 2020). I understand this contextualisation as reactivation of a myth which has been mobilised again and again for imagining the city's future—for example after German reunification. In the 1990s, as the German Democratic Republic (colloquially: East Germany) became part of the Federal Republic of Germany (colloquially: West Germany), the formerly divided Berlin was reunited into a single city. Back then, the revival of an imagined Berlin was guided by the idea of becoming a world city and service sector metropolis (Farías 2005). In the 2000s the successful future of the city was connected with the idea of becoming a cosmopolitically diverse—and therefore competitive—metropolis (Lanz 2011). Regarding the contextualisation of tourism conflicts, the world city comeback myth has a particular function. Following Farías (2005), one could claim that the "myth works [...] as a kind of urban memory which does not consist of verifiable historical events and dates, but of images which are used for the apprehension and assessment of current processes" (Farías 2005, p. 22, transl. by the author). In the case at hand, the world city comeback myth serves as a narrative which naturalises ever-growing tourism and turns concomitant conflicts into a minor side effect.

Moreover, the narrative of Berlin's world city comeback resonates with a reading of tourism conflicts which substantially builds on a dominant 'public opinion' in favour of tourism. From the outset, growing tourism and concomitant conflicts haven been dubbed a matter of customisation. According to this claim, Berliners will have to get used to the city's growing popularity as "the place to be" (visitBerlin 2010, visitBerlin 2019). As myths naturalise a certain status quo as irrevocably given (Assmann 1992), the world city comeback seems to leave no choice; "nothing else remains to be done but to accept the role [as world city]" (visitBerlin

2015, transl. by the author). Being proud of so many visitors, a former mayor of Berlin thwarted those who criticised tourism conflicts like noise, residential and commercial gentrification, etc.: "All attempts of some parts of town to talk problems into happening are disconnected from reality" (Wowereit 2012, transl. by the author). Thus, regarding the interpretation of the debate about tourism conflicts, destination managers and politicians (at least at the beginning of the past decade) appealed to the relaxed attitude of a cosmopolitan majority. Hence, the conclusion that no further destination governance action is required to solve tourism-related problems is at least implied.

A further problem interpretation still to be found in media statements until recently is the *argumentum ad hominem* expounding tourism critique as something which feeds on resentment. Corresponding with the results of Holm (2015) one can observe that in public discourses tourism critique is often tied to xenophobia and unwillingness to move ahead. So, critical voices regarding tourism externalities are attributed to a 'backward' minority which aims to stick with one's kind (visitBerlin 2019). Or alternatively, tourism critique is directly accused of being intolerant and xenophobic (visitBerlin 2011a).

Taking into consideration the above-mentioned problematisations, the question is how they resonate with the conclusions. I claim that the myth of the world city comeback, as well as the arguments referring to cosmopolitan majorities and 'backward' minorities, narrowed down the possibility of discussing tourism conflicts. The need for a "debate on principles" (visitBerlin 2011b, transl. by the author) has been literally refused (ibid.). At least in the first half of the past decade, this relativisation of tourism problems and critique facilitated non-decision-making on how urban tourism conflicts might be addressed.

Knowing problem perceptions statistically

Measuring attitudes towards urban tourism represents an established market research practice of DMOs. As a variant of governance by numbers, the statistical monitoring of attitudes towards tourism is understood as a "technology of persuasion" (Heintz 2008, p. 117). The statistical knowledge produced by destination managers endows them with considerable interpretational power. Statistical data is assumed to be eminently robust as it is comparatively difficult to prove numbers wrong. The so-called tourist acceptance survey (commissioned yearly by the DMO visitBerlin among Berliners, from 2012 to 2019) consisted of six questions. Four questions address aspects of tourism commonly thought of as positive (e.g. advantages of tourism for Berliners), two questions ask if and why tourism is perceived as annoying. The document analysis rendered visible that numerous policy documents contain references to the statistically proven sympathy towards tourism. Accordingly, it is exposed, for example, that a vast majority is "proud that people from all over the world visit their city" (visitBerlin 2013, p. 13, transl. by the author); or it is stressed that "84 percent of the Berliners think that they are good hosts" (Senate Department for Economics 2016, p. 2). All in all, one could say that the statistically produced knowledge prompts an interpretation of tourism conflicts which is "zooming in approval towards tourism" while simultaneously "zooming out the perception of problems" (Sommer and Helbrecht 2017, p. 165). The results of the survey might be representative, but only for the favourable scope of analysis defined in advance by visitBerlin.

Recalling the conceptual assumption that problem interpretations and conclusions/solutions co-emerge, how does this work with statistic-based ways of representing tourism acceptance? Statistical knowledge is used to numerically underpin media statements arguing that only a small minority of residents has a problem with tourism—while on the other hand a vast majority of the Berliners is displayed as being proud of the visitors and sees the benefits of tourism. Highlighting the approval towards tourism implies that tourism critique is just a matter of some parochial reactionists. Moreover, the acceptance numbers helped to justify the course of action regarding the management of conflict-prone tourism. This could be exemplified with the way in which a distinct concept for dealing with increasing tourism pressure (proposed by an opposition party in 2013) was declared not necessary. A decision-maker of the Senate Department for Economics is cited in a memo of the House of Representatives in the following way: "As these measures [regarding acceptance maintenance] work, which is proven by the high acceptance of 87 percent, no new concepts are needed" (Senate Department for Economics 2015, p. 52, transl. by the author). Third, it can be observed that the statistical findings trigger idiosyncratic solutions. The statistical result that "only 62 percent of the respondents [...] assess tourism [...] to be an important economic factor" (Senate Department for Economics 2016, p. 11) was read by destination managers as a reason to better market the economic benefit of tourism towards Berliners. Assuming that knowing the benefits of tourism would increase acceptance towards tourism (e.g. Senate Department for Economics 2016) resulted in a media campaign. It contained advertisements in local newspapers depicting the economic effects of tourism in combination with an acknowledgement of gratitude addressing Berliners in their role of hosting so many visitors.

Regulating problem-conclusion-combinations on paper

As indicated above, statistical knowledge is actively produced to construct acceptance and goodwill towards tourism. This served to deny the need for a new concept paper explicitly dealing with conflict-laden tourism. Instead of a concept, just a short "report on tourism acceptance" (Senate Department for Economics 2014) was delivered in 2014. In the following, two telling examples regarding the way conflictive tourism was known in concept work, are depicted.

The first example concerns Berlin's official tourism strategy of 2011 (which was replaced in 2018). It is interesting to note that the conflictive side of tourism was already addressed within the 'Tourism Concept 2011+' (Senate Department for Economics 2011, transl. by the author). It reports that tourism partly leads to phenomena of overuse and conflicts compromising the quality of life of residents (Senate Department for Economics 2011). The conclusion regarding the future development of tourism was "that a tourism-friendly development in accordance with the interest of residents" (ibid., p. 23, transl. by the author) needs to be aspired to. However, besides the aim of better managing tour-bus traffic, no tangible measures were listed to mitigate conflicts. As Novy (2017, p. 64) points out, the "extensive references to issues such as overcrowding [...] and the need to address them" did not make it into the final version of the concept. Problematisation of conflict-prone tourism was at best indistinct and the accountability of the city's destination management regarding the handling of tourism problems remained vague.

The second example of regulating problem-conclusion-combinations on paper shows how concept work was replaced by a far more noncommittal *reporting work*. As part of the parliamentary resolution concerning the federal state's budget for 2014/15, in 2013 the municipal government was required to create a coordinated concept and recommendations for action—with the goal of preserving acceptance of tourism (Budget Committee of Berlin's House of Representatives 2013). Instead of developing a concept comprising a schedule of responsibilities, the Department for Economics produced a short report of ten pages. It concluded that a "distinct concept for acceptance preservation" is not "useful" (Senate Department for Economics 2014, p. 10). For not containing a schedule of responsibilities (saying who is going to do what until when) this report was, unsurprisingly, heavily criticised by the oppositional Green Party (Ludwig 2015). Their proposal to produce a more detailed concept was declined—as mentioned above—with the explanation that the existing measures did work—which was, according to the argument, proven by the high acceptance of tourism of 87% (Senate Department for Economics 2015). Finally, this so-called acceptance report was updated yearly until 2017. Taking this example into consideration, it can be recorded that the yearly production of an acceptance report needs to be understood as a way of knowing conflict-prone tourism—albeit without increasing the destination governance accountabilities.

Both examples show that procedures of producing strategy papers (or solely reports) are—in this case—used without entrenching clear conclusions regarding conflictive tourism. In the first example (Tourism Strategy 2011+) conclusions and problem descriptions have been wiped out in the course of the concept development. In the second example, a non-binding report replaced a concept paper, potentially more binding. It needs to be noted here that also the latest tourism strategy from 2018, which is explicitly committed to issues of sustainable urban tourism, is not making transparent which measures are pursued, until when, by whom.

The power of (not) knowing urban tourism conflicts

Berlin's destination governance combines strikingly different ways of knowing and contesting conflict-prone tourism. In this chapter, three empirically reconstructable techniques of arranging doable tourism problems and practicable solutions are depicted: media-specific, statistical and conceptual ('paper work') techniques of knowing conflictive tourism.[1]

Partly supplementing and contradicting each other, the mentioned techniques effectively structure the "political-epistemological space" (Lemke 2000, p. 2) in which tourism problems and adequate solutions can co-emerge. It has been revealed how the identified problems are intimately linked to the availability of solutions. Marketing the economic benefits of tourism (assuming thereby to increase acceptance of tourism), for example, only seems to be a plausible solution against the backdrop of distinct problem interpretations, in this case numerical acceptance data. At the same time, acceptance statistics justified (at least at the beginning of the debate about tourism conflicts) *not* governing tourism conflicts. The myths and the arguments discrediting tourism critique must be read as attempts to stifle the debate about how to deal with tourism and accompanying conflicts.

Taking the empirical findings into consideration, the destination management techniques of (not) knowing urban tourism are powerful. The inferred claim is that the latter proceedings are consequential to the extent that they paralyse a more sustainable urban tourism governance. Keeping down the issue of dealing with conflict-prone tourism requires active and influential work. Thus, the claim that ways of (not) knowing tourism conflicts are powerful is twofold.

First, it refers to the destination governance capability of simplifying tourism conflicts and matching them with solutions. Along the lines of Scott's (1998) notion of '*thin simplifications*', one could argue that '*thin problematisations*' of tourism conflicts facilitate what Bachrach and Baratz (1962, p. 952) dubbed active "nondecisionmaking". According to Holmes (1988, p. 22, cited from Hall 2009, p. 4) this "allows political actors, organisations and collectives to 'leave selected topics undiscussed for what they consider their own advantage'". Thus, the destination governance accountability to act upon the problematised conflicts remains vague.

Second, the claim that there are potent ways of *not* knowing tourism conflicts builds on the observation that the knowledge production about tourism conflicts does not seem to make any progress. However, I argue that a more detailed understanding of tourism conflicts represents a precondition to manage tourism conflicts better. To give some examples regarding persistent gaps in knowledge: at the time of writing, the federal state of Berlin did not possess data regarding the numerical and geographical development of hotels within the last decade (Senate Department for Urban Development and Housing 2019); the status quo of holiday apartments and short term rentals can only be estimated; the mapping of retail, gastronomy and service industries is still not more than a goal mentioned in Berlin's tourism strategy paper from 2018. Nevertheless, to preserve, for example, a balanced mix of uses or to build additional accommodation facilities within residential areas in a sustainable way—which is the alleged goal—requires better knowledge of the gradual change and development of tourism.

One explanation for the somehow 'fixed' ways of (not) knowing urban tourism conflicts can be hypothesised. Since the public–private DMO visitBerlin and the Senate Department for Economic Development are political organisations, they have "a bias in favor of the exploitation of some kinds of conflict and the suppression of others because *organization is the mobilization of bias*" (Schattschneider 1960, p. 71, quoted from Bachrach and Baratz 1962, emphasis in the original). Widely assumed to be Berlin's major destination governance actors, visitBerlin and the Senate Department for Economics are—willingly or not—in charge of addressing New Urban Tourism conflicts. However, regarding matters of tourism development, the organisational bias of visitBerlin and the Senate Department for Economics still seems to be in favour of destination marketing issues. This hypothesis can be underpinned by the observation that many attempts to handle tourism conflicts in terms of urban development approaches take years to get conducted or are abandoned. To give some examples: the hotel development plan first mentioned as badly needed in 2013 has not been realised until now; the idea of establishing a tourism advisory board constituted by residents was already conceptualised in 2015—the founding of the advisory board is still in the making; the conceptual ideas for a spatial tourism development guideline (developed for the first time in 2007 by the Senate Department for Urban Development) apparently have been abandoned.

Summarising this somewhat provisional explanation could look like this: the power of (not) knowing urban tourism conflicts turns up in the various thin problematisations mobilised by destination governance actors to approach tourism-induced problems. To align doable problems and viable solutions, the municipal destination governance builds on different knowledge formats (statistics, media statements, myths, concept work). Maintaining thin problematisations of tourism conflicts limits a more far-reaching understanding and governance of (possibly unsolvable) contradictions of New Urban Tourism.

How to better know tourism conflicts

For a further discussion of academic and more practice-related ways of knowing (New Urban) Tourism problems, it needs to be accepted that the complexity of New Urban Tourism inevitably needs to be reduced in order to be known/governed. This also implies reflecting upon critical stances often articulated in tourism research regarding the allegedly flawed governance of urban 'overtourism'. It does not move things forward to denounce deficits of urban tourism governance with abstract explanations (e.g. neoliberalism, post-politics) and most of the well-intentioned "good governance" advice articulated by tourism scholars fizzles out in the real world. It is also clear that more (detailed) knowledge about urban tourism externalities does not guarantee better governance.

Hence, it is worthwhile challenging thin problematisations of tourism governance. Equally, it seems interesting to ask how it is possible for researchers and destination managers to get in touch about the question of *how* we know New Urban Tourism and its problems. How is it possible to mutually increase the complexity and variety of problem constructions?

Especially against the backdrop of recent attempts to divide the discourse about the future of tourism in pro-growth vs. growth-limiting positions (critically discussed by Higgins-Desbiolles, 2020), it seems to be essential to strengthen the diversity of applied and academic perspectives. Without being limited from the outset by the imperative to find solutions, transdisciplinary research co-operations between (tourism) scholars, destination managers, planners, lawyers, politicians and residents could help to unlock analytical potentials between idiosyncratic ways of knowing New Urban Tourism conflicts. To jointly venture into "co-laborative" (Niewöhner 2014) research co-operations with epistemic partners from tourism research and urban practice might help to gradually establish a more complex understanding of tourism frictions. Participatory discussion formats, as recently developed by Berlin's destination managers for example, could serve as excellent starting points. Gaining multiperspectivity and jointly compiling knowledge seems to be a better option than relapsing again and again into simplifications regarding (New Urban) Tourism conflicts.

Note

1 Regarding further destination management techniques for knowing conflictive tourism, see my PhD thesis, which this chapter partly builds upon (Sommer 2021). Notably, continuously advanced participatory discussion formats to negotiate issues of conflictive tourism (i.e. creative workshops, planned resident advisory board) must be mentioned here. These widely non-binding participatory formats only seem to be reasonable when tourism conflicts are assumed to result primarily from "communication problems" (Senate Department for Economics 2014 p. 2, transl. by the author).

References

Assmann, J., 1992. Frühe Formen politischer Mythomotorik. Fundierende, kontrapräsentische und revolutionäre Mythen. *In*: D. Harth and J. Assmann, eds. *Revolution und Mythos*. Frankfurt am Main: Suhrkamp, 39–61.

AutorInnenkollektiv, 2010. Wissen und soziale Ordnung. Eine Kritik der Wissensgesellschaft. Mit einem Kommentar von Stefan Beck.: *Working Papers des Sonderforschungsbereiches 640, 1/2010* [online]. Available from: https://edoc.hu-berlin.de/bitstream/handle/18452/3891/1.pdf?sequence=1 [Accessed 6 November 2020].

Bachrach, P. and Baratz, M., 1962. Two faces of power. *American Political Science Review*, 56 (4), 947–52.

Budget Committee of Berlin's House of Representatives, 2013. *Beschlussempfehlung*. Available from: https://www.parlament-berlin.de/ados/17/IIIPlen/vorgang/d17-1400.pdf [Accessed 6 November 2020].

Dodds, R. and Butler, R.W., 2019. The enablers of overtourism. *In*: R. Dodds and R.W. Butler, eds. *Overtourism–issues, realities and solutions*. Berlin: De Gruyter, 6–22.

Farías, I., 2005. Zukunft zum Greifen nah. Strukturelle Bedingungen, Semantiken und Verortung des Berliner Stadtmarketing. *Berliner Blätter*, 37, 22–31.

Füller, H., et al., 2018. Manufacturing marginality. (Un-)governing the night in Berlin. *Geoforum*, 94, 24–32.

Hall, C.M., 2009. *Power in Tourism: Tourism in Power* [online]. Available from: https://www.researchgate.net/publication/283694446_Power_in_tourism_Tourism_in_power [Accessed 6 November 2020].

Hall, C.M., 2011. A typology of governance and its implications for tourism policy analysis. *Journal of Sustainable Tourism*, 19 (4–5), 437–457.

Heintz, B., 2008. Governance by numbers. Zum Zusammenhang von Quantifizierung und Globalisierung am Beispiel der Hochschulpolitik. *In*: G.F. Schuppert and A. Voßkuhle, eds. *Governance von und durch Wissen*. Baden-Baden: Nomos, 110–129.

Higgins-Desbiolles, F., 2020. The "war over tourism": challenges to sustainable tourism in the tourism academy after COVID-19. *Journal of Sustainable Tourism*, 29 (4), 551–569.

Holm, A., 2015. Welche Stadt sehen wir? Die Urbanisierung des Tourismus. *Schader Stiftung* [online]. Available from: https://www.schader-stiftung.de/themen/kommunikation-und-kultur/fokus/kunst-und-gesellschaft/artikel/welche-stadt-sehen-wir-die-urbanisierung-des-tourismus-als-simulacrum-des-staedtischen [Accessed 6 November 2020].

Lanz, S., 2011. Berliner Diversitäten: Das immerwährende Werden einer wahrhaftigen Metropole. *In*: W.-D. Bukow, et al., eds. *Neue Vielfalt in der urbanen Stadtgesellschaft*. Wiesbaden: VS, 115–131.

Lemke, T., 2000. Neoliberalismus, Staat und Selbsttechnologie. Ein kritischer Überblick über die governmentality studies. *Politische Vierteljahresschrift*, 41 (1), 31–47.

Ludwig, N., 2015. Statement in a meeting of the budget committee of Berlin's House of Representatives, 28 January [online]. Cited from *meeting memo*. Available from: https://www.parlament-berlin.de/ados/17/Haupt/protokoll/h17-072-ip.pdf [Accessed 6 November 2020].

McCann, E., 2017. Governing urbanism: urban governance studies 1.0, 2.0 and beyond. *Urban Studies*, 54 (2), 312–326.

Müller, C., et al., 2019. Städtische Quartiere in der Tourismusfalle? Zur Wahrnehmung von Tourismus und Airbnb in Berlin–Ein Werkstattbericht. *In*: T. Freytag and A. Kagermeier, eds. *Touristifizierung urbaner Räume*. Mannheim: MetaGis-Verlag, 49–58.

Niewöhner, J., 2014. Ökologien der Stadt. Zur Ethnografie bio- und geopolitischer Praxis. *Zeitschrift für Volkskunde*, 110 (2), 185–214.

Novy, J., 2017. The selling (out) of Berlin and the de- and re-politicization of urban tourism in Europe's 'Capital of Cool'. *In*: C. Colomb and J. Novy, eds. *Protest and resistance in the tourist city*. New York: Routledge, 32–52.

Novy, J. and Colomb, C., 2017. Urban tourism and its discontents: an introduction. *In*: C. Colomb and J. Novy, eds. *Protest and resistance in the tourist city*. New York: Routledge, 1–30.

Novy, J. and Colomb, C., 2019. Urban tourism as a source of contention and social mobilisations: a critical review. *Tourism Planning & Development*, 16 (4), 358–375.

Oskam, J., 2020. Introduction. *In*: J. Oskam, ed. *The overtourism debate: mimby, nuisance, commodification*. Bingley: Emerald, 1–7.

Scott, J.C., 1998. *Seeing like a state: how certain schemes to improve the human condition have failed*. New Haven: Yale University Press

Senate Department for Economics, 2011. *Tourismuskonzept Berlin. Handlungsrahmen 2011+* [online]. Available from: https://digital.zlb.de/viewer/metadata/33656385/1/ [Accessed: 21. December 2020].

Senate Department for Economics, 2014. *Konzept zum Akzeptanzerhalt des Tourismus* [online]. Available from: https://www.parlament-berlin.de/ados/17/Haupt/vorgang/h17-1598-v.pdf [Accessed 6 November 2020].

Senate Department for Economics, 2015. *Statement of state secretary in the Berlin's budget committee meeting on 28.01.2015*. Cited from meeting memo [online]. Available from: https://www.parlament-berlin.de/ados/17/Haupt/protokoll/h17-072-ip.pdf [Accessed: 6. November 2020].

Senate Department for Economics, 2016. *Bericht zur Weiterentwicklung des Konzepts und der Handlungsempfehlungen zum Akzeptanzerhalt des Tourismus*. Berlin.

Senate Department for Urban Development and Housing, 2019. *Answer to parliamentary request by Katalin Gennburg (Document Nr. 18 / 17 858)*. Berlin.

Sommer, C., 2021. *Seeing like a tourist city. Governance-Techniken der (Ent-) Problematisierung stadttouristischer Konflikte–das Fallbeispiel Berlin*. Thesis (PhD). Humboldt-Universität zu Berlin. DOI: 10.18452/22587.

Sommer, C. and Helbrecht, I., 2017. Seeing like a tourist city: how administrative constructions of conflictive urban tourism shape its future. *Journal of Tourism Futures*, 3 (2), 157–170.

Sommer, C. and Kip, M., 2019. Commoning in new tourism areas. Co-performing evening socials at the Admiralbrücke in Berlin-Kreuzberg. *In*: T. Frisch, et al., eds. *Tourism and everyday life in the contemporary city*. Oxon: Routledge, 211–231.

Valverde, M., 2011. Seeing like a city: the dialectic of modern and premodern ways of seeing in urban governance. *Law & Society Review*, 45 (2), 277–312.

visitBerlin, 2010. Statement of visitBerlin-CEO in an interview "Da bilden sich touristische Ameisenstraßen". *taz – die tageszeitung*, 6 August.

visitBerlin, 2011a. Statement of visitBerlin-CEO in an article "Kein Ballermann". *taz – die tageszeitung*, 13 May.

visit Berlin, 2011b. Statement of visitBerlin-CEO in an interview "Provinzielle Diskussion". *Zitty*, 17 May.

visit Berlin, 2013. *Jahresbericht*. Zahlen und Fakten 2012. Berlin.

visit Berlin, 2014. *Kurzkonzept Akzeptanzerhaltung*. Berlin.

visit Berlin, 2015. Statement of visitBerlin-CEO in an article "Berlin erwartet einen neuen Ansturm von Touristen". *Die Morgenpost*, 17 May.

visitBerlin, 2016. *Jahresbericht 2015* [online]. Available from: https://about.visitberlin.de/sites/default/files/MAM//asset/2017-05/Geschäftsbericht%202015.pdf [Accessed 6 November 2020].

visit Berlin, 2019. Interview with visitBerlin-CEO in an article "Eine Großstadt kann nicht unter sich bleiben". *Der Tagesspiegel*, 6 October.

visit Berlin, 2020. Statement of visitBerlin-CEO in an interview. *RBB-Inforadio*, 2 February. Available from: https://www.inforadio.de/programm/schema/sendungen/int/202001/02/silvester-tourismusmarketing-visit-berlin.html [Accessed 2 February 2020].

Wowereit, K., 2012. Statement in visitBerlin press-release, 14 September.

Part III

Representations and identities

Hopes and challenges for New Urban Tourism

12 Shock of the new

The rhetoric of global urban tourism in the rebuild of Christchurch, New Zealand

Alberto Amore and C. Michael Hall

Introduction

Policy and governance are an important feature in the design, implementation and management of urban tourism spaces. Sound urban tourism governance can lead to "improved experiences for the visitors; reduction of negative impacts and greater net benefits for the host community; and improved functioning of the total, interdependent industry within the urban environment" (Edwards *et al.* 2009, p. 102). Contemporary approaches to decision-making and delivery of destination strategies has led to greater community involvement and the associated establishment of so-called 'smart' urban tourism governance networks. These, in turn, have fostered new forms of urban experience that "are characterised by the increasing quest by tourists for contact with mundane life in ordinary residential quarters" (Dirksmeier and Helbrecht 2015, p. 276). However, the full implications of the so-called New Urban Tourism are still to be evaluated as, over the last decade, the emphasis on market-driven urban tourism policies have only exacerbated the urban socio-economic impacts of the 2007–2008 Global Financial Crisis. Not surprisingly, debates on the future of urban tourism policy stress the further gearing of Destination Management and Marketing Organisations (DMMOs) to commercial goals (Amore and Hall 2017a) and the rise of new global corporate players in urban tourism, such as Airbnb and Uber (Amore *et al.* 2020, Hall *et al.* 2017, Oskam 2019).

Urban spaces are the reflection of policies and development practices. From a spatial planning perspective, the legitimisation of public–private partnerships and urban development corporations, reflect the fourth wave of urban neoliberal praxis (Olesen 2014). Cities embarked on flagship projects between the late 1970s and the 1990s to better position themselves as tourism destinations (Mugerauer 2009) as part of the discourse and practice of place competition. The blending of architecture with place branding (Hall 2008) and the emphasis on culture and creativity (Richards and Wilson 2006) have provided urban redevelopment alternatives to hierarchical steering modes of urban tourism governance. Cities have therefore shifted "from corporatist urban governance to adaptive urban metagovernance" (Amore 2019, p. 73), with urban regeneration strategies utilising both demand and supply perspectives to compete in a globalised urban tourism market, with cities redesigning spaces to attract and retain key city users and desirable temporary urban populations (Spirou 2010).

Such a shift is far from straightforward. Urban regeneration and tourism are inextricably tied to capital circulation, globalisation and the position of cities in the global urban hierarchy (Ashworth and Page 2011). This is particularly evident in the implementation of urban tourism recovery strategies in cities following a major disaster (Amore 2019, Gotham and Greenberg 2014). In such contexts, "national governments handle very large disaster at the top political tier" (Johnson and Olshansky 2013, p. 16), with substantial rearrangements of urban governance allegedly aiming at expediting recovery. Command-and-control steering modes of hierarchical governance are further legitimised, with a concentration of executive powers in the hands of selected few and a systematic side-lining of tourism relevant stakeholders and local residents. The latter, ultimately, have become more vocal against their minimised role in urban policy processes in recent years, with grassroot movements in post-disaster cities (e.g. New York) (Hardt and Negri 2011) and cities in general (e.g. Barcelona, Venice) (Caccia 2016).

The following chapter provides a longitudinal narrative of the redevelopment of Christchurch as a 'New' Urban Tourism destination following the 2010–2011 Canterbury earthquakes. The aim of the chapter is to provide an account of the stakeholders' roles and the policy discourses that emerged with regard to the rebuild of the city centre, which is the focal point for tourism in the city. Ultimately, the chapter sheds light on key episodes of governance that have shaped the city of Christchurch as an urban tourism destination over the last decade.

Divergent urban visions

Urban tourism development takes different forms with differing sets of consequences for the diverse sets of stakeholders involved (Searle 2009). As a result, tourism-related urban development projects often lead to socio-political conflicts "in which some stakeholders are more equal than others and have greater influence in determining the outcomes of development" (Searle 2009, p. 203). For instance, urban tourism developments are often tied to area-based regeneration projects as part of hallmark cultural events (Ferrari and Guala 2017, Johnson 2016) and sport events (Amore and Hall 2017b, Hall and Wilson 2011). These are often portrayed as a 'once in a lifetime' opportunity to redefine spaces and cities for leisure, tourism, hospitality, retail and public space and generate employment (Wise and Harris 2017). However, in such developments, stakeholders with decisional powers can be distinguished representatives from civic society with tokenistic participatory roles. These stakeholders coalesce around an urban development agenda that reiterates the rhetoric of urban competitiveness, economic growth and international tourism appeal (Hall 2006, Searle 2009). The latter, instead, find themselves in an increasingly depoliticised vacuum in which the rhetoric and logic of global urban tourism overrules community-grounded opinions and expectations regarding the future of cities.

These socio-political dynamics occur in hundreds of cities around the world (Amore 2019), particularly in the development of brownfield sites for leisure and tourism amenities and infrastructure (Saxena 2013). For example, in the redevelopment of Darling Harbour in Sydney, Australia, urban growth coalitions favoured the building of shopping and cultural amenities as a way to foster agglomeration economies and attract hotels and other tourism-related services (Ritchie 2009). Government authorities established a dedicated development corporation (the Darling Harbour Authority) that "overrode the state planning act with its various investigatory, participatory and environmental impact requirements" (Searle 2009, p. 207). Throughout its short lifespan, the Darling Harbour Authority oversaw and implemented projects and plans with little public participation or scrutiny (Hall 1998). Similarly, the organisation of sport events like the Olympic Games emphasises redevelopment visions and short-term economic objectives for urban tourism that are often questionable (Amore 2019).

The rapid physical development of urban tourism spaces, as opposed to more incremental development, become contentious between stakeholders when there are divergent urban visions (Amore *et al.* 2017, Doorne 1998). Particularly in post-crisis redevelopment, this can lead to "an explosion of spaces in which a multiplicity of meanings of and visions for the city erupt and enter into conflict" (Gotham and Greenberg 2014, p. 96). This was observed, for example, in post-Katrina New Orleans, with grassroot movements and advocacy groups fighting for transparency and inclusiveness in the redevelopment of the city and its future as a tourist destination (Airriess *et al.* 2008, Gotham and Greenberg 2014). Similarly, neighbourhood associations in Brooklyn opposed the expedited refurbishment of the Gowanus Canal area into a prime retail and real estate development for new residents and city users following hurricane Sandy (Glück 2013, Gould and Lewis 2018). Nevertheless, there are cases where both residents and real estate developers aligned in the top-down vision for the city, as in the case of Shenzhen, China (Sonn *et al.* 2017).

In extremely deprived or derelict areas, consensus among multiple stakeholders for urban tourism redevelopment is reasonably common because of the sheer urgency of development action (Amore and Hall 2016, Searle 2009). However, in some cases, the concentration of decision-making process in the hands of a restricted group of non-elected officials and consultants has eroded the ability of community organisations to meaningfully participate in decision-making processes (Nel 2015). As result, the use of authoritative modes of urban governance has tended to encourage urban development models and modes of governance that are closely aligned to the interests of international financial groups and developers (Amore 2019, Porter 2009). This raises questions as to whether the urban tourism policies underpinning the New Urban Tourism discourse can be actually implemented in post-disaster contexts or be instrumentalised to support the hegemonic urban development rhetoric.

Whose New Urban Tourism?

The notion of New Urban Tourism was originally introduced by Roche (1992) in the study of the impact and functions of mega events. In his view, it was necessary to move towards a broader account of "contexts of structural change and of discontinuity and reorganization in local economies" (Roche 1992, p. 565). From a spatial perspective, New Urban Tourism is associated with the mobility of tourists and their interaction with residents outside purpose-built amenities and attractions in the city. This has long been an element of the use of informal accommodation in cities as well as VFR tourism, however the significance of such spatial shifts was first observed in the inner-city neighbourhoods of Islington and Southwark in London (Long 2000, Maitland 2006), and has become a distinctive feature of regional capitals in Europe and North America (Hutton, 2009) — which may also suggest that urban researchers may need to get out more as the dispersal of visitors in urban areas has long been recognised (Page and Hall 2003). More recently, New Urban Tourism has been referred to in almost Romantic terms as being "characterised by the increasing quest by tourists for contact with mundane life in ordinary residential quarters" (Dirksmeier and Helbrecht 2015, p. 276). The trope of New Urban Tourism therefore acts as a catalyst for deeper entanglements of city users, spaces and identities (Gibson 2010) and what could be described as the suburbanisation of tourism. As Pappalepore *et al.* (2014, p. 228) suggest, New Urban Tourism has "the potential to challenge traditional views of tourism consumption by blurring the boundaries between tourists, day visitors and residents". Supposedly, the result is the rise and establishment of New Urban Tourism areas (Pappalepore *et al.* 2010) beyond traditional urban tourism precinct experience (Hayllar and Griffin 2005).

Research on New Urban Tourism and the relations between tourists and local residents in cities is gaining momentum. There is an argument that New Urban Tourism is a flywheel for gentrification and the further displacement of low-income residents away from inner city areas (Dirksmeier and Helbrecht 2015). Pappalepore *et al.* (2010) suggest that visitor experiences in New Urban Tourism occurs in gentrified areas in which the global and highly mobile population of creatives and professionals (Florida 2009) are part of the tourist product (Deery *et al.* 2012) and contribute to cosmopolitanism and diverse urban identities (Maitland 2007, Selby 2004). The rise of temporary letting platforms and the promotion of heritage conservation projects are seen as instruments in the gentrification of urban spaces (Amore *et al.* 2020, Spirou 2010), particularly in world cities (Lees 2003). Awareness of how New Urban Tourism is framed within current urban development discourses destinations is relatively recent (Amore 2019) and focuses predominantly on the debates around overtourism and resident perception (Adie and Falk 2020, Milano *et al.* 2019). Yet relatively little is known about crisis-driven urban redevelopment and the New Urban Tourism entangle when it comes to post-disaster rebuilding of cities.

The emergence of New Urban Tourism raises some interesting questions in terms of finding (and naming) urban tourism processes that were already occurring. For example, the embeddedness of short-term accommodation provision in New Urban Tourism is not a new phenomenon and neither is the development of hospitality, heritage, cultural and retail space that attracts visitors outside of the purpose-built tourism attractions and facilities. Such areas of more informal experiential urban tourism, that lay outside of the interest of tourism academics for many years, were often tied up with the (initially) low-rent, inner city, migrant, artistic, student and gay and lesbian spaces, that have served as the focal point for consumer-driven gentrification (Doan and Higgins 2011, Faro and Wotherspoon 2000, Hughes 1991, Nash and Gorman-Murray 2015, Ruting 2008), within which tourism has acted as a further catalyst. However, two things have arguably contributed to their description as 'new'. First, the stretching of spatial and temporal relations as the result of globalisation processes, particularly tourism mobilities and Information Communications Technology (ICT), has linked such spaces with international demand so as to accelerate gentrification. This is where the role of platforms, such as Airbnb, become especially significant as conduits that encourage and enable both international supply and demand (Gössling and Hall 2019). Second, destination marketing organisations have come to promote New Urban Tourism locations off the beaten track in order to diversify their product (Hall 2013). This, in turn, has led some scholars to question New Urban Tourism as part of the ongoing academic fetish of putting old wine in new bottles.

Undeniably, the notion of New Urban Tourism is far from neutral. Looking at the development policies over the last years, the promotion of alternative and creative forms of urban leisure and tourism consumption underpin the market-driven agenda of pioneering regeneration (and gentrification) that has been reshaping cities around the world over the last 40 years (Amore 2019). Those stressing the increasing socio-economic disparities between visitors and vulnerable resident populations recognise that tourism per se is not the problem and instead suggest that policy-decisions and policy-related actions, especially related to housing, are the root cause of increased urban inequality (Milano *et al.* 2019, Spirou 2010). Legitimate questions arise as to whose interests, values and believes urban tourism redevelopments actually respond to. Addressing this question is key in the study of socio-economic vulnerabilities and resilience of urban destinations (Amore 2019, Hall *et al.* 2017).

Context and source material

Christchurch is the second largest city of New Zealand and the main urban tourism destination in the country's South Island. Prior to the earthquakes, the visitor economy in Christchurch represented 10.2% of total spending in the city, for a combined total of more than NZ$1.7 billion in 2010 (NZMBIE 2010). The 2010–2011 earthquakes and the scale of the damages to hospitality infrastructure and attractions in the city resulted in a sharp downturn of tourism in the city and the closure of most of the main urban tourism area for more than two years (Amore 2017). Christchurch's share of the national visitor spend was 9.2% pre-earthquake, dropping to 6.9% in 2012, and only recovered to 7.8% by 2016 (C&CC 2016).

The redevelopment of Christchurch following the earthquakes has been a mix of fast-tracked, delayed and oversized projects (Amore and Hall 2016, 2017b). The Christchurch City Council (CCC) did gather input from a series of community engagement workshops "to create an environment which people will want to visit and shop in" (CCC 2011b, p. 10). However, the legislatively and economically dominant central-government–led Canterbury Earthquake Recovery Authority (CERA) developed the Greater Christchurch Recovery Strategy with input from international consultancy companies (Amore 2017). The governance arrangements therefore meant that contestation between different stakeholders was inherent to the redevelopment of the city centre. CERA officials and the dedicated Christchurch Central Development Unit (CCDU) framed the redevelopment strategy of the city centre on a series of anchor projects (CCDU 2012), while the CCC envisioned the creation of a low-rise central city with urban precincts acting as clusters for organic growth (CCC 2011a).

The following is based on longitudinal research that began in 2013 to identify and understand the governance of urban tourism spaces in Christchurch. The findings encompass the early recovery phase under CERA (2011–2016) and the current phase under Regenerate Christchurch and Ōtākaro Limited. The redevelopment of Christchurch as a tourist destination is strongly tied to how the national recovery authorities saw the rebuild "as a once in a lifetime opportunity to radically change the highly parcelled ownership of land in the CBD and sell allotments to attract major international developers" (Amore and Hall 2016, p. 190). Data were retrieved from grassroots organisations, as well as from the CCC and CERA's archives. These were triangulated with findings from a two-round series of semi-structured interviews with 52 stakeholders. The names of the participants indicated below are fictional and their affiliations are not reported.

Christchurch between recovery and innovation

The rebuild of the city has been highly contested, with residents' and local stakeholders' inputs on the draft central city recovery being systematically sidelined or silenced by the CERA and the CCDU. For example, the delivery of refurbished cultural and performing arts amenities in the Christchurch city centre favoured a narrative that aligned with those of project developers and influential lobbying groups rather than the Christchurch arts community. Sport, event and business tourism stakeholders also reinforced the visions and strategies promoted by the recovery authorities, leaving small developers and tourism businesses at arm's length on key decisions. Most of the projects launched between 2012 and 2017 had almost no public input or consultation. As one of the participants observed:

> Groups like the City Council and the Government picked people, not because they have a real knowledge or skill, but just because of who they were. [...] The lack of collaboration ... and the attitude of the government with regards to the acquisition of land for the anchor projects eventually led privates and developers to work outside the CBD.
>
> (Frank)

Stakeholders representing the creative industry were prioritised over local artists and craft vendors in the vision for the refurbished Arts Centre. Prior to the earthquakes, this complex was one of the key tourist attractions in the city, with an estimate of 500,000 visitors (Amore 2017). In defining the new vision for the site, tourism was considered a key management priority (Nahkies 2012) along with the inclusion of new tenants and leisure activities. The then Director for the Arts Centre, Andre Lovatt, reiterated the importance of "enabling the Arts Centre to reclaim its vitally important place in the lives of the people of Christchurch, and those who visit" (ACC 2013a, p. 7) and of pursuing the attraction of creative entrepreneurs to the site (ACC 2013a, 2013b). He further reiterated that it was important "to focus our efforts on making the Arts Centre a place [where] domestic and international visitors will value the authenticity of the Arts Centre's offer and become patrons themselves" (ACC 2013b, p. 4). As one respondent further observed:

> If the residents themselves love the site and actually want to spend time here and are part of making it, in actual fact that becomes exciting for tourists, because it's a place with a real vibe and sense of community and participation and activity.
>
> (Jim).

The relationships between arts, culture and tourism were highlighted in other redevelopment projects across the city. For example, the draft Performing Arts Precinct and Town Hall project stated that:

> Events hosted at a performance venue provide a focal point for a reason to visit Christchurch. A large performance venue will bring in headline performances that will attract visitors from throughout the South Island. Such visitors may stay on after a show, increasing the tourism spend in the city.
>
> (Deloitte 2015, p. 12).

The idea of the arts as an activator for hospitality-related businesses was acknowledged by stakeholders during the interviews. "Cafes and restaurants tend to group around where people are coming in and out" (Amanda) and arts facilities represented a key asset in the redevelopment of urban tourist spaces. The experiential dimension was also reiterated by Anna: "there was always the view that the performing arts, and the broad arts in general, bring vibrancy to the central city". Notwithstanding a consensus among different stakeholders on the vision for Christchurch as a cultural and creative destination, the actual role of the arts in the recovery was very generic and often geared towards leisure and entertainment (Amanda). As participants stated:

> It was the Convention Centre Precinct first and foremost. Absolutely dominated. [They] also asked the Performing Arts to be close by so the partners of people going to a convention could easily go be entertained.
>
> (Edward)

> [They] liked the area for the Performing Arts Precinct was going to be close to the Convention Centre. So, people from the convention would logically have all that stuff going on in that part of town.
>
> (Gabriel)

Focusing on the future of Christchurch as a business tourism destination, the divergence between stakeholders was significant. Recovery authorities were vocal on the importance of building an international conference centre to boost hospitality development in the city centre (Elton, Sam and Thomas). This echoed the Greater Christchurch Visitor Recovery Plan view of the conference centre anchor project as "essential to re-instate high value business tourism in a city that has previously excelled in the conference and convention sector" (C&CT 2012, p. 7). Similarly, the local DMMO regarded the project as a much-needed priority "to a vibrant city centre with major hotel development attached to it" (CINZ 2014, n.p.).

During the early recovery phase (2011–2016), the quest of the national government "to identify priorities, prepare plans and set directions for recovery activities" (CERA 2011, p. 1) overruled the local government draft city centre recovery plan. Although there were commonalities, the local government emphasised more the need to redevelop for the local market, while the central government focused on the importance of international tourism and markets, critically with attention to the experiential aspects of New Urban Tourism.

The announcement by central government of a dedicated regeneration unit "to progressively pass governance and management of the rebuild to the Canterbury community" (Brownlee 2015, n.p.) marked the end of the first stage of the city redevelopments. While some projects and precincts announced in the first stage are now completed or almost completed, the rebuild of the city has only moved slowly since. Construction of a new stadium has not yet started, while the Arts Centre has not been as successful as had been hoped, although Airbnb properties in the inner city have placed pressure both on existing hospitality businesses and housing demands (Prayag *et al.* 2020). As a result, in late 2020 the Christchurch City Council announced the planned placement of restrictions on short-term holiday rentals (Law 2020). But, like many destinations, the future is on hold because of COVID-19.

Conclusions

Site contestation over specific projects and the rhetoric of New Urban Tourism to legitimise the redevelopment agenda have been central in the design and delivery of key anchor projects in Christchurch. The experience of Christchurch shows how the cultural and creative stakeholders envisioned a destination in which visitors and residents interacted as part of the authentic tourist experience, one of the elements of the New Urban Tourism literature (Pappalepore *et al.* 2014). The case of the Arts Centre is of particular interest, as the new vision for the site fostered a process of gentrification and displacement of former tenants of what had previously been a location deliberately geared to the local community (Amore 2016). This also echoes the literature on New Urban Tourism perceptions among residents in inner city areas (Dirksmeier and Helbrecht 2015, Panton and Walters 2018). Importantly, such a vision, as well as the deliberate development of specialised precincts, rather than build on existing organic development initiatives, was top-down driven by central government as a means to accelerate the

Christchurch rebuild in a fashion that was thought to be attractive to the international tourism market.

The rhetoric regarding the future of Christchurch as a business destination is also an emblematic example of fierce neoliberal urban governance in post-disaster contexts (Glück 2013, Gotham and Greenberg 2014, Gould and Lewis 2018). Although convention and business tourism might not seem to fit under the New Urban Tourism umbrella, it needs to be noted that the arts and cultural dimensions were seen as critical for the attraction of the market because of the "vibrancy" that they helped bring. However, as Amore (2019, p. 33) suggested, such a "neoliberal rebuild agenda is legitimized through no or tokenistic community involvement [...] that heightens the detachment between institutions and their citizens". Indeed, the local people who comprise the vibrancy were ignored, or when they were consulted, then the results of such a process were not included in the redevelopment plan.

The Christchurch post-earthquake experience provides several insights into New Urban Tourism. First, the experiential aspects of New Urban Tourism helped provide a vision for post-disaster urban tourism redevelopment. Second, local communities had little input to the vision that was top-down driven and designed to accelerate development and pre-existing regeneration processes. This highlights the importance of governance and metagovernance as an (ignored) driver for New Urban Tourism. Third, the explicit emphasis on precincts and sites as part of the packaging of place experiences in top-down–led redevelopment reinforces questions as to just how new the New Urban Tourism really is. The previously available informal accommodations have now been commoditised by online platforms but otherwise the various elements of tourism-led regeneration trajectories have changed little over time. Significantly, the New Urban Tourism emphasis has not necessarily been fully successful, in part because of the paradoxes it provided, that something which is authentic because it is not-commoditised and is seen as representing the local community then becomes packaged and promoted to help sell a city as a destination. This situation reflects the possibility that more organic regeneration and gentrification trajectories that involve local communities may well be better at providing positive outcomes for both tourists wanting something of the inner-city local experience as well as the residents and businesses of such locations. Finally, from a wider perspective it should be noted that in academic terms the features of what comprises the so-called New Urban Tourism were already well established in the Christchurch context before the term came along, with visitors dispersed through the city for accommodation (though not available online), events, markets, greenspace, festivals and shopping experiences (Hall 2008, Hart 1992, Pearce 2011). What has changed arguably is the promotion and packaging of such experiences together with the integration of short-term guest accommodation, some of which is now purpose-built, into selling the city. However, this should not be described as New Urban Tourism as a result of the penchant of academics to invent and use terms in order to appear relevant. Instead, it should be called what it really is: tourism-related capitalism that serves only limited interests.

Acknowledgements

The authors would like to thank the editors for the valuable comments on this chapter.

References

ACC, 2013a. *The Arts Centre of Christchurch. Annual report 2012*. Christchurch: Arts Centre of Christchurch.

ACC, 2013b. *A hub for creative entrepreneurs: the Arts Centre of Christchurch vision*. Christchurch: Arts Centre of Christchurch.

Adie, B.A. and Falk, M., 2020. Residents' perception of cultural heritage in terms of job creation and overtourism in Europe. *Tourism Economics*. https://journals.sagepub.com/doi/10.1177/1354816620943688

Airriess, C.A., et al., 2008. Church-based social capital, networks and geographical scale: katrina evacuation, relocation, and recovery in a New Orleans Vietnamese American community. *Geoforum*, 39 (3), 1333–1346.

Amore, A., 2016. I do (not) want you back! (Re)gentrification of the Arts Centre, Christchurch. *In*: C.M. Hall, et al., eds. *Business and post-disaster management: business, organisational and consumer resilience and the Christchurch earthquakes*. Abingdon: Routledge, 79–96.

Amore, A., 2017. *Before/after: the governance of urban tourist spaces in the context of the Christchurch CBD recovery*. Thesis (PhD). University of Canterbury, Christchurch.

Amore, A., 2019. *Tourism and urban regeneration: processes compressed in time and space*. Abingdon: Routledge.

Amore, A., de Bernardi, C., and Arvanitis, P., 2020. The impacts of Airbnb in Athens, Lisbon and Milan: a rent gap theory perspective. *Current Issues in Tourism*. https://www.tandfonline.com/doi/full/10.1080/13683500.2020.1742674

Amore, A. and Hall, C.M., 2016. 'Regeneration is the focus now': anchor projects and delivering a new CBD for Christchurch. *In*: C.M. Hall, et al., eds. *Business and post disaster management: business, organisational and consumer resilience and the Christchurch earthquakes*. Abingdon: Routledge, 181–199.

Amore, A., & Hall, C.M., 2017a. National and urban public policy agenda in tourism. Towards the emergence of a hyperneoliberal script? *International Journal of Tourism Policy*, 7 (1), 4–22.

Amore, A., & Hall, C.M., 2017b. Sports and event-led regeneration strategies in post-earthquake Christchurch. *In*: N. Wise and J. Harris, eds. *Regeneration, events, journal of sport and tourism*. Abingdon: Routledge, 100–118.

Amore, A., Hall, C.M., and Jenkins, J.M., 2017. They never said "Come here and let's talk about it": exclusion and non-decision making in the rebuild of Christchurch, New Zealand. *Local Economy*, 32 (7), 617–639.

Ashworth, G.J. and Page, S.J., 2011. Urban tourism research: recent progress and current paradoxes. *Tourism Management*, 32 (1), 1–15.

Brownlee, G., 2015. Regeneration the focus of Chch Governance [Press release]. [online]. *Beehive.govt.nz*. The official website of the New Zealand Government. Available from: https://www.beehive.govt.nz/release/regeneration-focus-chch-governance [Accessed 10 April 2020].

C&CC, 2016. *Christchurch visitor strategy–setting the direction 2016*. Christchurch: Christchurch & Canterbury Corporation.

C&CT, 2012. *Greater Christchurch visitor recovery plan*. Christchurch: Christchurch & Canterbury Tourism.

Caccia, B., 2016. Dalle piattaforme civiche alle citta ribelli. *EuroNomade*, 23 September [online]. Available from: http://www.euronomade.info/?p=7979 [Accessed 25 May 2020].

CCC, 2011a. *Central city plan. Draft central city recovery plan for ministerial approval–December 2011*. Christchurch: Christchurch City Council.

CCC, 2011b. *Draft central city plan–August 2011, Volume 2: regulatory changes to the Christchurch City Council City Plan 2005 and global stormwater consent*. Christchurch: Christchurch City Council.

CCDU, 2012. *Christchurch central business district blueprint–implementing anchor projects*. Christchurch: Christchurch Central Development Unit.

CERA, 2011. *Draft recovery strategy for greater Christchurch*. Christchurch: Canterbury Earthquake Recovery Authority.

CINZ, 2014. Convention Centre good news for tourism. [Press release]. *Scoop Independent News* [online]. Available from: https://www.scoop.co.nz/stories/AK1408/S00149/convention-centre-good-news-for-tourism.htm [Accessed 2 May 2018].

Deery, M., Jago, L., and Fredline, L., 2012. Rethinking social impacts of tourism research: a new research agenda. *Tourism Management*, 33 (1), 64–73.

Deloitte, 2015. *Christchurch Town Hall. Strategic and economic case*. Christchurch: Deloitte.

Dirksmeier, P. and Helbrecht, I., 2015. Resident perceptions of new urban tourism: a neglected geography of prejudice. *Geography Compass*, 9 (5), 276–285.

Doan, P.L. and Higgins, H., 2011. The demise of queer space? Resurgent gentrification and the assimilation of LGBT neighborhoods. *Journal of Planning Education and Research*, 31 (1), 6–25.

Doorne, S.M., 1998. Power, participation and perception: an insider's perspective on the politics of the Wellington waterfront redevelopment. *Current Issues in Tourism*, 1 (2), 129–166.

Edwards, D., Griffin, T., and Hayllar, B., 2009. Urban tourism precincts: an overview of key themes and issues. *In*: B. Hayllar, T. Griffin and D. Edwards, eds. *City spaces-tourist places: urban tourism precincts*. Amsterdam: Elsevier, 95–106.

Faro, C. and Wotherspoon, G., 2000. *Street seen: a history of Oxford street*. Carlton South, VIC: Melbourne University Press.

Ferrari, S. and Guala, C., 2017. Mega-events and their legacy: image and tourism in Genoa, Turin and Milan. *Leisure Studies*, 36 (1), 119–137.

Florida, R.L., 2009. *Who's your city?: how the creative economy is making where to live the most important decision of your life*. New York: Basic Books.

Gibson, C., 2010. Geographies of tourism: (un)ethical encounters. *Progress in Human Geography*, 34 (4), 521–527.

Glück, Z., 2013. Race, class, and disaster gentrification. *Tidalmag* [online]. Available from: http://tidalmag.org/blog/race-class-and-disaster-gentrification/ [Accessed 26 June 2017].

Gotham, K.F. and Greenberg, M., 2014. *Crisis cities: disaster and redevelopment in New York and New Orleans*. Oxford: Oxford University Press.

Gould, K.A. and Lewis, T.L., 2018. From green gentrification to resilience gentrification: an example from Brooklyn. *City & Community*, 17 (1), 12–15.

Gössling, S. and Hall, C.M., 2019. Sharing versus collaborative economy: how to align ICT developments and the SDGs in tourism? *Journal of Sustainable Tourism*, 27 (1), 74–96.

Hall, C.M., 1998. The politics of decision making and top-down planning: Darling Harbour Sydney. *In*: D. Tyler, Y. Guerrier and M. Robertson, eds. *Managing tourism in cities: policy, process and practice*. Chichester: John Wiley & Sons, 9–24.

Hall, C.M., 2006. Urban entrepreneurship, corporate interests and sports mega-events: the thin policies of competitiveness within the hard outcomes of neoliberalism. *The Sociological Review*, 54 (2), 59–70.

Hall, C.M., 2008. Servicescapes, designscapes, branding, and the creation of place-identity: south of Litchfield, Christchurch. *Journal of Travel and Tourism Marketing*, 25 (3/4), 233–250.

Hall, C M., 2013. Regeneration and cultural quarters: changing urban cultural space. *In*: M.K. Smith, ed. *The Routledge handbook of cultural tourism*. Abingdon: Routledge, 355–361.

Hall, C.M., Le-Klähn, D.-T., and Ram, Y., 2017. *Tourism, public transport and sustainable mobility*. Clevedon: Channel View Publications.

Hall, C.M., Prayag, G., and Amore, A., 2017. *Tourism and resilience: individual, organisational and destination perspectives*. Clevedon: Channel View Publications.

Hall, C.M. and Wilson, S., 2011. Neoliberal urban entrepreneurial agendas, Dunedin Stadium and the Rugby World Cup: or 'if you don't have a stadium, you don't have a future'. *In*: D. Dredge and J.M. Jenkins, eds. *Stories of practice: tourism policy and planning*. Burlington, VT: Ashgate, 133–152.

Hardt, M. and Negri, A., 2011. The fight for 'real democracy' at the heart of Occupy Wall Street: the encampment in Lower Manhattan speaks to a failure of representation. *Foreign Affairs*, 1 October [online]. Available from: https://www.foreignaffairs.com/articles/north-america/2011-10-11/fight-real-democracy-heart-occupy-wall-street?page=show [Accessed 25 May 2020].

Hart, A.J., 1992. *Planning for tourism in Christchurch: a comparative study*. Thesis (Master). University of Canterbury, Christchurch.

Hayllar, B. and Griffin, T., 2005. The precinct experience: a phenomenological approach. *Tourism Management*, 26 (4), 517–528.

Hughes, R., 1991. *The shock of the new*. Revised edition. London: Thames and Hudson.

Hutton, T.A., 2009. *The new economy of the inner city: restructuring, regeneration and dislocation in the 21st century metropolis*. London: Routledge.

Johnson, L.A. and Olshansky, R.B., 2013. The road to recovery: governing post-disaster reconstruction. *Land Lines*, 25 (3), 14–21.

Johnson, L.C., 2016. *Cultural capitals: revaluing the arts, remaking urban spaces*. Burlington, VT: Ashgate.

Law, T., 2020. Proposed plan to restrict Christchurch's Airbnb-style accommodation 'anti-tourism'. *Stuff*, 7 September [online]. Available from: https://www.stuff.co.nz/business/property/122680121/proposed-plan-to-restrict-christchurchs-airbnbstyle-accommodation-antitourism [Accessed 20 September 2020].

Lees, L., 2003. Super-gentrification: the case of Brooklyn heights, New York city. *Urban Studies*, 40 (12), 2487–2509.

Long, P., 2000. Tourism development regimes in the inner city fringe: the case of Discover Islington, London. *Journal of Sustainable Tourism*, 8 (3), 190–206.

Maitland, R., 2006. Culture, city users and the creation of new tourism areas in cities. *In*: M.K. Smith, ed. *Tourism, culture and regeneration*. Wallingford: CABI, 25–34.

Maitland, R., 2007. Tourists, the creative class, and distinctive areas in major cities. *In*: G. Richards and J. Wilson, eds. *Tourism, creativity and development*. Abingdon: Routledge, 73–86.

Milano, C., Cheer, J.M., and Novelli, M., eds., 2019. *Overtourism: excesses, discontents and measures in travel and tourism*. Wallingford: CABI.

Mugerauer, R., 2009. Architecture and urban planning: practical and theoretical contributions. *In*: T. Jamal and M. Robinson, eds. *The SAGE handbook of tourism studies*. London: SAGE, 290–313.

Nahkies, G., 2012. *The Christchurch Arts Centre: a review of governance structure issues–May 2012*. Wellington: Boardworks International.

Nash, C.J. and Gorman-Murray, A., 2015. Recovering the gay village: a comparative historical geography of urban change and planning in Toronto and Sydney. *Historical Geography*, 43, 84–105.

Nel, E., 2015. Evolving regional and local economic development in New Zealand. *Local Economy*, 30 (1), 67–77.

NZMBIE, 2010. *Spend by visitors in Christchurch for the year to March 2010*. Wellington: New Zealand Ministry for Business, Innovation and the Employment.

Olesen, K., 2014. The neoliberalisation of strategic spatial planning. *Planning Theory*, 13 (3), 288–303.

Oskam, J.A., 2019. *The future of Airbnb and the sharing economy: the collaborative consumption of our cities*. Clevedon: Channel View Publications.

Page, S.J. and Hall, C.M., 2003. *Managing urban tourism*. Harlow: Pearson.

Panton, M. and Walters, G., 2018. It's just a Trojan horse for gentrification': austerity and stadium-led regeneration. *International Journal of Sport Policy and Politics*, 10 (1), 163–183.

Pappalepore, I., Maitland, R., and Smith, A., 2010. Exploring urban creativity: visitor experiences of Spitalfields, London. *Tourism Culture & Communication*, 10 (3), 217–230.

Pappalepore, I., Maitland, R., and Smith, A., 2014. Prosuming creative urban areas: evidence from East London. *Annals of Tourism Research*, 44, 227–240.

Pearce, D.G., 2011. Tourism, trams and local government policy making in Christchurch: a longitudinal perspective. *In*: D. Dredge and J.M. Jenkins, eds. *Stories of practice: tourism policy and planning*. Burlington, VT: Ashgate, 57–78.

Porter, L., 2009. Whose urban renaissance? *In*: L. Porter and K. Shaw, eds. *Whose urban renaissance? An international comparison of urban regeneration strategies*. New York: Routledge, 241–252.

Prayag, G., et al., 2020. Integrating MLP and 'after ANT'to understand perceptions and responses of regime actors to Airbnb. *Current Issues in Tourism*. https://www.tandfonline.com/doi/full/10.1080/13683500.2020.1768226.

Richards, G. and Wilson, J., 2006. The creative turn in regeneration: Creative spaces, spectacles and tourism in cities. *In*: M.K. Smith, ed. *Tourism, culture and regeneration*. Wallingford: CABI, 12–24.

Ritchie, B.W., 2009. Contribution of urban precincts to the urban economy. *In*: B. Hayllar, T. Griffin and D. Edwards, eds. *City spaces-tourist places: urban tourism precincts*. Amsterdam: Elsevier, 151–182.

Roche, M., 1992. Mega-events and micro-modernization: on the sociology of the new urban tourism. *The British Journal of Sociology*, 43 (4), 563–600.

Ruting, B., 2008. Economic transformations of gay urban spaces: revisiting Collins' evolutionary gay district model. *Australian Geographer*, 39 (3), 259–269.

Saxena, G., 2013. Cross-sector regeneration partnership strategies and tourism. *Tourism Planning & Development*, 11 (1), 86–105.

Searle, G., 2009. Conflicts and politics in precinct development. *In*: B. Hayllar, T. Griffin and D. Edwards, eds. *City spaces-tourist places: urban tourism precincts*. Amsterdam: Elsevier, 203–222.

Selby, M., 2004. *Understanding urban tourism: image, culture and experience*. London: IB Tauris.

Sonn, J.W., et al., 2017. A top-down creation of a cultural cluster for urban regeneration: the case of OCT Loft, Shenzhen. *Land Use Policy*, 69, 307–316.

Spirou, C., 2010. *Urban tourism and urban change: cities in a global economy*. New York: Routledge.

Wise, N. and Harris, J., 2017. Introduction: framing sport, events, tourism and regeneration. *In*: N. Wise and J. Harris, eds. *Regeneration, events, journal of sport and tourism*. Abingdon: Routledge, 1–8.

13 New Urban Tourism in the post-conflict city

Sharing experiences of violence and peace in West Belfast

Henriette Bertram

Introduction

New Urban Tourism is usually regarded as a challenge to residential neighbourhoods, which is certainly justified in many cases (Maitland 2007, Novy 2013). In cities emerging from protracted intrastate conflict, however, tourism may become a bearer of hope for economic recovery as well as physical and social regeneration. Tourism in formerly conflict-ridden destinations that supports peacebuilding and helps to restructure social relations between the conflict parties has been conceptualised as "Phoenix Tourism" (Causevic and Lynch 2011). This chapter discusses New Urban Tourism in post-conflict Belfast where touristification has long been the dominant strategy of urban renewal (Neill *et al.* 2014, Bertram 2017b).

The focus of this chapter is on the residential area of West Belfast which has emerged as a popular sub-destination in Northern Ireland in recent years, mostly attracting visitors interested in Political or Dark Heritage (Wiedenhoft Murphy 2010, Skinner 2016, Bertram 2017). West Belfast is a nationalist stronghold and characterised by a close-knit, politically and culturally active community, and severe economic deprivation (Carden 2011). West Belfast is the only part of the city where the conflict is recognised as an important element of the city's heritage that is worth sharing with visitors.

The following section is dedicated to understanding the concepts of New Urban Tourism and Dark or Political Heritage Tourism in the context of post-conflict destinations. I will then give an overview on the strategies of urban renewal and tourism development in Belfast to provide the context on tourism in West Belfast. I will focus on the narratives of the conflict as well as actors in the field of tourism. Neighbourhood tourism in West Belfast is, on the one hand, a combination of New Urban Tourism and (Dark) Heritage Tourism. On the other hand, and even though it lacks certain features generally associated with New Urban Tourism, I argue that the touristic offerings aim to create an 'authentic' sense of life in the neighbourhood during and after the conflict for tourists and can therefore also be seen as a special case of New Urban Tourism.

I will show that neighbourhood tourism—which is criticised and protested against in many other cities—is seen as the preferred remedy for deprivation and an opportunity to share the experiences of ordinary people during the conflict in West Belfast. The economic turnout for the community is, however, questionable (Bereskin 2017). To a certain extent, tourism has indeed played a role in improving social relations between communities. Nevertheless, as the most important actors are local individuals and organisations, there is a possibility for it to be instrumentalised in a way that emphasises one communities' narrative over the other.

Tourism in post-conflict destinations: phoenix from the ashes?!

For many tourists, it is important to experience 'authenticity' and the everyday life of a destination (Leonard 2011), highlighting the significance of New Urban Tourism. Visitors increasingly explore (former) working-class neighbourhoods (Dirksmeier and Helbrecht 2015) which are perceived as vibrant, with a rich cultural life and strong evening economy, but remain somewhat 'edgy' and with a feel of novelty (Füller and Michel 2014). Tourist and resident activities can thus be difficult to differentiate, calling into question the "assumption that we can distinguish, isolate and examine a distinctive urban tourist" (Ashworth and Page 2011, p. 7). Residents, in this setting, become "part of the tourist product" (Dirksmeier and Helbrecht 2015, p. 277). In many traditional tourist cities, a strong critique of New Urban Tourism has emerged, blaming the influx of visitors for a range of "perceived evils" (Dirksmeier and Helbrecht 2015, p. 280, cf. Novy 2013, Novy and Colomb 2017).

Likewise, Heritage Tourism has gained popularity following travellers' quest for authenticity and uniqueness. It has been defined as the commodification and staging of "elements of the past" (Chhabra *et al.* 2003, Timothy and Boyd 2003). This means that authenticity is actively produced and there is always more than one possible version (Leonard 2011). The way heritage sites are dealt with can therefore be seen as a—conscious or not—way of showing how a society and its groups see themselves and how they want to be seen from outside (Ashworth and Graham 2005, Friedrich *et al.* 2018).

Heritage is not always 'dark', but many sites of Heritage Tourism, especially those dealing with dissonant heritage, can be considered sites of Dark Tourism: places associated with death, suffering, atrocity or tragedy (Lennon and Foley 2010, Light 2017). Sites of Dark Tourism "probably constitute the largest single category of tourist attractions in the world" (Smith 1998), encompassing a wide range of places from battlefields to fun-factories (Causevic and Lynch 2011, Farmaki 2013), especially, but not exclusively, in post-conflict settings (Tunbridge and Ashworth 1996, Bertram 2017). Visitors of these places are said to pursue remembrance, education and entertainment (Light 2017, Seaton 2018).

During protracted intrastate conflict, tourism is hardly a primary concern. Once a conflict has officially ended, it can, however, help to "overcome the conflict-ridden image" (Isaac *et al.* 2019). Post-conflict destinations can be classified as places of Political Tourism, "made interesting for reasons of political dispute" (Warner 1999). Simone-Charteris and Boyd (2010, p. 109) define Political Tourism as

> travel to sites, attractions, and events associated with war, conflict, and political unrest for educational, commemorative, or diplomatic reasons, showing solidarity or empathy, curiosity, in search of authenticity; enjoying the thrill of political violence or a combination of these.

Political and Dark Tourism overlap significantly both in terms of motivations and places visited. For the purpose of this chapter, I will conceptualise Dark Tourism and Political Tourism as specific forms of Heritage Tourism.[1]

The development of tourist attractions in cities and regions scarred by decades of violent conflict is certainly ambivalent. It has been criticised as creating "Potemkin villages" of the "neoliberal project" (Nagle 2009) or superficial "oases of normality" for visitors and well-off residents only (Guttal 2005) that bear little relevance for the majority of residents, creating a contrast to the 'authenticity' tourists are looking for. Patrick (2019) concludes that "projects aimed primarily at attracting tourists rarely seek to improve the quality of life for permanent residents".

Regarding the elements of heritage associated with the conflict, many aspects remain unagreed between the former conflict groups, making it difficult to find a common narrative. Post-conflict situations are "highly unregulated settings" (Patrick 2019) in which it is relatively easy for "memorial entrepreneurs" (Dwyer and Alderman 2008) like former combatants to put themselves in charge of the tourism sector and communicate their views of past events. If memorial entrepreneurs in one former conflict group are more active or more skilful, a result can be the symbolic disinheritance of the other group, meaning that its view of past events is not as visible and well-documented (Friedrich *et al.* 2018).

Another possible result is that tourism contributes to peace and provides opportunities to reflect the conflict and its heritage (Boyd 2019, Patrick 2019). Especially for ex-combatants and other people heavily involved in the conflict, talking about it, for example as tour guides, creates an opportunity to gain "empathy, self-reflection, and personal catharsis" (Causevic and Lynch 2011, p. 792). For this kind of touristic experience, Causevic and Lynch (ibid.) have coined the term 'Phoenix Tourism'. According to them, it is not a label for a subsector of the touristic market like Dark or Heritage Tourism, but "rather resembles the role which tourism has in the process of social renewal through the transformation between the emotions of sorrow and codification of the heritage" (ibid., p. 796).

Cities are "intensive microcosm[s] for the wider societal tensions and fragmentations, and their diverse related discourses" during conflict (Gaffikin and Morrissey 2011, p. 79). Within them, certain neighbourhoods are often more affected than others in terms of casualties, destruction and general deprivation. It is precisely in these neighbourhoods that the concepts of New Urban Tourism and Dark Heritage Tourism, contradictory as they may seem, intertwine. They create opportunities for peacebuilding and catharsis as implied in the concept of Phoenix Tourism and at the same time potential for new conflict depending on the narrative chosen.

Urban development and tourism in post-conflict Belfast

Even though the Northern Ireland conflict, locally and somewhat euphemistically referred to as 'the Troubles', ended over 20 years ago, there only have been rudimentary efforts to come to terms with the past.[2] Some of the most violent incidents have been investigated, but there is neither an all-encompassing strategy of dealing with the heritage of conflict nor an agreed narrative (Bertram 2019). This makes memorialising the conflict and related events "a tricky business", as there are two "distinct and opposing sets of communal social memories and heritages" (Hodson 2019, p. 227) that are still "actively intertwined" with present collective identities in the city of Belfast (Leonard 2011, p. 114).

Belfast as the capital city of the region was deeply affected during the conflict in terms of social and physical development. Almost 75% of all casualties of the conflict originated from the deprived residential neighbourhoods of North and West Belfast (Morrissey and Smyth 2002). Many neighbourhoods are still perceived as either 'nationalist/catholic' or 'unionist/protestant' spaces, resulting in more or less rigid segregation between the main societal groups in terms of housing, work and leisure (Mac Ginty *et al.* 2007, Bereskin 2017). Barriers, colloquially called 'peace lines', separate some of them. The term refers to walls or fences, usually made of brick or metal and up to six meters high that prevent traffic (motorised and sometimes pedestrian) between two neighbourhoods. Peace lines were introduced in 1969 as temporary measures to stop rioting in the early days of the conflict, but have proven quite permanent in many areas up to the present (Jarman 2008). In spite of the government strategy "Together: Building a United Community" (T:BUC) that has been in place since 2013 and aims to dismantle all interface barriers until 2023, a high number of peace lines and other elements of security architecture remain intact (Jarman 2008, Byrne 2011, The Newsroom 2020).

At the same time, the physical appearance as well as the image of Belfast have changed significantly during the last 20 years. Urban development and planning in the city has mostly been concerned with creating a positive and renewed image to visitors, investors and inhabitants. This approach has been labelled putting "lipstick on the gorilla" (Neill 2006) because it avoids dealing with the past and its spatial remnants in meaningful ways and side-lines the residential areas that have been most affected by the conflict (Nagle 2009). It has nevertheless been very successful in attracting tourists, creating new and exciting spaces for a growing middle class and 'normalising' the city's image (Neill 2004, Bertram 2017b, Boyd 2019). When visiting these spaces, religious or political affiliations hardly play a role anymore. The revitalisation of the city centre has taken place since the 1980s and was seen as a "non-military battle with political violence in showcasing the advantages of normalisation" (Neill *et al.* 2014, p. 5), creating shared or at least neutral space (Bereskin 2017, Hodson 2019).

City marketing has identified seven culturally significant quarters, all but one (Gaeltacht Quarter, see next section) situated in the inner city or on 'neutral ground' (see Figure 13.1). Queen's quarter refers to Belfast's most established

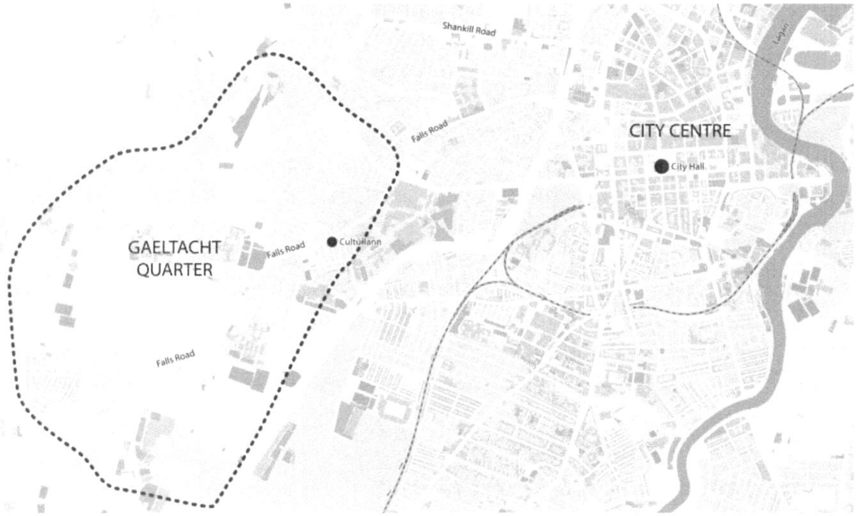

Figure 13.1 Map of Belfast City © OpenStreetMap, edited by Henriette Bertram

university and is situated in South Belfast where people have experienced a laid-back academic lifestyle and attended independent bars and cafés even during the height of the conflict. Arguably, South Belfast is the area of the city where a more 'conventional' kind of New Urban Tourism takes place. Titanic Quarter and Linen Quarter make references to the industrial heritage of Belfast. A spectacular and iconic Titanic museum was built at the heart of the new Titanic Quarter in 2012, the first major touristic development that references local history (Bereskin 2017). Cathedral Quarter is named after its most famous landmark, St. Anne's Cathedral, and celebrates independent cultural initiatives and nightlife. The last two quarters are Market Quarter und Library Quarter. Only Cathedral, Titanic, Queen's and Gaeltacht Quarter, however, have elaborated marketing concepts that go further than putting up street signs with the quarter's name on them (Carden 2011).

Eventisation efforts include the establishment of a "diverse cultural infrastructure" of high-quality theatres and museums (Neill *et al.* 2014, p. 11), the attraction of sporting and other big events and the exploitation of touristic interest in the series *Game of Thrones*, parts of which were filmed in studios and outdoor locations in Northern Ireland. Belfast won the Guardian and Observer Travel Award as "Best UK City" in 2016 (Belfast Telegraph 2016). In the same year, the Lonely Planet noted on its website that the city had transformed "from bombs-and-bullets pariah to a hip-hotels-and-hedonism party town", and strongly recommending a visit (Lonely Planet 2016). According to Belfast City Council, visitor numbers have reached almost 9.5 million per year (Belfast City Council 2020), most of which travel from other places within Northern Ireland and the Republic of Ireland (Bereskin 2017).

O'Dowd and Komarova (2009) have given urban development in post-conflict Belfast the label "consumerist Belfast" (Bereskin 2017, p. 157). They observe a stark contrast to the still-existing 'Troubles Belfast' which includes the places scarred by the conflict that have hardly changed in the post-conflict period. The conflict that has shaped all aspects of daily life in the city for so long and continues to do so in certain areas does not make an appearance in the newly developed places of 'consumerist Belfast'. Events related to conflict are not officially recognised as part of the city's heritage. 'Troubles Belfast' does not appear in urban planning documents or tourism strategies for the city except for when reference is made to the perceived lack of security some tourists experience when wandering beyond the city centre (Belfast City Council and Northern Ireland Tourist Board 2014, Bertram 2018a). The only part of Belfast where the heritage of the conflict is allowed to play a role in its presentation to outsiders is West Belfast. I will describe and discuss the way this is carried out in the following section.

Neighbourhood tourism in West Belfast

Belfast is not a large-scale city; some of the physically fragmented and socially deprived quarters of 'Troubles Belfast' are situated only a few hundred yards away from the new places of 'consumerist Belfast' (O'Dowd and Komarova 2009). West Belfast[3] is unequivocally part of 'Troubles Belfast', traditionally a catholic/republican stronghold and the location of some of the most severe fighting during the conflict.

West Belfast is one of the areas of Northern Ireland where the repercussions of the conflict most strongly shaped everyday life. The Irish Republican Army (IRA) recruited massively in the neighbourhood from the late 1960s onwards. To date, many residents are in one way or the other connected to nationalist/republican political and military organisations. During the most violent period of the 'Troubles', parts of the neighbourhood became a no-go area for the police, and public transport was suspended. It was replaced by so-called Black Cabs, taxis that followed fixed routes (Hurley Dépret 2007). West Belfast remained the "centre of the war" until 1998 (Carden 2011, p. 9), priding itself on its resistance against British "oppression".

Like most of the residential quarters surrounding the inner city, West Belfast has not participated greatly in the so-called 'peace dividend': neither its physical appearance nor its socioeconomic status changed greatly after the end of the conflict. During the 'Troubles', unemployment figures rose to up to 40% (Hart 1990). West Belfast is separated from the adjacent protestant-unionist neighbourhood Shankill—which was in many ways just as badly affected by the conflict and remains equally deprived—through a peace line.

> This line is not a continuous structure, but a jumble of concrete walls, metal fences, barbed wire, and steel gates that are physical representations of past violence and current sources of contention between local residents.
>
> (Wiedenhoft Murphy 2010, p. 543)

Additional urban design features as well as the period of relative isolation during the 'Troubles' have worked together and produced a very close-knit, politically and culturally active community that strongly identifies with the neighbourhood, the nationalist cause and the Irish culture (Carden 2011).

At the same time, West Belfast is the only deprived residential neighbourhood generating significant interest from outsiders. It had cultivated its own touristic profile focusing on the community's situation during the conflict and on Irish culture long before the end of the conflict, attracting smallish numbers of visitors interested in political history, especially those sympathising with the nationalist cause (Adams 2014, Bereskin 2017). Neill (1999, p. 276) has described it as "arguably the most distinctive cultural quarter in the city", possessing a "unique place identity almost as a city within a city". In the words of Paul Maskey, Sinn Féin politician and then Member of the Local Assembly—now Member of the British Parliament—promoting tourism in the neighbourhood helps "to lift those communities from areas of social need into areas that have great employment and socio-economic opportunities that will become must-see areas for tourists" (Skinner 2016, p. 35). Although the content of the tourist strategy for West Belfast differs immensely from its counterpart in the 'neutral' parts of the city, they are not so different with regard to form and activities: the most important aspect in West Belfast is the staging of certain elements of local heritage (see Figure 13.2).

Unsurprisingly, the most important tourism actors in West Belfast are civil society and private economy actors, some of them associated with nationalism and in many cases qualifying as memorial entrepreneurs following Dwyer and Alderman (2008). Public actors like Belfast City Council or the Northern Ireland Tourist Board (NITB) were reluctant to actively promote 'Troubles Tourism' because of its potential to spark controversy (Wiedenhoft Murphy 2010, Leonard 2011). One of the most active tourism organisations in West Belfast is Fáilte Feirste Thiar/Welcome to West Belfast. According to their website, their mission is to "market the uniqueness of West Belfast as a tourist attraction, re-branding the area as a place to visit" (Fáilte Feirste Thiar/Visit West Belfast 2020). They advertise events, activities and accommodation in West Belfast and bring together companies and potential employees. The accommodation section, however, remains relatively meagre with only ten facilities in total, three of which being private places advertised on Airbnb (cf.Bereskin 2017).

Fáilte works in close partnership with Coiste na n-I-archimí, the republican Ex-Prisoners' Committee who offer "unique walking tours delivered by former political prisoners who share their personal experiences of the British/Irish conflict" (Coiste na n-Iarchimí 2020). Coiste's most popular tour is the Falls Mural Tour that attracts visitors "in their thousands" to West Belfast (see Figure 13.3) (Coiste na n-Iarchimí 2020). Tour guiding gives ex-prisoners an opportunity to find their place in the post-conflict society, earn a living and narrate their version of past events. Tour guides are unapologetic about the role they themselves and their community played during the 'Troubles' and declare their political point of view openly at the beginning of the tour. They welcome questions about their background but reserve their right not to answer some of them (Skinner 2016).

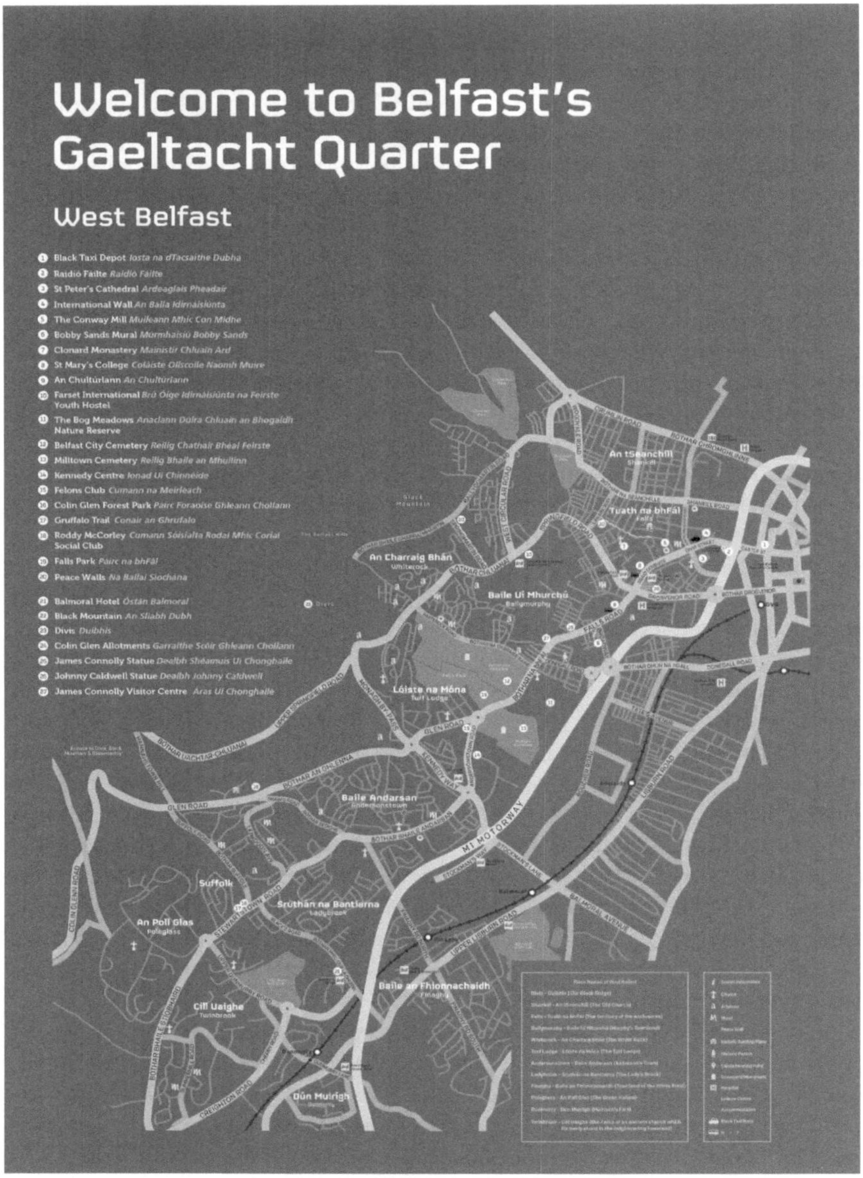

Figure 13.2 Map of Gaeltacht Quarter © Fáilte Feirste Thiar/Visit West Belfast

Figure 13.3 Mural in West Belfast © Henriette Bertram

They are very successful in combining their own personal experiences during the conflict into a narrative of 800 years of colonialism and injustice against nationalists in Northern Ireland (Leonard 2011, Skinner 2016). Some tours are carried out in co-operation with the corresponding organisation, EPIC, in Shankill (Ex-Prisoners' Interpretative Centre 2013, Bereskin 2017). The guides explain that their unionist counterpart will see things differently and leave visitors to make sense of the two narratives for themselves, trusting, as Leonard (2011, p. 123) writes, "that their meta-narrative will outshine the fragmented Loyalist narrative on offer". The narrative promoted by nationalists indeed seems to be much more coherent and easier to 'sell' to outsiders than the one promoted by unionists who, after the end of the conflict, had to cope with the loss of privileges, deindustrialisation and internal feuding (Leonard 2011).

Other, usually local, operators offer tours in the famous Black Taxis. Some of them do the tours in addition to their normal taxi-driving, others fully concentrate on tourism. Some of them include the neighbouring Shankill Road, others cooperate with taxi drivers in Shankill, handing tourists over to their counterpart at the peace line. Whether ex-prisoners or not, the taxi guides also admit to their biased point of view and explain to tourists that they will tell the story of the community from this point of view only (Wiedenhoft Murphy 2010).[4]

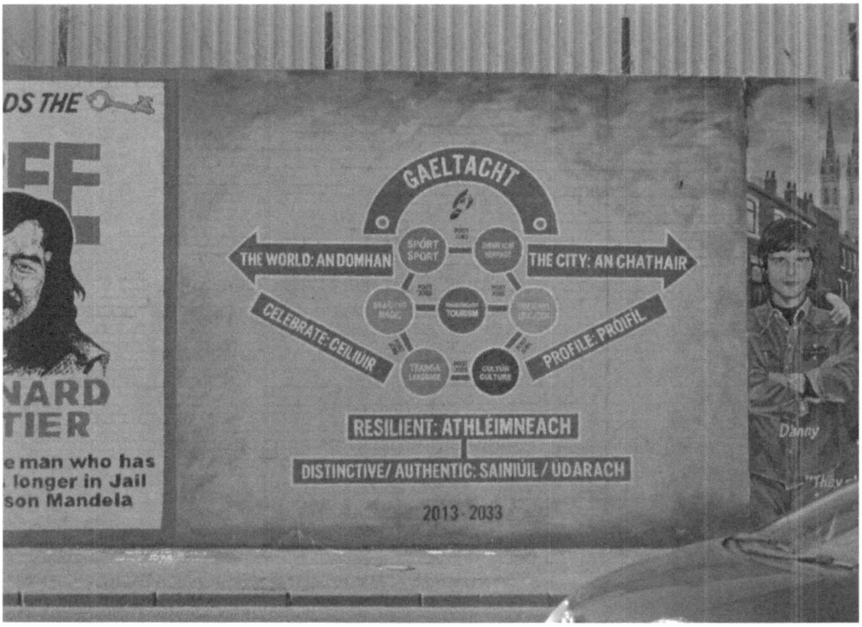

Figure 13.4 Mural of Gaeltacht Quarter. © Henriette Bertram

Traditionally, tourism in West Belfast was mostly geared towards visitors interested in the local heritage of the conflict. As the demand grew, local tourism actors have extended and depoliticised the focus to include Irish language and culture, marketing the area as Gaeltacht Quarter since 2006 (Bereskin 2017). *Gaeltacht* is an Irish-language term referring to a predominantly Irish-speaking community or region (see Figure 13.4). The term is not entirely correct for West Belfast, given that the official as well as the routinely spoken language is English, even though many inhabitants have at least some command of the Irish language. Driver and promotor of the Gaeltacht Quarter is the company An Cheatrú Ghaeltachta Teo/ The Gaeltacht Quarter Ltd. Behind the company is a group of "business people, Irish language enthusiasts and tourism and economic development professionals who almost all originate from West Belfast or have long term links with the area" (Carden 2011, p. 12) —again, an association of local civil society and commercial actors operating largely independently from city-wide or regional political entities. Given that the Irish culture and language in Northern Ireland are closely associated to the nationalist community, it is the only one of the seven quarters that could be seen as not politically neutral. As a matter of fact, when Gaeltacht Quarter was announced, it immediately sparked protest from a unionist councillor who insisted that the same amount of money should be spent on the (protestant-unionist) cultural tradition of Ulster Scots (Canning 2006).

Several cultural institutions work together to fill the structure of the Gaeltacht Quarter with Irish cultural life: Forbairt Feirste, an "Irish Language development agency which strives to unleash the economic power of Belfast's Irish speaking community to the benefit of the entire city" (Forbairt Feirste 2020) was founded in 1994 and has campaigned for the designation of the Gaeltacht Quarter for a long time. Another long-standing institution is the Irish-speaking cultural centre Cúlturlann McAdam O'Fiaich on the arterial Falls Road. It includes an Irish language school, a book and crafts shop as well as a café and restaurant. In addition to that, it serves as a venue for performances, discussions, readings or exhibitions (Kabel 1999, Cultúrlann McAdam Ó Fiaich 2020). The second (and older) Irish language centre in the area is Cumann Chluain Ard. It is regarded as the breeding ground for the revival of the Irish language and culture in Northern Ireland (Kabel 1999, Carden 2011), but is nowadays not as well known as Cultúrlann. The tourism strategy based on promoting heritage is flanked by substantial physical regeneration efforts including Conway Mill and Cultúrlann on the Falls Road.

Conclusion: New Urban Tourism?

This chapter has shown that there is indeed a special quality to neighbourhood tourism in post-conflict settings. It bears elements of (Dark) Heritage Tourism as well as Phoenix Tourism. The two themes of tourism, conflict—as framed and narrated by ex-combatants and other residents—and Irishness, on the one hand strongly resonate with the notion of Phoenix Tourism as conceptualised by Causevic and Lynch. Combined, they give the opportunity of looking back, achieving catharsis and provide a positive narrative viable for the future. As organisations of ex-combatants on both sides of the peace line work together on walking tours, tourism even has a certain power of reconciliation. The fact that tourism in Shankill plays a negligible role highlights the importance of an appealing and convincing narrative that visitors can relate to.

The question remains whether West Belfast can be seen as New Urban Tourism as discussed in this book. Given that there are only a few tourist accommodations and only a handful of stylish bars and restaurants, the negative effects of New Urban Tourism like gentrification and overcrowding are not (yet) noticeable in West Belfast. The shift from 'sightseeing' to 'life-seeing' is therefore quite selective. It is safe to say, however, that West Belfast is certainly an 'edgy' part of the city that despite its emphasis on tradition and heritage provides 'new' and unique experiences for tourists. It is true that guided tours are usually not an element of New Urban Tourism and tourists taking the tours will never actually feel like a local, even for a short time. It can be argued though that the tours do aim to satisfy the demand for authenticity and immersion into the local culture by combining local cultural heritage, a coherent narrative of oppression and resistance and at the same time foregrounding personal experiences. All relevant tourism actors are locals or have long-standing ties with the area; their stories of the conflict are told from the point of view of the community. The tours have been created to mediate the feeling of community life and its struggles during conflict;

the locals and their stories become the most important selling point of the quarter and are therefore a significant feature of the 'tourist product'. With the launch of the Gaeltacht Quarter, the focus of tourism has widened in recent years, now enabling tourists to immerse into the Irish language and culture, to hear the locals speak their (supposed) native language; again, an experience that—due to its political implications—is only possible in West Belfast. The Gaeltacht is used to showcase an element of heritage that is positive, lively and unconnected to violence and creates a picture of a community that has overcome hard times. Given that visitors interested in the political history of Northern Ireland have strolled the streets of West Belfast even before the ceasefires, one might even say that New Urban Tourism was a part of the neighbourhood before it became a marketed tourist destination.

In this regard, neighbourhood tourism in Belfast not only provides opportunities for catharsis and co-operation, but also has the potential to produce new kinds of tensions: due to the disputed heritage of the conflict, city and regional authorities in Belfast and Northern Ireland refrain from prescribing narratives or approaches to memorialisation in fear of losing their neutral reputation. The prerogative of interpretation of the conflict is therefore largely left to the communities and their memorial entrepreneurs who seize the chance to disseminate their point of view. An important lesson here is therefore that without a general and more global process of dealing with the past and/or a neutral institution facilitating the development of New Urban Tourism in post-conflict cities, it is quite likely that one community's narrative will informally win recognition over the other and impede reconciliation in the long run.

Notes

1 For a detailed discussion of the relationship between Dark and Heritage Tourism see Light 2017.

2 Parties involved in the Northern Ireland conflict were (largely catholic) nationalists and republicans who advocate for a united Ireland and (mainly protestant) unionists and loyalists who are in favour of maintaining the Union with Great Britain, and the British state. In 1998, the governments of the United Kingdom and the Republic of Ireland as well as major political parties and groupings of Northern Ireland signed the Good Friday or Belfast Agreement. The Agreement marked the end of the Northern Ireland conflict and the start of a lengthy peace process including a range of deadlocks, setbacks and renewed outbreaks of violence. Power-sharing between unionists and nationalists in the Northern Ireland Assembly remains difficult: in January 2020, the Assembly came together for the first time in three years, finally ending a political stalemate that arose after the elections in 2017 (McCormack 2020). For a detailed analysis of the conflict see for example McGarry and O'Leary (2004), O'Leary and McGarry (2016), Tonge (2013). A more detailed account and analysis of the peace process can be found in an edited volume by Gilligan and Tonge (2019) or in Mac Ginty (2019).

3 West Belfast contains Lower and Upper Falls with five electoral wards each. The Gaeltacht quarter, where most of the tourist attractions can be found, is situated in Lower Falls (Falls and Clonard wards).

4 There are other, 'neutral', commercial tour operators like City Sightseeing who include West Belfast and Shankill in their 23-stop Hop-on Hop-off tour of Belfast (Belfast City

Sightseeing 2020). These tours are not relevant in regard to New Urban Tourism in the quarter, which is why I will not discuss them in further detail.

References

Adams R., 2014. *Crumlin road gaol/Girdwood Park*. Belfast. Interview with author.

Ashworth, G.J. and Graham, B., 2005. Introduction. *In*: G.J. Ashworth and B. Graham, eds. *Senses of place: senses of time*. Aldershot: Ashgate, 3–11.

Ashworth, G.J. and Page, S.J., 2011. Urban tourism research: recent progress and current paradoxes. *Tourism Management*, 32 (1), 1–15.

Belfast City Council, 2020. *Tourism* [online]. Available from: https://www.belfastcity.gov.uk/tourism-venues/tourism/tourism-about.aspx [Accessed 28 February 2020].

Belfast City Council and Northern Ireland Tourist Board, 2014. *Belfast: integrated tourism strategy 2015–2020* [online]. Available from: https://minutes3.belfastcity.gov.uk/documents/s2976/Appendix%203%20-%20Belfast%20Integrated%20Tourism%20Strategy.pdf [Accessed 19 December 2020].

Belfast City Sightseeing, 2020. *Hop-on/hop-off bus stop locations* [online]. Available from: https://belfastcitysightseeing.com/hop-on-hop-off-bus-stops/ [Accessed 29 February 2020].

Belfast Telegraph, 2016. Belfast is named the best UK city at guardian and observer travel awards. *Belfast Telegraph*, 8 March. Available from: https://www.belfasttelegraph.co.uk/news/northern-ireland/belfast-is-named-the-best-uk-city-at-guardian-and-observer-travel-awards-34520514.html [Accessed 17 December 2020].

Bereskin, E., 2017. Tourism provision as protest in 'post-conflict' Belfast. *In*: C. Colomb and J. Novy, eds. *Protest and resistance in the tourist city*. London: Routledge, 152–170.

Bertram, H., 2017. Re-imaging the post-conflict quarter. Tourismus in Westbelfast zwischen Konflikt und kulturellem Erbe. *Geographische Zeitschrift*, 105 (2), 167–189.

Bertram, H., 2018a. Discourses and practices of dealing with the spatial legacy of conflict in Belfast: the cases of Crumlin road gaol and Girdwood Park. *In*: E. Crooke and T. Maguire, eds. *Heritage after conflict: Northern Ireland*. London: Routledge, 117–131.

Bertram, H., 2017b. *Schattenorte in Belfast Stadterneuerung nach dem Ende des Nordirlandkonflikts*. Bielefeld: Transcript.

Bertram, H., 2019. Dealing with the spatial remnants of conflict in Belfast: the Andersonstown Barracks site in West Belfast. *In*: M. Ristic and S. Frank, eds. *Urban heritage in divided cities: contested pasts*. London: Routledge, 210–224.

Boyd, S., 2019. Post-conflict tourism development in Northern Ireland: moving beyond murals and dark sites associated with its past. *In*: R. Isaac, E. Çakmak, and R. Butler, eds. *Tourism and hospitality in conflict-ridden destinations*. London: Routledge, 226–239.

Byrne, J., 2011. *The Belfast peace walls: problems, politics and policies of the troubles architecture*. University of Ulster.

Canning, M., 2006. Belfast gets a 'Gaeltacht' quarter. *The Irish Times*, 11 September. Available from: https://www.irishtimes.com/news/belfast-gets-a-gaeltacht-quarter-1.1001043 [Accessed 17 December 2020].

Carden, S., 2011. Post-conflict Belfast sliced and diced: the case of the Gaeltacht Quarter. *Conflict in Cities and the Contested State. Working Paper No.* 20 [online]. Available from: http://www.conflictincities.org/PDFs/Working%20Paper20_1.3.11.pdf [Accessed 17 December 2020].

Causevic, S. and Lynch, P., 2011. Phoenix tourism. *Annals of Tourism Research*, 38 (3), 780–800.

Chhabra, D., Healy, R., and Sills, E., 2003. Staged authenticity and heritage tourism. *Annals of Tourism Research*, 30 (3), 702–719.

Coiste na n-Iarchimí, 2020. *Coiste* [online]. Available from: http://coiste.ie/ [Accessed 20 February 2020].

Cultúrlann McAdam Ó Fiaich, 2020. *About us* [online]. Available from: http://www.culturlann.ie/en/about-us/ [Accessed 11 February 2020].

Dirksmeier, P. and Helbrecht, I., 2015. Resident perceptions of new urban tourism: a neglected geography of prejudice. *Geography Compass*, 9 (5), 276–285.

Dwyer, O. and Alderman, D., 2008. Memorial landscapes: analytic questions and metaphors. *GeoJournal*, 73 (3), 165–178.

Ex-Prisoners' Interpretative Centre, 2013. *EPIC tours* [online]. Available from: http://www.epic.org.uk/ [Accessed 31 August 2020].

Fáilte Feirste Thiar/Visit West Belfast, 2020. *Home* [online]. Available from: http://visitwestbelfast.com/ [Accessed 11 February 2020].

Farmaki, A., 2013. Dark tourism revisited: a supply/demand conceptualisation. *International Journal of Culture, Tourism and Hospitality Research*, 7 (3), 281–292.

Forbairt Feirste, 2020. *Forbairt Feirste - here comes the New Belfast* [online]. Available from: http://www.forbairtfeirste.com/ [Accessed 28 February 2020].

Friedrich, M., Stone, P., and Rukesha, P., 2018. Dark tourism, difficult heritage, and memorialisation: a case of the Rwandan Genocide. *In*: P. Stone, et al., eds. *The Palgrave handbook of dark tourism studies*. London: Palgrave Macmillan, 261–289.

Füller, H. and Michel, B., 2014. 'Stop Being a Tourist!' New dynamics of urban tourism in Berlin-Kreuzberg. *International Journal of Urban and Regional Research*, 38 (4), 1304–1318.

Gaffikin, F. and Morrissey, M., 2011. *Planning in divided cities: collaborative shaping of contested space*. Hoboken, NJ: Wiley-Blackwell Publishing.

Gilligan, C. and Tonge, J., 2019. *Peace or war? Understanding the peace process in Northern Ireland*. London: Routledge.

Guttal, S., 2005. The politics of post-war/post-conflict reconstruction. *Development*, 48 (3), 73–81.

Hart, M., 1990. Belfast's economic millstone? The role of the manufacturing sector since 1973. *In*: P. Doherty, ed. *Geographical perspectives on the Belfast region*. Newtownabbey: Geographical Society of Ireland, 37–53.

Hodson, P., 2019. Titanic struggle: memory, heritage and shipyard deindustrialization in Belfast. *History Workshop Journal*, 87, 224–249.

Hurley, Dépret, M., 2007. 'Troubles tourism': debating history and voyeurism in Belfast, Northern Ireland. *In*: P. Scranton and J. Davidson, eds. *The business of tourism: place, faith, and history*. Philadelphia, PA: University of Pennsylvania Press, 137–162.

Isaac, R., Çakmak, E., and Butler, R., 2019. Introduction. *In*: R. Isaac, E. Çakmak, and R. Butler, eds. *Tourism and hospitality in conflict-ridden destinations*. London: Routledge.

Jarman, N., 2008. Security and segregation: interface barriers in Belfast. *Shared Space*, 6, 21–34.

Kabel, L., 1999. Das Irische als kulturelle Zweitsprache in Belfast. *In*: A. Wigger, R Ködderitzsch, and S. Zimmer, eds. *Akten des Zweiten Deutschen Keltologen-Symposiums: (Bonn, 2.-4. April 1997)*. Tübingen: Niemeyer, 96–104.

Lennon, J. and Foley, M., 2010. *Dark tourism*. Andover: Cengage Learning.

Leonard, M., 2011. A tale of two cities: 'Authentic' tourism in Belfast. *Irish Journal of Sociology*, 19 (2), 111–126.

Light, D., 2017. Progress in dark tourism and thanatourism research: an uneasy relationship with heritage tourism. *Tourism Management*, 61, 275–301.

Lonely Planet, 2016. *Introducing Belfast* [online]. Available from: http://www.lonelyplanet.com/ireland/northern-ireland/belfast/introduction [Accessed 11 March 2016].

Mac Ginty, R., 2019. Northern Ireland. *In*: A. Özerdem and R. Mac Ginty, eds. *Comparing peace processes*. London: Routledge, 211–236.

Mac Ginty, R., Muldoon, O., and Ferguson, N., 2007. No war, no peace: Northern Ireland after the agreement. *Political Psychology*, 28 (1), 1–11.

Maitland, R., 2007. Culture, city users and the creation of new tourism areas in cities. *In*: M. Smith, ed. *Tourism, culture and regeneration*. Wallingford: CABI Publishing, 25–34.

McCormack, J., 2020. *Stormont: what is it and why did power-sharing collapse in Northern Ireland?* Available from: https://www.bbc.com/news/uk-northern-ireland-politics-50822912 [Accessed 28 February 2020].

McGarry, J. and O'Leary, B., 2004. *The Northern Ireland conflict: consociational engagements*. Oxford: Oxford University Press.

Morrissey, M. and Smyth, M., 2002. *Northern Ireland after the Good Friday agreement: victims, grievance, and blame*. London: Pluto Press.

Nagle, J., 2009. Potemkin village: neo-liberalism and peace-building in Northern Ireland? *Ethnopolitics*, 8 (2), 173–190.

Neill, W., 1999. Whose city? Can a place vision for Belfast avoid the issue of identity? *European Planning Studies*, 7 (3), 269–281.

Neill, W., 2004. *Urban planning and cultural identity*. London: Routledge.

Neill, W., 2006. Return to Titanic and lost in the maze: the search for representation of 'post-conflict' Belfast. *Space and Polity*, 10 (2), 109–120.

Neill, W., Murray, M., and Grist, B., 2014. Introduction: Titanic and the New Belfast. *In*: W. Neill, M. Murray and B. Grist, eds. *Relaunching Titanic: memory and marketing in the New Belfast*. London: Routledge, 3–13.

Novy, J., 2013. "Berlin Does Not Love You": notes on Berlin's "Tourism Controversy" and its discontents. *In*: M. Bernt, B. Grell, and A. Holm, eds. *The Berlin reader: a compendium on urban change and activism*. Bielefeld: Transcript, 223–237.

Novy, J. and Colomb, C., 2017. Urban tourism and its discontents: an introduction. *In*: C. Colomb and J. Novy, eds. *Protest and resistance in the tourist city*. London: Routledge, 1–30.

O'Dowd, L. and Komarova, M., 2009. Regeneration in a contested city: a Belfast case study. *Conflict in Cities and the Contested State. Working Paper No. 10* [online]. Available from: http://www.conflictincities.org/PDFs/WorkingPaper10_14.4.10.pdf [Accessed 19 December 2020].

O'Leary, B. and McGarry, J., 2016. *The politics of antagonism: understanding Northern Ireland*. London: Bloomsbury Publishing.

Patrick, N., 2019. Memorial entrepreneurs and dissonances in post-conflict tourism. *In*: R. Isaac, E. Çakmak, and R. Butler, eds. *Tourism and hospitality in conflict-ridden destinations*. London: Routledge.

Seaton, T., 2018. Encountering engineered and orchestrated remembrance: a situational model of dark tourism and its history. *In*: P. Stone, et al., eds. *The Palgrave handbook of dark tourism studies*. London: Palgrave Macmillan, 9–31.

Simone-Charteris, M. and Boyd, S., 2010. Developing dark and political tourism in Northern Ireland: an industry perspective. *In*: G. Gorham and Z. Mottiar, eds.

Contemporary issues in Irish and global tourism and hospitality. Dublin: Dublin Institute of Technology, 106–123.

Skinner, J., 2016. Walking the falls: dark tourism and the significance of movement on the political tour of West Belfast. *Tourist Studies*, 16 (1), 23–39.

Smith, V., 1998. War and tourism. *Annals of Tourism Research*, 25 (1), 202–227.

The Newsroom, 2020. The Duncairn peace wall has not been demolished — it is being rebuilt with a cosmetic makeover. *Belfast News Letter*, 12 February. Available from: https://www.newsletter.co.uk/news/opinion/letters/duncairn-peace-wall-has-not-been-demolished-it-being-rebuilt-cosmetic-makeover-1483978 [Accessed 28 February 2020].

Timothy, D. and Boyd, S., 2003. *Heritage tourism*. Harlow: Pearson Education.

Tonge, J., 2013. *Northern Ireland: conflict and change*. Hoboken, NJ: Taylor and Francis.

Tunbridge, J. and Ashworth, G., 1996. *Dissonant heritage: the management of the past as a resource in conflict*. Chichester: Wiley.

Warner, J., 1999. North Cyprus: tourism and the challenge of non-recognition. *Journal of Sustainable Tourism*, 7 (2), 128–145.

Wiedenhoft Murphy, W., 2010. Touring the troubles in Westbelfast: building peace or reproducing conflict? *Peace & Change*, 35 (4), 537–560.

14 The race, class and gender of websites

Marketing and mythologising urban Africa online

Annie Hikido

Introduction

A red flower made of recycled aluminium spans the top of the website for Lungi's B&B in Cape Town, South Africa. The site reads, "STAY IN OUR LOVELY GUESTROOMS. Lungi's B&B is a friendly ran B&B, where you will feel immediately at ease. It offers you the unique experience of family life and the culture in a South African township ...". Below the cover photo, the site shows two colourful bedrooms priced at R450 per night ($27 USD). Further down, Lungi, a middle-aged Black woman, stands in the doorway of a large sea-green structure. She wears a headwrap, a pale blue long-sleeved shirt, and a long, patterned skirt. The caption tells viewers that they will "experience the township inside-out" by staying with Lungi and her family. Testimonials from Western European guests exclaim the uniqueness and comfort of their stay. A cluster of young Black boys look up at the viewer in a photo next to a solicitation for feedback.

Lungi's B&B is part of a growing township hospitality sector within the South African township tourism market. She and other older Black women have turned their township homes into tourist accommodations where guests can stay overnight, take walking tours and engage with township residents. The invitation to 'live like a local' points to a New Urban Tourism paradigm that prizes tourists' integration into places through mundane interactive experiences (Maitland 2007). Selling South African township experiences, however, confronts the challenge of territorial stigma. The townships of Cape Town, collectively referred to as the Cape Flats, are often referred to as risky places that tourists should avoid. Those who operate township tourism businesses must prove to tourists that the 'everyday' township is safe and comfortable. Lungi's website aims to do just this. Through images and text, the site projects her home as a welcoming place where Western tourists enjoy memorable township stays.

In this chapter, I analyse how marketing websites present township accommodations to draw tourists to the margins of Cape Town. I argue that the websites anticipate and recode the racial, classed and gendered meanings of Black townships. A global myth frames Black urban neighbourhoods as dangerous, impoverished and hypermasculine by way of racialised criminality. Websites create counter-myths that frame Black townships as safe, entrepreneurial and maternal through communal caretaking. By focusing on the township hospitality market in South Africa, this chapter illuminates how New Urban Tourism in the Global South involves the strategic negotiation of neighbourhood representations and the possibility of doing so online for an international audience. For Lungi and township hostesses, the stakes are high. Township tourism provides them income in a state where Black women have been historically excluded from attaining wealth and stable livelihoods.

Black women and the geography of South African inequality

Black women's entrepreneurial township accommodations grew from South Africa's history of racial segregation and gendered labour market exclusion. In 1948, following centuries of Dutch and British colonialism that displaced and impoverished Black persons, the white state institutionalised what racial segregation that already existed through the apartheid regime. Blacks were to live in designated rural 'homelands' and stay only in the cities to work. Influx controls that required Blacks to carry passbooks regulated Black urban entry and residency. Black men who provided cheap labour lived temporarily in worker hostels. Black women, who were denied employment in many formal labour markets, were largely ineligible for urban tenure (Lee 2009). The imperatives of employment and family unification, however, pushed Black women to migrate to the cities.

In Cape Town, Black women were especially barred from jobs and targeted for removal. The Western Cape province's 1955 Coloured Labour Preference Policy privileged Coloured over Black labour, pushing Black women into informal jobs.[1] Some Black women secured employment as domestic or factory workers, but without valid passbooks, many sold goods and services in the townships (Berger 1992). Staying in the home offered protection from police scrutiny, but home raids became frequent during the 1970s (Lee 2009). Police aggressively targeted women in order to curb Black family settlement in the Cape (West 1982). But as rural destitution continued to push Black urban migration, the number of 'illegal' Black residents increased and the state was forced to establish additional townships. The scattered settlements became known as the Cape Flats, the sandy flatland east of the city.

Cracks in the apartheid regime became visible by the mid-1980s. In the Western Cape, pass requirements and the Coloured Labour Preference Policy were abolished in 1986. As apartheid legislation crumbled, waves of Black migrants established homes in the Black townships and informal settlements across the peninsula. Since apartheid was dismantled in 1994, Cape Town's Black population has increased dramatically (Western 1996). The racial and class geography of apartheid, however, remains entrenched in the post-apartheid landscape. The especially high level of unemployment among Black women indicates how racial and gender restrictions structurally persist in the post-apartheid labour market (Skinner and Valodia 2001).

While the number of persons searching for work in the Cape Flats has redoubled, the city's white urban core has become a global tourist destination. According to one account, the total number of international tourist arrivals to South Africa increased by 23.3% in 1995, the year after apartheid was dismantled, and foreign tourism grew by roughly 10% for the next four years (Cornelissen 2005b). The overwhelming majority of overseas tourists stay in Cape Town (Cornelissen 2005a). Black townships have become integrated into the city's tourism itineraries. In the 1990s, following the fall of apartheid, bus tours based in the central business district began transporting tourists to see the 'other side' of South Africa. But as the township tourism market expanded, this predominantly white Western class of tourists has started to fear that the 'safari' model constitutes unethical voyeurism (Butler 2010). Alternatively, Black township residents who operate township tourism businesses are perceived to provide a more respectful township experience. Thus, while older Black women remain structurally disadvantaged in the formal labour market, they can capitalise on symbolic advantages in tourism as the 'real' voices of the township. Tourists' hunger for authenticity opens up not only opportunities for much-needed capital but also new township representations.

Township hospitality as New Urban Tourism in Cape Town

Cape Town did not become a tourism destination with the fall of apartheid. 'The Mother City' has provided a sun-soaked holiday for European tourists since the 19th century (Bickford-Smith 2009). Today, much of Cape Town still seems to be marketed as a paradisaical getaway for white elites. Racial segregation, a highly developed central business district and the suffusion of white Dutch and British culture (Hikido 2019a) make it "easy to forget that this is a city on the African continent" (McDonald 2008, p. 269). But since South Africa's post-apartheid aggressive agenda to insert itself into the global economy, Cape Town has become an emergent 'global city' (Sassen 2005) with a significantly expanded tourism sector (Cornelissen 2005a). Cape Town is also a 'world tourism city', given its integration into multiple global circuits and frequent mention as "one of the most beautiful and compelling places to visit on the planet" (Bickford-Smith *et al.* 1999). In these cities, large tourism industries compel some tourists to actively seek experiences that appear to be "off the beaten track" (Maitland and Newman 2009).

Townships provide tourists such "alternative public spaces" (Richards 2011). Similar to the low-income 'gritty' neighbourhoods studied in post-industrial Western contexts and other so-called 'poverty tours' of the Global South (Frenzel *et al.* 2012), South African townships intrigue tourists because they represent residential and therefore more 'real' tourism experiences. The quest for authenticity vis-à-vis the everyday lives of locals is the hallmark of the New Urban Tourism paradigm (Maitland 2007). But unlike Berlin's Kreuzberg (Füller and Michel 2014), Brazil's *favelas* (Freire-Medeiros 2013), or India's slums (Meschkank 2011), the apparent authenticity of Cape Town's townships also stems from the racial segregation borne of South Africa's colonial and apartheid history. Black townships exist in stark contrast to the white central business district's "first world shoppertainment" (Pirie 2007). Township tourism's 'backstage' authenticity

(MacCannell 2013) is rooted in the Western imaginary of 'real Africa' as Black poverty.

Apartheid's social landscaping created a distinct geography for township tourism development. Capital has been consistently channelled away from townships in favour of the central city (McDonald 2008), conditioning the concentration of unemployment, crime and inadequate housing in these homogenously Black neighbourhoods. Mainstream media and laypeople frame townships as "no-go zones" (George and Booyens 2014). The stigma poses a major hurdle to township tourism development. In order to dissipate perceptions of township danger, Black women as township hostesses shift the locus of authenticity from street poverty to domestic settings and neighbourhood activities. By embracing the racialised, gendered and classed connotations embedded in the home and the community, hostesses assuage Western tourists' inhibitions of crime and exploitative voyeurism (Hikido 2019a). Those who advertise their businesses through websites begin reframing townships in this way online, affording communication with pre-departure tourists.

Analysing counter-myths online

Tourism scholars have analysed text and images in various types of travel promotion material in order to unpack the ideologies that inform their construction (e.g. Nelson 2005, Wulff 2007). The method, sometimes referred to as 'discourse analysis' or 'semiotic analysis', draws upon Roland Barthes' concept of 'mythologies' (1984) to distil the message that producers intend to convey to consumers. It recognises two 'layers' of meaning signification. The surface-level 'denotation' communicates objective value-neutral information. The underlying 'connotation' draws upon ideologies and symbolically reproduces them, thereby naturalising cultural meanings in representations and making them myths. By extracting ideologies, scholars "piece together the type of world that such a text [or image] presupposes, the world it calls once again into being each time it is read [or seen]" (Parker 1994, p. 95).

In this chapter, I apply the same techniques to analyse websites that advertise township accommodations in Cape Town. My focus on attractions that are not yet considered 'mainstream', however, provides a unique angle of analysis. Previous research of promotional tourism materials points to how popular guidebooks and state websites reproduce dominant ideologies. For example, the presentation of tourists as white and people of colour as objects of tourist interest in advertisements reflects hierarchies of race, gender and nation (Bhattacharyya 1997, Burton and Klemm 2011, Patil 2011). These constructions not only assume normative consumer ideologies but also reinforce them, thereby sustaining the myths that prop up inequitable social orders. By turning to websites that showcase accommodations in disreputable locations, I show how social actors anticipate dominant myths and strategically deploy symbols to create counter-myths in order to attract tourists.

Following Nelson's example (2005), I first conducted web searches as a tourist-cum-researcher in 2013. By typing in "Cape Town township bed and breakfast",

I located six popular township accommodations. Four sites were independent domains and two were linked to popular travel platforms. On each, potential guests could learn about the hostess, her business and her neighbourhood. The sites also allowed guests to make reservations and provided testimonials from former guests. Of the six, only three are still functional at the time of this writing.[2] These sites constitute the first part of my sample. Kopanong B&B, established in 1999, and Lungi's B&B, established in 2010, are both in Khayelitsha and have independent websites. Liziwe's Guesthouse, established in 2005, is based in the Gugulethu township and has a page supported by a travel hub called SA-Venues.

Airbnb was established in Cape Town in 2015. The online platform enables locals to rent out vacant rooms or entire properties. Most of the Airbnb locations in Cape Town are in the white business district and surrounding suburbs, but the company has sought to enlist Black township residents through training and recruitment sessions (Hikido 2019b). The second part of my sample includes the Airbnb sites of three township hostesses who participated in one or more of the programs. Majoro's B&B and Malebo's B&B in Khayelitsha have been hosting international guests in their township homes since the late 1990s. Mikana's B&B, also in Khayelitsha, opened in 2017. Hostesses established their B&Bs as their primary source of income, distinguishing their enterprises from the home-sharing model that Airbnb is known for. In 2017, at an international tourism convention based in the central business district, I observed an Airbnb representative announce that the site would help hostesses and other township residents attract more guests and generate more revenue (Hikido 2019b).

The independent and third-party sites operate differently, but as marketing websites they are analytically comparable. Independent sites require direct payments for a unique domain, whereas third-party sites such as SA-Venues and Airbnb take a cut of the profits for providing a marketing platform. Yet all types share techniques of symbolic construction and the goal of attracting unstuffy middle-class Western tourists. The websites are not solely the hostesses' creations. During my fieldwork between 2014 and 2019, I learned that white Europeans who were former guests typically assisted the construction of independent websites (Hikido 2018). I also learned that Airbnb's white American training team advised hostesses regarding how to present their sites. Therefore, each site represents a collaboration between Black South African hostesses and the white Western persons who aided their site construction.[3]

Creating urban counter-myths through race, class and gender

In his comparative study of "poverty tourism" in South Africa, India and Brazil, German geographer Manfred Rolfes found that tourists, prior to visiting, associated the townships with "poverty", "Black/African inhabitants", "crime" and "poor housing" (2010, p. 430). The tourist imaginary thus explicitly racialises and classes townships as Black and poor and implicitly genders them as masculine through connotations of criminality. Websites that market township hospitality anticipate this myth. Using carefully selected images and text, they create counter-myths that reconfigure the intersections of race, class and gender to signal the positive aspects of everyday township life. I identify four themes: comfort and cleanliness, mothers and families, community vibrancy and industriousness and African authenticity.

Comfort and cleanliness

Given tourists' presumptions of substandard township housing, the websites must first and foremost show guests that they will not be staying in dilapidated shacks. All township accommodation sites include at least one photo of their rooms. Airbnb provides a template that enables hostesses to promise comfort immediately by uploading cover photos as page headliners. Majoro's and Malebo's pages display their bedrooms, living rooms and dining areas. Their tidy lounges with couches and coffee tables portray middle-class domesticity. Appliances such as televisions, refrigerators and washing machines point to the conveniences of modern lifestyles. Crisply made beds topped with artful arrangements of pillows show where guests will sleep. The cover photo for Mikana's site features a queen-sized bed with polka-dotted linens. Site visitors learn that they will sleep in clean, comfortable conditions.

Most of the websites feature white people as guests in and around the homes. Liziwe's SA-Venues site shows photos of an opulent bedroom with striped walls and paisley curtains, a large dining room, and Liziwe opening up heated food trays for older white guests. Kopanong's and Lungi's also feature photos of white people at their places. On Kopanong's site, white persons gather around the hostess and her daughter in front of the house and on the street. On Lungi's, a video clip shows young white women helping Lungi prepare a meal and taking photographs outside with a Black man. These images confirm township accommodations' safety and suitability for the white middle-class. They also reify the assumption that such people are synonymous with "tourists" (Burton and Klemm 2011). Black South Africans also frequent some township accommodations (Hikido 2019b), but they are not featured on the business websites.

Text that describes the homes emphasises interior comforts and amenities, speaking directly to racialised and classed assumptions that townships are dirty, cramped and backwards. Kopanong's site states that, "all [rooms] have showers with hot running water. The rooms are clean and well-lit" (Kopanong Bed & Breakfast n.d.). Lungi's invites guests to experience "real township life from a shack house with all the luxury of a 3-star B&B, that's what we call 'Shack Chic'" (Lungi's Bed and Breakfast 2016). Malebo's Airbnb site informs guests that, "Offcourse [*sic*] you can use our Wifi (wireless internet), just ask for the WiFi code. My rooms are an absolute pleasure to sleep in and offer a wonderful resting environment after guests experience the dynamic Khayelitsha environment" (Malebo's B&B n.d.). Majoro's Airbnb site tells guests, "My place is a home away from home with a friendly atmosphere nice and cosy" (Majoros homestay n.d.). Disentangling associations of Black urban poverty and social distress, each site shows that a township overnight stay includes middle-class living standards

Figure 14.1 The lounge in Majoro's Homestay promises a comfortable stay that evokes middle-class domesticity.[4] © Annie Hikido

approved by white guests (see Figure 14.1). These racialised and classed material comforts come wrapped in gendered caretaking services.

Mothers and families

Narratives that link urban poverty, Blackness and crime direct tourists away from visiting townships. Township hospitality websites counter these ideologies by drawing upon discourses that link femininity, caretaking and the home. They interrupt Western tourists' fear of violence, which is associated with masculinity, aggression and streets. Race, class and nation intensify these gendered discourses. 'Controlling image' frame Black men, especially those in poor urban neighbourhoods, as dangerous threats. Motherly African women, on the other hand, evoke a 'mammy' figure who is inherently nurturing (Collins 2005). Websites portray hostesses as the mothers of township children and the adoptive mothers of their guests, suggesting that visitors will be taken care of and need not be afraid.

Lungi's website includes a photo gallery that communicates her maternal role and the township as a family-oriented environment. It shows the hostess baking over her kitchenette and sharing a meal with a group of white young adults, clusters of smiling boys and girls playing and dancing outside and women shopping at fruit stalls. Black men feature less often. The few that do are a church reverend, a craftsman and a man standing in front of a house that functions as a tuck shop. In tandem with bourgeois gender roles, the gallery frames Black men as dignified through their breadwinning activities in the community. As Black American anthropologist Michael Hanchard learned in Brazil, when Black men are positioned as husbands and fathers, onlookers' inhibitions are tempered (2000, p. 179).

Text also weaves stories of Black townships as familial places. Several sites explain how each hostess started her place. Given that their entrepreneurial projects turn their private homes into public accommodations, women's families play central roles in their narratives. Malebo's site (Malebo's B&B n.d.) explains that her husband plays an active paternal role in the B&B.

> After working for many years as a cook, I went back to school and became a chef. When my husband Alfred retired, we decided to join our talents and turn our home into a bed and breakfast. We have four grown children and are able to give our guests our attention.

Liziwe's site (Liziwe's Guesthouse n.d.) similarly reads,

> My husband and I have come a long way, from staying in a shack on the plot and receiving a donation of bricks from Sappi Craft where my husband has been working for 25 years. He would come home and clean bricks, and today that has paid off, we have beautiful walls and a home that we decided to turn into a Guest house, This has been a thrilling journey for us.

By explaining how their businesses rose from and serve their families, women's stories trace how their positions as mothers and wives inform their roles as hostesses. They additionally underscore a middle-class work ethic that further protects their reputations against dominant ideologies that intertwine poverty and Black women as 'handout' recipients (Collins 2005).

Other sites suggest that women's families will be an integral part of the tourist experience. Majoro's Airbnb site (Majoros homestay n.d.), for example, states,

> My place is a home away from home with a friendly atmosphere nice and cosy. You'll meet my daughter and my two grandchildren, a boy and a girl who are very friendly. African cuisine, warmth, fun and emotional connection makes my stay memorable!

Mikana's Airbnb site (Mikana's B&B n.d.) also informs guests that,

> Mikana's B&B offers a homely, safe and warm stay. If you stay with me, you will be welcomed by myself and my family who promise to make you feel right at home. I will be delighted to share my story with you and the history

> of the surroundings. All I ask in return, is that you share yours and we form a memorable friendship.

Both Majoro's and Mikana's sites suggest that the appeal of visiting the township is not seeing the 'dark' side of Cape Town. Instead, hostesses promise that staying in their homes will be a worthwhile experience because guests spend personalised time with themselves and their families. Guests effectively become, for a short time, a part of the family. Therefore, through maternal invitations, inflected by Blackness and middle-class domesticity, site visitors learn that they will be safe and taken care of during their township visit.

Community vibrancy and industriousness

Websites further disrupt assumptions about poor Black urban places by framing Black locals as responsible citizens enacting grassroots development. They suggest that tourists should come not to gape at urban depravation, but to engage with a vibrant community. These excursions, offered in the form of walking tours, allay tourists' fears of crime and generate impressions of poor but hard-working Black women and men.

Website photos that feature hostess' neighbourhoods emphasise the daily functionality of township life. Having established middle-class comforts in hostesses' home interiors, photos of the streets show glimpses of underdevelopment. But they focus on Black people participating in economic activities, Black children at play and infrastructural improvements. On Lungi's website, a page dedicated to the Khayelitsha township includes three images. One shows two Black adults standing at an ATM in front of a hand-painted building that reads "MAKHAYA CASH STORE", the next one shows a Black woman in an outdoor produce stall and the last shows a hand-painted mural that reads "KHAYELITSHA". The accompanying text states that, "While many people struggle to make a living, a lot of people have begun to take control of their lives and the entrepreneurial spirit of the people is evident to visitors" (Lungi's Bed and Breakfast 2016).

Photos of tourists on walking tours demonstrate the industriousness of township communities and their apparent welcome of white outsiders. On both Lungi's and Kopanong's websites, white guests are shown strolling through the streets, posing for photos with Black persons, observing Black persons cook and make crafts and playing with Black children. Mingling with neighbours proves to be a central part of the township experience. Lungi's home page (Lungi's Bed and Breakfast 2016) states,

> Don't forget to 'Meet the street' and be fascinated by the peoples' stories while being in the convenient environment of Lungi's Township B&B. Be surprised by the neighbourhood's unexpected creativity and entrepreneurship as well as the welcoming smiles.

Claiming that tourists will be 'surprised' by 'unexpected' township activities evidences the anticipation of tourists' negative stereotypes about under-resourced

Black urban places. The website asserts that visiting the townships will disabuse tourists of such beliefs. Mikana's Airbnb site similarly assures viewers that, "There is a strong sense of community where everyone knows each other, there is no doubt that you will definitely make some new friends".

Some websites frame hostesses as gendered agents of community uplift. Kopanong's site (Kopanong Bed and Breakfast n.d.) explains that Thope, the hostess, operates her B&B as an endeavour to generate skills and resources among her community.

> Thope trains and mentors local women in guesthouse keeping, small business management and catering skills. Dedicated to the economic betterment of the community, she is an active participant and advocate for local NGOs and helps match guests interested in volunteering to organizations that need the particular skills of the guest.

Thope is presented as an entrepreneur who assumes the responsibility of job creation and tackling poverty. She and Lungi in particular are displayed as feminised drivers of development. Their caretaking labours within the family are presumed to extend to that of the impoverished community (Keating *et al.* 2010). The texts imply that by staying with them, tourists contribute to townships' economic and social health.

Kopanong's guestbook page (Kopanong Bed and Breakfast n.d.) suggests that the portrayal of vibrant township communities enhances consumer satisfaction. One couple remarks,

> Thank you so much for taking the time to share your Khayelitsha with us. It was great to hear your upbeat optimism and desire to see more entrepreneurship to bring up the hopes of the Black people of Capetown. Take care and good luck.
>
> Ian and Leslie (California, USA)

Observing the "optimism" and "the hopes of the Black people of Capetown" are integral parts of the tourist experience. In other words, the presentation of vibrant and industrious township communities not only deflects stigmas associated with poor Black areas, such as crime, isolation and laziness; it also makes an experiential consumer product of daily township life.

African authenticity

South Africa's international tourist 'source pool' comes from Europe (Cornelissen 2005a). They arrive in the country with preformed conceptions of what 'Africa' looks like. It is not white people enjoying beaches and mountain hikes, as most Cape Town tourism sites advertise. Township hospitality websites play into tourists' belief that Black neighbourhoods represent the 'real' Cape Town (Rolfes 2010). But instead of highlighting poverty, hostesses use emblems of pan-African culture as racialised markers of their township's African authenticity.

Photos of hostesses' home interiors display African aesthetics. Since many Western tourists equate Africa with poor housing, the middle-class comforts of hostesses' accommodations risk being read as unreal. Supplanting tourists' economic idea of Africa with a cultural one smooths out what tourists might perceive to be 'Western' vis-à-vis middle-class homes. Hostesses decorate their rooms with African patterns, artwork and artefacts. Malebo's Airbnb site includes numerous photos of her living room, dining room and guestrooms. Illustrations of African people and woven baskets line the walls, gourds and sheaths of straw line the countertops and the bed linens feature stylised prints of African women. Kopanong's home page shows a guest room where trees and conical huts feature on the bed's duvet cover and pillowcases while a flower crafted from recycled materials stands on a nightstand. Lungi's brightly painted rooms feature buffalo, antelope and elephant masks with paintings of penguins and zebras (see Figure 14.2). Liziwe's dining room shows a large mural of safari animals and paintings of shirtless African men and African women in headwraps.

Websites also spotlight communal interaction, especially around food, as a source of African authenticity. Township life is distilled not into impoverished struggle but the oft-cited "*ubuntu*". This word roughly translates as "humanity" in multiple Nguni languages spoken in South Africa. In the context of township hospitality, it is exhibited through bringing tourists into the fold of township families and communities. Majoro's Airbnb site is titled "Majoro's B&B-Experience

Figure 14.2 Lungi's "Red Room" displays an African aesthetic through wall decorations and bed linens. © Annie Hikido

the true spirit of Ubuntu". In the description, it details that "African cuisine, warmth, fun and emotional connection makes my stay memorable!". Like the rooms described above, Majoro's living room includes touches of leopard print, hanging artefacts and other symbols of African culture among modern appliances and Victorian furniture. Mikana's Airbnb site, on the other hand, does not include any photos of African-themed rooms. Yet it promises tourists that, "it is inevitable that you will experience township lifestyle, indulge in authentic African cuisine and encounter the ultimate South African hospitality". This suggests that African décor bolsters the ambience of cultural authenticity, but the essence of township authenticity is located in its communal practices.

Outdoor activities also market African authenticity through engagement with locals and food. During walking tours, tourists are shown observing street vendors and sampling their wares. Lungi's website shows tourists drinking a home-made brew (*umqombothi*) from a tin can and images of Black women preparing meat and washing clothes. Under a heading titled "Daily Lives", the site assures tourists that, "All these activities are in the direct area where the township B&B is situated so that the community can also benefit from the presence of the tourists as well. And of course they are all very excited to meet you!" (Lungi's Bed and Breakfast 2016). Liziwe's site also offers a walking tour that includes, "Mzoli's place (Shisa nyama)", "a local taxi (iphela)" and "a Sangoma, the traditional healer of the township". The last item is a visit to the Gugulethu Seven memorial, a site that commemorates seven people who were murdered in a Gugulethu bombing during apartheid. Unlike markers of pan-African culture, the memorial nudges tourists to think about townships as creations of white supremacy. Kopanong's is the only site to historicise the township experience, noting on the homepage that, "Here we have an opportunity to talk about South Africa's recent history and future hopes in an authentic setting" (Kopanong Bed and Breakfast n.d.). Political discussions, however, are not major selling points; experiencing timeless African traditions is.

Conclusion

The tourism industry operates through the mythologisation of places. Previous studies have discussed how guidebooks construct myths to draw tourists (e.g. Nelson 2005, Ogden 2019). This chapter analyses another layer of marketing by focusing on a stigmatised destination. I demonstrate how websites anticipate dominant myths that deter visitation and create counter-myths to attract tourists. Both myths and counter-myths tap into a repertoire of meanings attached to race, class and gender. Websites for township hospitality aim to untangle the intersectional knot that associates townships with Black poverty and reweave a narrative of African communalism. In other words, the websites recode race, class and gender to market welcoming neighbourhoods with stand-in mothers instead of risky forays among criminals. The counter-myth challenges the territorial stigma that the dominant myth plays up. But both myths rely on selling the 'everyday' excursions that lie at the heart of New Urban Tourism.

My focus on online advertising also points to how the internet enables individuals to propagate counter-myths in digital spaces. Large companies and state agencies compile more commonly studied print material, granting them control over its content and distribution. Small business owners (and their supporters) can use websites to independently construct marketing narratives and immediately communicate with overseas audiences. Internet marketing, however, has its limitations. During my fieldwork in township accommodations, some Western tourists had seen the website, while others had not. Many of them admitted that they arrived with lingering inhibitions. Would staying in the township actually be safe, permissible and enjoyable? The hostess had to convince them in the flesh. Nonetheless, the websites begin *a process* of re-mythologising that draws in tourists who may not otherwise consider visiting.

Counter-myths are a tool of New Urban Tourism. They facilitate the commoditisation of marginalised neighbourhoods by reframing disreputable places as tourist spaces. This is especially true in the Global South, where Western tourists interpret poverty to signal ‘Third World’ authenticity and danger. In the South African case, counter-myths preserve the allure of ‘realness’ while combatting the narratives of risk that stem from deep racial segregation. They symbolically transform township homes and streets into welcoming destinations, which generates income and reclaims dignity for township communities. The cost, however, is the depoliticisation of Black space. Framing townships as hardworking communities while covering the daily struggles minimises the problems of persistent inequalities. Kopanong’s nod to history suggests that encouraging critical perspectives can be incorporated into New Urban Tourism. But to develop a more trenchant collective consciousness, myths must be *deconstructed* rather than countered. That is, tourists must strive to understand how structural inequalities condition myths in the first place. Without this lens, tourists may learn that townships represent the unjust laws of the apartheid past, but counter-myths will obscure how townships are also creations of the post-apartheid present.

Notes

1 The apartheid regime categorised mixed-race persons as ‘Coloured’ and slotted them between Blacks and whites in its racial hierarchy.
2 The other three sites were taken down because the hostess passed away, became too old to operate the business full-time or decided to pursue a different line of business.
3 The level of collaboration varied among places, some hostesses authoring text and selecting photos, while others passed on much of the design to their white contacts. In a study of Cuba guidebooks, Ogden argues that textual sources be studied *in contrast* to fieldwork (2019). Since my time in the townships illuminated the nuanced construction of the websites, however, I implore researchers to analyse text and fieldwork *in conjunction*.
4 The photographs in this chapter were taken by the author during her fieldwork in Khayelitsha. They are not the images displayed on the websites but capture the same rooms.

References

Barthes, R., 1984. *Mythologies*. New York: Hill and Wang.

Berger, I., 1992. *Threads of solidarity: women in South African industry, 1900–1980*. Bloomington, IN: Indiana University Press.

Bhattacharyya, D., 1997. Mediating India: an analysis of a guidebook. *Annals of Tourism Research*, 24 (2), 371–389.

Bickford-Smith, V., 2009. Creating a city of the tourist imagination: the case of Cape Town, "The Fairest Cape of Them All". *Urban Studies*, 46 (9), 1763–1785.

Bickford-Smith, V., Van Heyningen, E., and Worden, N., 1999. *Cape Town in the twentieth century: an illustrated social history*. Claremont: D. Philip.

Burton, D., and Klemm, M., 2011. Whiteness, ethnic minorities and advertising in travel brochures. *The Service Industries Journal*, 31 (5), 679–693.

Butler, S.R., 2010. Should I stay or should I go? Negotiating township tours in post-apartheid South Africa. *Journal of Tourism and Cultural Change*, 8 (1–2), 15–29.

Collins, P.H., 2005. *Black sexual politics: African Americans, gender, and the new racism*. New York: Routledge.

Cornelissen, S., 2005a. *The global tourism system: governance, development and lessons from South Africa*. London: Routledge.

Cornelissen, S., 2005b. Producing and imaging "place" and "people": the political economy of South African international tourist representation. *Review of International Political Economy*, 12 (4), 674–699.

Freire-Medieros, B., 2013. *Touring poverty*. Oxford: Routledge.

Frenzel, F., Steinbrink, M., and Koens, K., 2012. *Slum tourism: poverty, power and ethics*. New York: Routledge.

Fuller, H. and Michel, B., 2014. "Stop being a tourist!": new dynamics of urban tourism in Berlin-Kreuzberg. *International Journal of Urban and Regional Research*, 38 (4), 1204–1218.

George, R. and Booyens I., 2014. Township tourism demand: tourists' perceptions of safety and security. *Urban Forum*, 25 (4), 449–467.

Hanchard, M. G., 2000. Racism, eroticism, and the paradoxes of a U.S. Black researcher in Brazil. *In*: F. W. Twine and J.W. Warren, eds. *Racing research, researching race: methodological dilemmas in critical race studies*. New York: New York University Press, 165–185.

Hikido, A., 2018. Entrepreneurship in South African township tourism: the impact of interracial social capital. *Ethnic and Racial Studies*. 41 (14), 2580–2598.

Hikido, A., 2019a. Of mountains and multiculturalism: the Cape Town tourist gaze. *In*: B. Camming and Z. Matebeni, eds. *Beyond the mountain: queer life in "Africa's gay capital."* Pretoria: UNISA Press, 33–41.

Hikido, A., 2019b. *The black township hospitality market: intimacies, authenticities, and postcolonial imaginaries*. PhD Dissertation. University of California-Santa Barbara.

Keating, C., Rasmussen, C., and Rishi P., 2010. The rationality of empowerment: microcredit, accumulation by dispossession, and the gendered economy. *Signs*, 36 (1), 153–176.

Kopanong Bed and Breakfast, n.d. *Kopanong bed and breakfast* [online]. Available from https://www.kopanong-township.co.za/ [Accessed 30 November 2020].

Liziwe's Guesthouse, n.d. *South Africa venues* [online]. Available from https://www.sa-venues.com/visit/liziwesguesthouse/ [Accessed 30 November 2020].

Lee, R., 2009. *African women and apartheid: migration and settlement in urban South Africa*. New York: Palgrave Macmillan.

Lungi's Bed and Breakfast, 2016. *Lungi's bed and breakfast* [online]. Available from: https://lungis.co.za/ [Accessed 30 November 2020].

MacCannell, D., 2013. *The tourist: a new theory of the leisure class*. Berkeley, CA: University of California Press.

Maitland, R., 2007. Tourists, the creative class and distinctive areas in major cities: the roles of visitors and residents in developing new tourism areas. *In*: G. Richards and J. Wilson, eds. *Tourism, creativity and development*. London: Routledge, 73–86.

Maitland, R. and Newman, P., 2009. *World tourism cities: developing tourism off the beaten track*. New York: Routledge.

Majoros homestay–two doubles [online]. Airbnb. Available from: https://www.Airbnb.co.uk/rooms/21123256?source_impression_id=p3_1591842313_N2gkJQp9hSttQ3Ud&guests=1&adults=1 [Accessed 30 November 2020].

Malebo's B&B [online]. Airbnb. Available from: https://www.Airbnb.com/rooms/21263384?source_impression_id=p3_1591842269_r4Hg6bJoP48dyv2E&guests=1&adults=1 [Accessed 30 November 2020].

McDonald, D. A., 2008. *World city syndrome: neoliberalism and inequality in Cape Town*. New York: Taylor and Francis.

Meschkank, J., 2011. Investigations into slum tourism in Mumbai: poverty tourism and the tensions between different constructions of reality. *GeoJournal*, 76 (1), 47–62.

Mikana's B&B [online]. Airbnb. Available from: https://www.Airbnb.com/rooms/21186969?source_impression_id=p3_1591842209_JHglBXKpaGSqoOH9 [Accessed 30 November 2020].

Nelson, V., 2005. Representation and images of people, place and nature in Grenada's tourism. *Geografiska Annaler, Series B – Human Geography* 87 (2), 131–143.

Ogden, R., 2019. Lonely planet: affect and authenticity in guidebooks of Cuba. *Social Identities*, 25 (2), 156–68.

Parker, I., 1994. Discourse analysis. *In*: P. Banister, et al., eds. *Qualitative methods in psychology: a research guide*. Buckingham: Open University Press, 92–107.

Patil, V., 2011. Reproducing-resisting race and gender difference: examining India's online tourism campaign from a transnational feminist perspective. *Signs*, 37 (1), 185–210.

Pirie, G., 2007. Urban tourism in Cape Town. *In*: C.M. Rogerson and G. Visser, eds. *Urban tourism in the developing world: the South African experience*. New Brunswick, NJ: Transaction, 223–244.

Richards, G., 2011. Creativity and tourism: the state of the art. *Annals of Tourism Research*, 38 (4), 1225–1253.

Rolfes, M., 2010. Poverty tourism: theoretical reflections and empirical findings regarding an extraordinary form of tourism. *GeoJournal*, 75 (5), 421–442.

Sassen, S., 2005. The global city: introducing a concept. *Brown Journal of World Affairs*, 9 (2), 27–43.

Skinner, C. and Valodia. I., 2001. Globalisation and women's work in South Africa: national and local approaches to economic transformation. *Agenda*, 16 (48), 75–89.

West, M., 1982. From pass courts to deportation: changing patterns of influx control in Cape Town. *African Affairs*, 81 (325), 463–477.

Western, J., 1996. *Outcast Cape Town*. Berkeley, CA: University of California Press.

Wulff, H., 2007. Longing for the land: emotions, memory, and nature in Irish travel advertisements. *Identities*, 14 (4), 527–544.

15 New Urban Tourism and the right to complain

Tourism as a catchall for urban problems

Emily Kelling and Annika Zecher

In tourism studies, the concept of New Urban Tourism describes the interest (and practice) of tourists to *experience* the life of locals (Maitland 2010). It is used in contradistinction to merely 'seeing' important sights in the city, as passive consumers. While others have counterposed these two forms of tourism as 'sight-seeing' and 'life-seeing' (e.g. Wöhler 2011, Sommer and Helbrecht 2017), we want to add the differentiation between *life-seeing* and *life-experiencing*. From our media analysis, we observe that with an increasing interest for *life-experiencing*, old elements from sight-seeing—large groups, coaches, hotels—transform this new kind of individual tourism *back* into a sort of mass tourism—so that we prefer to call it *life-seeing*. While New Urban Tourists seek to experience "the everyday and mundane activities of city residents [which for the tourists] take on significance as markers of the real, and off the beaten track areas" (Maitland 2010, p. 176), what in fact they are producing is very beaten tracks. The neighbourhoods are no longer untrodden territory but due to references in travel guides and blogs, they become yet again "recognized tourist attractions" (Maitland 2010, p. 176).

Our data[1] consists of more than 925 newspaper articles from 6 daily newspapers published in Berlin from 2008 to 2019.[2] We assume that the media discourse constructs a social reality that may influence the perception of residents and politicians and therefore deserves attention. In a content analysis (Mayring 2015, Gläser and Laudel 2010), we coded and reconstructed common themes that form the public narratives on tourism in Berlin. In this, we distinguish neither between the newspapers nor between single journalists and their opinions. Specifically, we ask, what is the overall media discourse on New Urban Tourism in Berlin and how do the argumentations change over time?

Our data shows that Berlin's inner-city neighbourhoods become shaped by New Urban Tourism in ways that life-experiencing becomes such a dominating feature of the neighbourhoods' spatial constitution that it attains the quality of life-seeing in the perception of the residents. Their own life becomes a *mise-en-scène*: the residents become actors and their houses and neighbourhoods the scenery in the scripts of the tourists and the tourist industry. With this different perception of neighbourhood spaces, we see the emergence of a fierce political debate about urban development and the political priority setting of residents' needs versus the city's need for income through tourists. The urban development dynamics—rising housing prices and the perceived growing difficulty to find one's space in the city, with a displacement of residents, local shops and traders—are intense and the discontent with it finds expression in the debate about tourism, particularly New Urban Tourism. This negotiation goes as far as creating a moral conflict among Berliners about who has the right to complain and at whose expense.

Urban neighbourhoods as tourist spaces: from life-experiencing to life-seeing

From the perspective of Löw's (2016) relational sociology of space, a space is constituted not only by material objects but also by people. Moreover, spaces only exist through people connecting these spatial elements (people or material objects) to form spaces—spaces do not exist out there. As such, also a given street is not always the same space, because who is present and what happens on the street changes over time. Even in the same moment, different people may perceive the given elements in different ways; some may feel safe and others unsafe. Nonetheless, despite these differences, some core elements such as the path for the cars, the pedestrian way and street signs make all of us recognise the street as a street: the street is an institutionalised space with recurring arrangements.

Similarly, urban neighbourhood spaces, and all the spaces within, are created including all people present, tourists or not. Our analysis shows that a crucial question for the constitution of space and for the satisfaction of the residents in the context of New Urban Tourism concerns whether tourists are recognisable as such. If New Urban Tourists blend into the familiar environment, the neighbourhood space remains the same. But if New Urban Tourists become recognisable as tourists, the neighbourhood space changes: it is no longer the place for mundane, everyday residential activities but also—or even primarily—for tourist activities. Here, not the individual tourist, and not even high numbers of individual tourists, but large groups, coaches, guided tours and hotels and hostels as the tourism-facilitating infrastructure mark the critical turning point. The visibility of tourists and tourism has such an influence on the perception of space that the neighbourhood is perceived as dominated by tourism. In contrast to hotels, holiday homes, even if in high numbers, are much less visible on the streetscape.

Turning to why it is a problem if neighbourhood spaces are perceived as dominated by tourism, we draw on Wöhler's (2011) research on the touristic culturalisation of spaces. Wöhler (2011) states that any tourist visiting a place turns the everyday life of the residents into something special: they create something *other* compared to their own everyday life. Any touristic practice thus produces what is to be seen. If New Urban Tourists become recognisable as tourists, this *gazing* on the everyday life of the residents itself becomes noticeable—and the residents generally dislike being objectified. As pinpointed by the journalist Schneider (2017), "The neighbourhoods change, entire streets decay into sceneries for tourists, characterised by bars and restaurants" (Schneider 2017). Another journalist quotes the mayor of Berlin's district Friedrichshain-Kreuzberg Herrmann stating that "'some visitors think they are in a sort of Disneyland' and that the locals resemble 'background actors'" (Prösser 2014). This illustrates that the presence of tourists shapes the space to such an extent that it is no longer the familiar neighbourhood, with its institutionalised spatial arrangements, but has become *another space* altogether. In the words of sociologist Goffman (1956, preface), "the audience constitutes the third partner to the interaction—one that is essential and yet, if the stage performance were real, one that would not be there". Applied to our case, reality only exists without tourists. If tourists are present, it is no longer reality. This seems to capture the sentiment of residents in affected neighbourhoods: they feel their own everyday life constrained. The problem seems to be not the presence of strangers per se, given that urban life is characterised by heterogeneity, but that the residents and their neighbourhoods become touristic objects.

Illustrating this with newspaper findings, people are reported to have moved out of their neighbourhood once tourist coaches started frequenting the area (Treichel 2015). Richter (2014) considers tourism to massively impact areas so that "the population feels threatened; where spaces overcrowded by tourists turn into a plague for the residents". More specifically, residents are reported to be worried about their children walking on streets covered with broken glass (Haas *et al.* 2011); others as afraid of hostel tourists for their vandalising behaviour (Jacobs 2013). Shops are reported to change, accommodating the demands of New Urban Tourists. Grocery stores are replaced with bars (Haas *et al.* 2011). While in the early years, people are said to feel displaced from certain spots like their favourite pub (Pezzei 2010), over the years the tenet develops to describe tourism as causing the complete transformation of neighbourhoods: "it has little similarity with back in the time"—as rents for shops are rising, the old ones are driven out—"The shops were basically taken over by tourists" (Walde 2012). Even some of the restaurant owners, who profit from the proximity of other restaurants and tourists, are annoyed by tourists because, as residents, they have difficulties navigating the neighbourhood with strollers (Keseling 2013). By 2018, many journalists describe the displacement of small traders, non-profit organisations and social infrastructure with international chain stores, causing Hönicke (2018) to entitle his article "Federal government shall save the Kieze"—*Kiez* is the Berlin word for a residential neighbourhood. Messmer (2018) illustrates a situation in which "Many who formerly identified with their neighborhood, today would not even enter most of the new stores".

The extent of the constraints posed on residents by tourism is further emphasised by the founding of initiatives such as "Reclaim Your Kiez", which Schwilden (2014) poignantly comments, "rhetorically they have arrived at war". Even the head of the city's destination management organisation Visit Berlin, Kieker, stated that the growth of tourism has consequences for urban life: "tourist hotspots emerge where the visions of tourists displace normal life"; they develop into "tourist ant trails" (Pezzei 2010). Herrmann is quoted saying that Friedrichshain-Kreuzberg has become a "party laboratory" for tourists (Abel and Fahrun 2015). Generally, Friedrichshain-Kreuzberg can be considered one of the districts most impacted by New Urban Tourism. From 2010 onwards, politicians recognise the risk and deliberate measures to curb the negative impact. We find statements that there is a limit to how many hotels and hostels a *Kiez* can withstand (Nowakowski 2014). Even Kieker states, tourists in residential areas are such a problem that a third hostel in a single residential street is unacceptable (Fahrun 2014). In 2019, the rise in hotels causes a left-wing politician to warn about the "'uninhabitability of certain areas'" (Bünger 2019).

Tourism as a vehicle to discuss urban problems

With the shift to identifying residential neighbourhoods as dominated by tourism, we can observe a conflation of topics, problems and causal relationships: other problems in the *Kiez* and in the city are attributed to tourism. Examples for such larger problems are parties, the leisure industry, litter, noise and rising rents. This dynamic is obvious with the problem of 'party tourism'. A journalist writes,

> "You are entering the tourism sector" [orig. in English], is sprayed on the elevated railway viaduct at the Berlin Oberbaumbrücke. But that's not true, because on both sides of the inscription you are already right in the middle of it, in a kind of party-Ballermann [a party area in Mallorca, Spain] in the district of Friedrichshain-Kreuzberg. There are shards of broken glass on cycle paths, green spaces stink of urine, entrances to buildings smell of vomit, and unnerved residents don't sleep at night because of the noise.
>
> (Hannemann 2014)

Friedrichshain-Kreuzberg is the home to many clubs, frequented for party-associated practices by Berliners as well as by tourists. This is well known, as media reports evince, yet the common narratives work as such that partying Berliners become part of the disturbing tourist crowd. By engaging in these leisure activities, they qualify as tourists. The causality dominates to an extent that although its flaws are obvious, everything else is subsumed under the header of tourism. One journalist contends:

> It's not about sweepingly accusing tourists, but against the leisure industry! [...] The economy is heard, residents do not want to be just a backdrop scenery, authorities do not manage to punish violations of the rules! However, according to the Senate, 60 percent of the people who get noticed for their bad behaviour in the neighbourhoods come from Berlin and the surrounding area. Well, says one resident: Let's say "visitors" instead of "tourists".
>
> (Lackmann 2016)

We assume that the concept of tourism works so well as a catchall exactly because it cuts right to the chase of the matter: tourism inherently produces its own object. In the case of New Urban Tourism, this object is the everyday life of the city's residents. The home and its environment become a sort of fetish—something special for others—although most residents just want to live their customary life. With the inconspicuous tourist masses the disturbances of the leisure industry are attributed to the general party crowd. In contrast, where tourists are recognisable, they are made responsible. Even politicians reproduce this discursive move. Exemplified by the quote above, Berlin's Senate states that the majority of people noticed for their bad behaviour in the neighbourhoods are locals. Nonetheless, the first experiment to ease the tensions in 2015 was a pantomime performance explicitly targeting tourists.

> The initiative is led by an alliance of the district, the Club Commission, the Dehoga Hotel and Restaurant Association and Visit Berlin. A study was first carried out to determine which approaches cities with similar [tourism] problems, such as Barcelona, are pursuing. The result: Mime artists are to make life easier again for noise polluted residents.
>
> (Pohlers 2015)

Negotiating the right of residents to complain about and profit from tourism

We have thus shown that the narratives of tourists being responsible for *Kiez*-problems have obvious flaws. Nonetheless, these claims also enjoy legitimacy—evinced by being voiced frequently. We interpret this circumstance as an expression of the emotional charge of the root problem: people feel threatened in their ability to live in Berlin how they want and have been used to. This is due to an experienced inaccessibility of affordable housing, and the perceived change of the *Kieze*, which comes to the fore in the quotes above about the replacement of shops and residents with tourists and tourist infrastructure. In this wake, two moral questions are discussed. First, is it morally just to make tourists responsible for the problems in the city and the neighbourhoods? And second, is it ethically justifiable to let residential apartments to tourists via platforms like Airbnb given that this may jeopardise livelihoods and neighbourhood relations? In the following, we illustrate these debates.

Regarding the first, there is a strong voice in the media marking the absurdity of making tourists responsible for everything that annoys Berliners, for instance littering; rising prices in cafes (Rennefanz 2011); "the transformation of useful grocery stores into noisy amusement-mob-lounges" (Höge 2014); the ruining of Berlin's club culture (Leber 2012); or that, generally, "it destroys Kieze" (Prösser 2014). This is analysed as working with stereotypes, which makes tourists the scapegoats for urban problems (e.g. Kuhn 2014). These journalists criticise the observed aggression against people on rental bikes (Hüttl 2012) and those who place stickers against tourists in public spaces, judging this unjustified and damaging to the city (Meyer-Gatermann 2013). They call those criticising tourism "tourist haters"—one article is even entitled "Tourist haters get out!" (Leber 2012).

Gennies (2011) describes the way in which residents of Wrangelkiez expressed their discontent with local dynamics as 'Wutbürgertum'—'angry-citizen-dom'—which is a ridiculing reference to right-wing protesters. In the same vein, complaints about tourism are said to be entangled with xenophobia (Rennefanz 2012).

However, this voice of anti-tourist-haters itself sometimes has a xenophobic touch, for instance when Leber (2012) states, "It is not acceptable that a non-German in Berlin has to have a proper refugee background to be welcome". Some journalists also blame politicians: "The mood towards tourists is heated up and is still fired up by politicians. Tourism has been recognised as a plague" (BZ 2014).

Even the residents' complaints about tourists are interpreted as bizarre and absurd (Kögel 2011). Kittan (2014) states that partying tourists simply are loud, and "You have to put up with a certain amount of noise if you live directly in the city". Furthermore, the complaints are attributed to specific groups. "Salonek, a politician from the liberal party, expresses that it is absurd that especially supporters of the Green party, who like to travel authentically and self-determined as backpackers, now wanted to guide tourists on predetermined routes" (Kögel 2011). Others quote a survey, which identifies provincial newcomers to Berlin as typically complaining; they "would only dislike tourists because they need to strengthen their own fragile feeling of belonging by such malicious means" (Lenz 2012).

This critique is so effective that others feel the need to justify complaining. Maroldt (2012), for instance, writes, those who endure all the evil and complain, "do not need lecturing about lacking tolerance". Especially those directly experiencing nuisance in their building "have a right to complain" (Loy 2014). Some articles evince the emotional charge of the debate without taking sides (e.g. Stuckmann 2012). Overall, tourists and tourism become a reason to negotiate the own rules of conduct.

Turning to the second moral question:

> The fact that more and more private flats are rented out to temporary visitors is, in the view of many residents, one of the dark sides of the tourism boom that the capital has been experiencing for years.
>
> (Henneke 2011)

> Tenants' associations are concerned that the already scarce supply of housing in inner-city locations is shrinking even further due to the increasing number of holiday homes. Hotel and restaurant associations are worried about competition from private individuals. Residents worry about their night's sleep, politicians about their mandates.
>
> (Hädicke 2013)

These newspaper article quotes indicate the variety of voices against the letting of residential apartments as holiday homes in Berlin. They coalesce into one force, successfully bringing about a law aiming to constrain this market—Zweckentfremdungsverbot von Wohnraum (prohibition of misappropriating residential units). It builds on a variety of claims.

First, owners of residential units are condemned for earning more by letting to tourists than to regular tenants (e.g. Schönball 2018). Politicians also engage in the shaming of this practice (e.g. van Bebber 2013). Second, the operators of private holiday homes are claimed as having an unfair advantage vis-à-vis hotels because different standards apply, for instance regarding fire safety (Flatau 2011). Perhaps the most prominent argument is that the use of residential units as holiday homes means a reduction in available flats for Berliners to rent (e.g. Strauß 2010)—which is supposed to cause rent increases (e.g. Maroldt and Voss 2018). Fülling and Lange (2012) even call residential holiday homes "the speculative destruction of housing".

Residential holiday homes are perceived as an existential threat—they "endanger" "the social diversity of the population" (taz 2013). Their operation becomes a moral question: "It is unacceptable that entire buildings [...] are being converted and the last old lady living there is pushed out through annoying behaviour" (Rada 2010). Tenants in such situations suffer from the "tourist terror" (e.g. Loy 2013). Holiday homes are also framed as illegal. Operators are accused of hiding the numbers of holiday homes in a building on purpose; of breaking rent regulations and causing security issues (Klemp 2008), or of evading tax by not registering the income (Jacobs and Rink 2013). After the implementation of the legal prohibition, we find many complaints, also by politicians, about operators not applying for the obligatory permission, thus acting illegally (Fleischmann *et al.* 2015).

As a reaction to this discourse and the legal prohibition, many voices justify the operation of residential holiday homes. Some authors criticise common narratives positioning tourists as the scapegoat for rising rents (e.g. Rennefanz 2011); while others emphasise that tourists are not responsible for housing policy decisions (e.g. Petersen 2013). However, the most prominent claim, expressed throughout our research period, is that ordinary Berliners are thus empowered to profit from tourism (e.g. Schlösser 2018). Moreover, from 2014 onwards it is said, many would no longer be able to afford their rising rents and living expenses without subletting a room to tourists (e.g. Hasel 2014). In this opposition, Krone (2015) and others also accuse the state of using methods of the former German Democratic Republic's state security in their prohibition efforts by asking the public to report about potentially illegal rentals. Overall, the criticism was so strong that in 2013, during the drafting of the legal prohibition, a politician from the Green Party felt the need to justify their action by stating that "'The common good takes precedence over individual interests. If I can earn three or four times as much money through short-term rentals and thereby withdraw housing, the state has the right to intervene'" (Loy 2013). At this point, it is necessary to add that economists indeed have shown rents to increase with every nearby unit rented on Airbnb (Duso 2020). Nonetheless, as stated before, we interpret the above as such that tourism becomes a major discursive outlet through which the pressures of urban development, especially regarding housing prices, can be articulated.

Prioritising tourism over the needs of residents

The above description shows that the debate is emotionally loaded, touching the core of being able to live in the city. Nonetheless, we find politicians constructing narratives that place the residents' quality of life and general needs as secondary to the need of attracting tourists. The residents' satisfaction becomes a means to that end: massive inconveniences for the residents are to be avoided, not for their own well-being, but to prevent them leaving or becoming angry, which may be detrimental for the continuous rise in tourist numbers.

The popular narratives of 'tourism bringing money'—prominent in the media throughout (e.g. Maroldt 2018)—and 'the city relying on this revenue' (e.g. Richter 2011) is reinforced even when the criticism rises. Already in 2011, the Senate is quoted aiming for rising numbers and further investing into tourism despite the growing discontent among residents (Haas *et al.* 2011). In 2012, the idea emerges to address the growing discontent by distributing the tourists throughout all of Berlin, not just central areas (Dobberke 2012).

> Noise, garbage, streets clogged with coaches: Even the city's official tourism concept lists 'congestion' as a barrier to development. It states, 'In the tourist core zones, conflicts with the resident population sometimes occur at peak times, which impair the quality of [...] life'.
>
> (Kreller 2012)

By 2017, according to the senator responsible for tourism and the economy, Pop, the aversion against tourism assumes such an extent that maintaining its acceptance becomes the main task for politicians (van Bebber 2017). In 2018, the Senate therefore allocated a budget for district-specific advertisement to lure tourists into the outskirts (Beikler 2018a). "The goal is no longer just a quantitative increase in the number of tourists, but 'sustainable tourism that is compatible with the city'" (taz 2018). According to Pop, distributing tourists would allow keeping the high numbers and associated revenue, while responding to "the 'negative consequences'" such as "the overuse of public space in the centre" (Hering 2019). Furthermore, the idea is attracting specifically "quality tourists", interested in culture and the discovery of other *Kieze* (Beikler 2018b)—not just the famous "hotspots" (taz 2018).

Indeed, we find that the residents are somewhat instrumentalised for securing the continuous rise in tourist numbers. A flattening in the rising curve is already perceived as a failure and loss, as happened in 2017 (Thomsen 2018). In 2011, city advertisers identify the growing animosity expressed against tourists as a threat to "'Berlin's image as hospitable metropolis'" (Beikler *et al.* 2011), thus prioritising the image over the expression of discontent. In 2013, Kieker argues that residential holiday homes should not proliferate further, anticipating the population to consequently stop accepting tourism altogether, which eventually would decrease the city's popularity among tourists (Hoffmann 2013). Similarly, the Dehoga argued that the city's charm would falter for tourists when unnerved residents move out (Pohlers 2015). Finally, by 2019, the problem with urban tourism had become so apparent that Kieker warned, "'We are losing what I think is any city's most important momentum of success: our authenticity'" (Strauß 2019), where obviously success refers to the growth of tourism.

The narrative justifying this approach is that Berlin's economy, culture, residents and even the city itself 'depend' on the tourists. "Because the city cannot do without the income from city trips and congresses", as Dobberke (2012) paraphrases Kieker. Therefore, neither politicians nor residents may do anything that might stop tourists from visiting Berlin. Martenstein (2011) applied this argument to levying a city tax from tourists, postulating that tourists would stop visiting altogether. Senator Pop, supported by Dehoga, reasons similarly by stating "'you cannot limit the number of visitors. You cannot tell people that they cannot come, because then nobody will come'" (Vossen 2018).

This mantra of the dependency assumption also transpires strongly in opinion statements by journalists and residents. For instance, they fear the negative commentary of the anti-tourist sentiments among Berliners in foreign that newspapers (Leber 2012). Nowakowski (2012) argues that tourism is good, and the city needs it because it provides low-skilled people with a job, therefore deeming all complaints about tourism inappropriate. An interviewed resident states that those expressing their aversion against tourists, for instance with stickers, want to harm the city (Haas *et al.* 2011). Martenstein (2011) argues that the noise and dirt originating from tourists, which one must endure, is the flipside of the fact that they bring money. While the above statements were published in the early years of our data selection, also in 2017 Fahrun (2017) argues that tourism is worthwhile "not being dried up", given its economic importance. However, it also shows a change in discourse: the complaints no longer need to be justified; there is ongoing support of tourism.

This merges into an even more pronounced critique of the dependency assumption among journalists in recent years. We find a first critique by van Bebber (2017) emphasising that it is not Berlin's economy per se but only specific sectors that profit. Nowakowski (2017) puts the dependency assumption in a nutshell: "Berlin can be self-confident because the times when you could be grateful for every single tourist are over". Berlin can now prioritise its residents' needs instead of solely focusing on economic interests. Notably, this is the same author who in 2012 criticised complaints because tourism secures low-skilled jobs. Similar to Nowakowski's recent remark, Lenz (2015) states that Berliners let themselves be dominated by tourists and calls them to stop being submissive—"tourists are no shy deer". Hönicke (2018) urges, "you can't treat tourists better than residents just because they bring more money". Finally, there is a growing consensus that mass tourism anywhere in the world but also specifically in Berlin is negative (e.g. Hönicke 2018).

Conclusion

To recap the development of the debate, throughout the years, we find a dominant narrative of tourism being good for the city because tourists bring money. This narrative builds on the assumptions that the city has little viable industry. From 2010, there is a rise in complaints about tourism. Core to this is the identification of a 'party tourism' in certain areas, which was considered a major problem for neighbourhood relations and housing quality. Soon, the growth of residential holiday homes in central areas was considered a threat to affordable housing, and a nuisance to residents in nearby properties because of noise, littering, etc. A variety of arguments and actors concur in the objective of restricting the market. Interestingly, while at no point did the argument about tourism bringing money to the city lose weight, towards the end of our research period what was then called 'mass tourism' received a consensual devaluation among journalists. Nonetheless, some of the influential decision-makers keep reproducing the narrative that Berlin must make sure not to offend tourists, because otherwise they may stop visiting. In our view, this argument implies a logic of constructing the city as dependent on tourists. Notably, some journalists identify this argumentative logic and retort that the city can, by now, be self-confident and attend to the needs of its residents and their housing quality. They understand that doing so does not necessarily incur a drop in tourism; and that even if the number of tourists stops rising, or even falls, this should not be perceived as a failure anymore.

At core, this is about the right to the city (Lefebvre 2006), about the primacy of economic interests versus those of the city's people—at least of those who do not profit from tourism. Even though it may appear differently at times from individual commentaries in the media, generally, it is not the tourists per se who are the object of discontent, but the perception of the sell-out of the city. Nonetheless, *tourism* is the means by which this discontent is articulated. We argue that it is a suitable vehicle that functions as an outlet for criticising general urban development dynamics. Drawing on Wöhler's (2011) argument that any touristic practice makes the ordinary life of the 'locals' something other than normal, thereby producing its own object, we understand the critique of tourism as an expression of the residents realising and disliking this dynamic. Adding to Wöhler's idea, when life-experiencing, itself not generally despised, turns into life-seeing and is also understood as such by affected Berliners, this has such an emotional charge because the residents turn into the main attraction and an advertising feature for the sell-out of the city. Moreover, while those with decision-making power continue to advertise for more and more visitors, the residents seem to feel left without control, neither over the city's development generally nor over their private lives directly.

Notes

1 Our data was conducted for the research project Kiez in der Tourismusfalle? Eine Untersuchung zur Veränderung von Wohnqualität durch touristische Übernachtungsmöglichkeiten in ausgewählten Berliner Wohnquartieren, funded by the German Research Foundation—DFG (GZ: FR 2522/5-1 & WE 5894/2-1).

2 The newspapers are: *Der Tagesspiegel*, *taz-die tageszeitung*, *Berliner Kurier*, *Berliner Zeitung*, *B.Z.*, *Berliner Morgenpost*. The selection criteria were one or more of the following keywords: tourist(s), tourism or touristification.

References

Abel, A. and Fahrun, J., 2015. Wir brauchen ein Sicherheitskonzept. *Berliner Morgenpost*, 21 September.

Beikler, S., 2018a. Berlin will Touristen in die Bezirke umleiten. *Tagesspiegel*, 11 June.
Beikler, S., 2018b. Spandau ist doch auch schön. *Der Tagesspiegel*, 10 June.
Beikler, S., Keilani, F., and Leber, S., 2011. Schmerzlich willkommen. *Der Tagesspiegel*, 3 March.
Bünger, R., 2019. Touristifizierung folgt auf Gentrifizierung. *Der Tagesspiegel*, 9 March.
BZ, 2014. Gegenüber Touristen brauchen wir eine neue Willkommens-Kultur. *BZ*, 27 May.
Dobberke, C., 2012. Berlin ist Millionen Reisen wert. *Der Tagesspiegel*, 28 August.
Duso, T., et al., 2020. Airbnb and Rents. Evidence from Berlin. *DIW Berlin Discussion Paper No.* 1890.
Fahrun, J., 2014. Auf dem Weg zum sanften Tourismus. Auch "Visit Berlin" hat erkannt, dass die Besucher manche Innenstadtkieze stark belasten. *Berliner Morgenpost*, 26 July.
Fahrun, J., 2017. Abschied von Masse als Ziel. Der Tourismus in Berlin braucht neue Leitlinien. *Berliner Morgenpost*, 23 November.
Flatau, S., 2011. Was tun gegen Kneipenlärm und Ferienwohnungen? *Berliner Morgenpost*, 1 March.
Fleischmann, M., Weeg, C., and Otto, V., 2015. Legal, illegal, scheißegal! *Berliner Kurier*, 30 July.
Fülling, T. and Lange, K., 2012. Kampf den Ferienwohnungen. *Berliner Morgenpost*, 27 June.
Gennies, S., 2011. Kreuzberg will seine Ruhe haben. Anwohnerwut auf Touristen bei Grünen-Diskussion. *Der Tagesspiegel*, 2 March.
Gläser, J. and Laudel, G., 2010. *Experteninterviews und qualitative Inhaltsanalyse als Instrumente rekonstruierender Untersuchungen*. 4th ed. Wiesbaden: VS Verlag.
Goffman, E., 1956. *The presentation of self in everyday life*. Edinburgh: University of Edinburgh.
Haas, B., Bauer, T., and Drachsel, C., 2011. Partytouristen ärgern die Kreuzberger. *Berliner Morgenpost*, 11 May.
Hädicke, G., 2013. Urlaub machen, wo andere wohnen. *Berliner Zeitung*, 4 November.
Hannemann, U., 2014. Kodex Kotze. Bezirksbürgermeisterin will Touristenhorden zivilisieren. *taz*, 20 August.
Hasel, V.F., 2014. Heimweh: Allein unter Freunden–im eigenen Haus. *Der Tagesspiegel*, 21 May.
Henneke, M., 2011. "Sie kommen"–Tür an Tür mit Touristen. *Der Tagesspiegel*, 20 August.
Hering, M.-M., 2019. Es kommen immer mehr Touristen. *Der Tagesspiegel*, 23 February.
Hoffmann, K.P., 2013. Trotz aller Pannen ist Berlin beliebt wie nie. *Der Tagesspiegel*, 21 February.
Höge, H., 2014. Der Hass auf Touristen, historisch betrachtet. *taz*, 30 August.
Hönicke, C., 2018. In zehn Jahren sind unsere Städte komplett zerstört. *Tagesspiegel*, 21 November.
Hüttl, T., 2012. AUFGETISCHT. Tina Hüttl war im Käfer. Wo das Kalb an die Forelle denken lässt. *Berliner Zeitung*, 17 November.
Jacobs, S., 2013. Nacht ohne Ruhe. *Der Tagesspiegel*, 2 October.
Jacobs, S. and Rink, T., 2013. Pankower Ferienwohnungen vor dem Aus. *Der Tagesspiegel*, 4 January.
Keseling, U., 2013. Insel der Träumer–und der Touristen. *Berliner Morgenpost*, 17 February.
Kittan, T., 2014. Nein, Touristen bringen Geld. *BZ*, 26 May.
Klemp, M., 2008. Hilfe, mein Haus ist ein Geheim-Hotel. Müll und Krach: Bei den armen Mietern liegen die Nerven blank. *Berliner Kurier*, 15 October.

Kögel, A., 2011. Sie liebt dich doch. *Der Tagesspiegel*, 19 June.
Kreller, A., 2012. Berliner Monopoly. Hoteliers und Investoren setzen auf einen anhaltenden Boom im Städtetourismus–das ist nicht ohne Risiko. *Der Tagesspiegel*, 8 April.
Krone, T., 2015. Am Ende hilft nur Petzen. *taz*, 3 February.
Kuhn, A., 2014. Touristen im Dienst. Statt mit Benimmregeln wollen zwei Studentinnen Besucher durch Freiwilligendienste integrieren. *Berliner Morgenpost*, 5 October.
Lackmann, T., 2016. Holzauge, rette die Stadt! *Tagesspiegel*, 1 March.
Leber, S., 2012. Touristenhasser raus! Zu laut, zu blöd und ständig im Weg: In unserer toleranten Weltstadt gehört es inzwischen zum guten Ton, gegen Besucher zu hetzen. *Der Tagesspiegel*, 7 April.
Lefebvre, H., 2006. *Writings on cities*. Oxford: Blackwell Publishing.
Lenz, S., 2012. Provinzielle Abneigung. *Berliner Zeitung*, 15 September.
Lenz, S., 2015. Touristen sind keine Rehe. *Berliner Zeitung*, 31 August.
Löw, M., 2016. *The sociology of space. Materiality, social structures, and action*. New York: Palgrave Macmillan.
Loy, T., 2013. Ärger mit den Kurzzeitnachbarn. *Der Tagesspiegel*, 20 January.
Loy, T., 2014. Ferienwohnungen in Berlin. *Der Tagesspiegel*, 10 May.
Maitland, R., 2010. Everyday life as a creative experience in cities. *International Journal of Culture, Tourism and Hospitality Research*, 4 (3), 176–185.
Maroldt, L., 2012. Gast, Freund, Schuft. Berlin–eine Stadt wehrt sich gegen ihre Besucher. *Der Tagesspiegel*, 2 March.
Maroldt, L., 2018. Tagesspiegel checkpoint, diverse. *Tagesspiegel Checkpoint*, 2 May.
Maroldt, L. and Voss, O., 2018. Wie Airbnb zum Amazon des Tourismus werden will. *Tagesspiegel*, 13 July.
Martenstein, H., 2011. Touristen machen Lärm. *Der Tagesspiegel*, 17 April.
Mayring, P., 2015. *Qualitative Inhaltsanalyse. Grundlagen und Techniken*. 12th ed. Weinheim: Beltz.
Messmer, S., 2018. Es knirscht in der Hood. *taz*, 27 October.
Meyer-Gatermann, A., 2013. Kreativ, lässig und luxuriös. Wer am hiesigen Hotelmarkt reüssieren will, braucht trotz steigender Touristenzahlen einen langen Atem–und Ideen, die dem kulturellen Ruf Berlins gerecht werden. *Berliner Zeitung*, 29 June.
Nowakowski, G., 2012. Millionen Touristen … // … können nicht irren: Berlin ist für Menschen aus aller Welt ein Faszinosum. *Der Tagesspiegel*, 11 August.
Nowakowski, G., 2014. Immer mehr Touristen. *Der Tagesspiegel*, 17 August.
Nowakowski, G., 2017. Tourismus in Berlin an der Grenze. *Tagesspiegel*, 20 November.
Petersen, L., 2013. Jetzt können Sie die Kommune I besetzen. *BZ*, 27 September, p. 17.
Pezzei, K., 2010. Mit den Massen kommen die Probleme. *taz*, 29 December.
Pohlers, A., 2015. Pantomime gegen Ballermannisierung von Berlin. *Tagesspiegel*, 9 May.
Prösser, C., 2014. Eine Stadt kommt unter die Rollen. TOURISMUS Partys unterm Schlafzimmerfenster, Kotze im Hauseingang: Vielen Berlinern sind Touristen ein Graus. Was kann man tun gegen die Auswüchse der Reisefreuden–oder müssen wir uns an die Eigenarten der zahlenden Besucher gewöhnen? *taz*, 22 November.
Rada, U., 2010. Reichtum macht bräsig und erstickt Kreativität. *taz*, 7 August.
Rennefanz, S., 2011. Zu Gast bei Nörgelfritzen. *Berliner Zeitung*, 12 March.
Rennefanz, S., 2012. Zum Nichtstun verführt. *Berliner Zeitung*, 14 July.
Richter, C., 2011. Immer mit der Ruhe. *Berliner Morgenpost*, 11 May.
Richter, J., 2014. Probleme entstehen da, wo Touristen die Stadtteile verändern, 17 November.

Schlösser, R., 2018. Vielen Berliner Vermietern drohen hohe Bußgelder. *Berliner Zeitung*, 24 July.

Schneider, J., 2017. Kassemachen auf dem Kiez. *Süddeutsche Zeitung*, 5 August.

Schönball, R., 2018. In fremden Betten. *Der Tagesspiegel*, 27 June.

Schwilden, F., 2014. Touris in Berlin–eine Hassliebe. *Berliner Morgenpost*, 12 October.

Sommer, C. and Helbrecht, I., 2017. Seeing like a tourist city: how administrative constructions of conflictive urban tourism shape its future. *Journal of Tourism Futures*, 3 (2), 157–170.

Strauß, S., 2010. Wechselnde Nachbarschaft. Das Geschäft mit der Vermietung von Wohnungen an Touristen boomt. *Berliner Zeitung*, 16 August.

Strauß, S., 2019. Das Dilemma. *Berliner Zeitung*, 23 February.

Stuckmann, S., 2012. ADEL berichtet FOLGE 30 Braune Bedrohung. Stefan Stuckmann erzählt, wie unser Redaktionspraktikant Cedric zu Guttenberg die Stadt erlebt. *Der Tagesspiegel*, 22 September.

taz, 2013. Schulz ist dagegen. *taz*, 7 February, p. 22.

taz, 2018. Touris raus–aus der Innenstadt? *taz*, 21 February.

Thomsen, J., 2018. Tourismus-Bilanz 2017: Warum immer weniger Gäste aus Europanach Berlin reisen. *Berliner Zeitung*, 26 Febuary.

Treichel, T., 2015. Prachtmädchen und Superschlüpfer. *Berliner Zeitung*, 28 January.

van Bebber, W., 2013. Die Milieuschützer. *Der Tagesspiegel*, 4 May.

van Bebber, W., 2017. Die Politik kuscht vor den Touristen, weil sie Geld in die Stadt tragen. Der Widerstand gegen die Touristifizierung wächst aber in Berlin. *Der Tagesspiegel*, 2 June.

Vossen, L., 2018. Die Stadt lebt davon, dass sie frei ist. *Berliner Morgenpost*, 20 May.

Walde, G., 2012. Der Berliner ist immer noch mufflig, aber er trainiert. *Berliner Morgenpost*, 28 July.

Wöhler, K., 2011. *Touristifizierung von Räumen*. Wiesbaden: VS Verlag für Sozialwissenschaften.

16 Science-driven mobility as a form of New Urban Tourism

Insights from student and research internationalisation in Lund, Sweden

Lena Eskilsson and Jan Henrik Nilsson

Introduction

During recent decades, higher education and scientific research have become increasingly important in modern societies, in terms of employment and value creation (Moretti 2013). Due to its dependence on international connections, this sector is mainly concentrated in urban areas. Universities generate considerable flows of international travel. There are many different visitors coming to university cities, for different reasons and for various lengths of stay. International scholars come for short visits, attending meetings and conferences, or taking part in other academic work within research and teaching. Researchers move to university cities for temporary career positions, e.g. post-docs. Students stay for shorter or longer periods—exchange students stay for a couple of months, master's students and PhDs for longer periods of time. Researchers, engineers and teaching staff move to university cities as semi-permanent visitors (Hannam and Guereño-Omil 2014, Maitland 2006). As research and education become increasingly internationalised, the impact of this process becomes highly relevant to study.

We have decided to call the phenomenon of visitor flows generated by universities 'science driven mobility'. It includes short-term, medium-term and long-term visitors to university cities, and in our study we focus on international travel flows. The phenomenon is related to a general tendency towards globalisation (Dicken 2015). International travel advanced by developments of new business models and technologies (e.g. low-cost aviation and collaborative economies) makes travel cheaper and more easily accessible (Nilsson 2020). In Europe, urban tourism has grown faster than other forms of tourism (Eurostat 2019). The increase of urban travel has not only had a quantitative impact on cities; recent developments have brought about such a qualitative change that we are speaking of New Urban Tourism (Maitland 2007, Maitland and Newman 2009), building on the notion of new tourists as being "flexible, independent and experienced" (Poon 1993, p. 114). In particular, the three following perspectives can be identified.

First, New Urban Tourism is linked to spatial change. In major cities, residential districts outside traditional tourist areas are increasingly affected by tourism. The resulting changes, often related to tourist-driven gentrification, have been the subject of extensive scientific debate (Gotham 2005, Nilsson 2020).

Recently, cities other than traditional tourist cities have also seen an increase in incoming tourism. University cities are examples of this phenomenon; places like Cambridge receive large flows of tourists (Hakeem and Khan 2018, Maitland 2006). Spatial change within university cities also includes processes of 'studentification'; different urban transformation processes resulting from students residing in neighbourhoods close to higher education institutions (Nakazawa 2017, Gregory 2020).

Second, traditional categories of tourists are increasingly questioned, since being physically and socially mobile constitutes the lives of many people today: for work, leisure and studies (Urry 2007). This is often not considered as tourism; the contemporary "urban traveller notion connects to the blurred boundaries between tourism and other forms of mobility" (Pasquinelli 2017, p. 34). Therefore, international students are interesting to study in the context of the debate:

> Beyond the studentification [...], international students become involved in broader urban processes such as the tourism industry, marginal gentrification or entrepreneurial creativity, thus becoming a new class of transnational urban consumers.
>
> (Calvo 2018, p. 2142)

Third, New Urban Tourism represents a cultural shift where visitors tend to seek local authentic experiences rather than iconic 'touristic' sights. The atmosphere of everyday life attracts visitors who often identify as temporary locals (Larsen 2019). New tourist districts have certain place-specific qualities in common; they are neo-bohemian and associated with high cultural capital among New Urban Tourists. Thus, New Urban Tourism is often associated with youth and alternative lifestyles. Students and visiting researchers are simultaneously actors in the knowledge economy and spreading new urban lifestyles (Calvo 2018). Yet, students, researchers and universities are, with few exceptions (e.g. Russo and Arias Sans 2007), absent or unproblematised in tourism literature.

In summary, this points at two possible research gaps on the relation between science-driven mobility and New Urban Tourism: first, students and researchers as actors in New Urban Tourism, and second, university cities as arenas where New Urban Tourism takes place. The notion of science-driven mobility also challenges traditional conceptions of tourism as a distinct social phenomenon, and tourism geographies as a research field. The purpose of this chapter is to analyse science-driven mobility as a part of New Urban Tourism, and discuss this phenomenon as a way to challenge common conceptualisations of tourism. The chapter is based on a case study of the university city of Lund in the south of Sweden. It is a small city with approximately 120,000 inhabitants, with a large and highly internationalised university and significant international research facilities. Thereby, we argue, the impact of science-driven mobility is likely to be relatively more important and visible than in larger cities.

Science-driven mobility and tourism

Analysing science-driven mobility as a form of New Urban Tourism is challenging. It poses essential questions about the nature of tourism. Generally, tourism is conceptualised temporally and spatially—a person staying away from her permanent residence for more than 24 hours, but for less than a year (UNWTO 2020). However, this is a statistical concept and, by nature, rigid. As shown by Hall (2005) there are forms of temporary mobility (e.g. sojourning, working holidays and study or working abroad) that clearly lie at the conceptual borders of tourism. A basic problem is the lack of definition of permanence, as opposed to temporality (Bell and Ward 2000). The new interest in mobilities questions the dichotomy between the static and the mobile place-bound moorings, and mobile flows are viewed as mutually interdependent. In social science, the focus on mobilities challenges borders between transport research, tourism studies and migration studies—making it a post-disciplinary field of study highly relevant for urban tourism studies (Cresswell 2006, Kannisto 2016).

Professional travel conducted by teachers and researchers falls, broadly speaking, into the business tourism category. It is however an under-researched field of study, despite its empirical significance. Professional travel accounts for 14% of all international arrivals in Europe (Beaverstock *et al.* 2010, UNWTO 2020). Other visitors coming to university cities as part of science-driven mobility are more challenging to analyse from a tourist perspective. According to Novy (2019), tourism has become gradually more diverse and variegated in its composition. There is hence a need to move beyond narratives that see tourism as a distinct, easily separable social phenomenon. In the context of the blurred boundaries between tourism and other forms of mobility, Novy (2019) talks about "temporary city users". Examples mentioned are artists in residence, academics on sabbaticals or students on exchange. The number of students studying abroad has increased significantly during the last century; a lucrative international market for higher education has developed (Hannam and Guereño-Omil 2014).

Another concept that challenges the general time–space notion of tourism is 'lifestyle mobilities'. Lifestyle mobilities cover a wide range of voluntary relocation to places that promise a different lifestyle or way of life (Cohen 2011). According to Kannisto (2016), the literature on the topic can be divided into three categories: studies on long-term travel, studies of lifestyle migration and studies on professional lifestyle travel. The last category uses their profession as a means of being mobile. Typical professional travellers appreciate flexibility, and are not tied to a particular place. We argue that the concept of 'professional lifestyle travellers' could be broadened to the context of science-driven mobility, especially when analysing mobility connected to career positions at the university and research facilities.

These new forms of travelling pose challenges to cities where temporary city users and professional lifestyle travellers set down for shorter or longer periods, being neither tourists nor permanent residents. Their choice of services and activities differ from those of other categories (Yu *et al.* 2018). Abdullateef and Biodun

(2014) argue for recognising foreign students as tourists. However, by concentrating on what students do outside studies, they adhere to the idea of tourism as being merely associated with leisure. Instead, international students ought to be viewed as a tourist niche market of semi-professional character; flows of students to university cities have been classified as "educational travel" and "academic tourism" (Rodríguez *et al.* 2012). In the context of urban transformation and New Urban Tourism, students are actors both in the knowledge economy (as students), in the housing market (as temporary inhabitants), as well as in the travel economy (as tourists) and the leisure economy (as youth) (see Calvo 2018, p. 2144). International students could in many ways be categorised as a form of "transnational new middle classes living (temporarily) in the contemporary city" (Calvo 2018, p. 2149).

In the context of New Urban Tourism, it is interesting to analyse spatial perspectives of the incoming temporary city users. Research shows that the general spatial behaviour of students in the university city change over time. When the students first arrive, they act more like tourists in the sense that they explore the city and consume the same services as other tourists/visitors (e.g. tourist sights, cultural events, nightlife venues, cafés). At the same time, they are in a short-term migratory context, meaning that they also use everyday facilities like grocery stores, sport facilities and health care. After some time, they often fall into routines, stop exploring and mainly use their favourite places in the city. In that sense, they then act more like a local citizen (Yu *et al.* 2018, Calvo 2018, Gregory 2020).

In our study we define and discuss science-driven mobility by the purpose of the travel it generates. Adding a time perspective, three broad visitor categories within science-driven mobility can be identified, based on length of their stay.

The first category is the short-term visitors, i.e. mainly professional travellers who come to the university city for meetings, or shorter research and teaching visits. These visitors mainly fall into the concept of business tourism when it comes to the length and purpose of their stay. Within this category we also see short-term urban visitors in the form of additional mobility; international students have a high capacity generating new visits. Studies in Spain have shown that international students receive an average of three visits per student from friends and relatives during their time of study abroad (Rodriguez *et al.* 2012).

The second category is the medium-term visitors, i.e. researchers, educational travellers like exchange students and other temporary city users who stay for a longer time period than conventional urban tourists. For master students we see a blurring of categories, as the duration of their stay often exceeds the general tourism time definition (UNWTO 2020). Hence, statistically the master students count as tourists if they take a one-year master, not if they follow a two-year programme. Nevertheless, in most cases it is a temporary move to a university city and not permanent migration.

The third category is the long-term visitors, including PhD students, post-docs and other types of professional lifestyle travellers moving to the university city for research and/or teaching opportunities. Given the time span, the purpose of travel and the fact that some of them stay more or less permanently in the city, it is clearly challenging to analyse these as visitors. We argue, however, that these should be included in the discussion about science-driven mobility and New Urban Tourism. An important argument is the mutual links between the different visitor categories. Medium-term and long-term visitors generate additional short-term visitors, such as people visiting friends and relatives (VFR segment), thus influencing the first category. Temporary city users tend to come back later in life to revisit the place. Especially student mobility tends to promote this form of mobility, new interpersonal networks are formed when students go abroad, which tend to increase future travel. Maintaining such networks requires substantial travel, which is likely to be sustained when students move on in their academic careers, possibly even affecting their future place of stay (Frändberg 2013, Calvo 2018). These findings suggest an interesting dynamic where international studies may act as a driver of additional travel; travel drives interpersonal networks which drive more mobility.

To sum up, by using the concept of science-driven mobility we point at the increasing importance of research and education as drivers of mobility, and the impact these forms of mobility have on university cities. It is a broad concept, capturing both the different categories of visitors that take part in science-driven mobility for various lengths of time, and the mutual links between them. By using this concept, we also emphasise the interdependence of different actors, facilities and institutions, for developing place-specific conditions favouring science-driven mobility. Hence, the discussion about science-driven mobility can contribute to our understanding of the complex dynamics of New Urban Tourism geographies.

Data

Based on a case study of Lund, we show a city where the impact of science-driven mobility is particularly visible. As rapidly increasing mobility of students and researchers is a widely spread phenomenon, such a case study may have implications for other cities as well. However, it is highly questionable to make generalised assumptions based on one case study; it merely points at interesting contemporary tendencies in urban development (cf. Flyvbjerg 2006). The work with collecting empirical material follows a mixed-methods approach, using document analysis, interviews and quantitative sources (Bryman 2016). The most substantial source of information comes from analysis of documents, e.g. reports, planning material and strategic policy documents. These come primarily from three sources: Lund University and its affiliated research facilities, Lund Municipality and the Greater Copenhagen region. The document analysis was supplemented by official statistics and information acquired through direct personal contacts with public sector and business representatives. Interviews were conducted with persons holding leading positions at Lund University, Lund Municipality and the local destination marketing organisation (DMO) (Visit Lund 2020).

Internationalisation at Lund University

The rapid internationalisation process in Lund can be traced back to two major events. In 1995, Sweden became a member of the European Union. The EU-membership reduced legal and institutional barriers to Denmark and the rest of Europe. Five years later, in 2000, the Öresund Bridge connecting Malmö and Copenhagen opened (see Figure 16.1). The bridge facilitated travel between the cities, and connected Copenhagen airport directly to Malmö and Lund. In the new millennium, efforts to build cross-border clusters intensified, focusing on physics, medical technology, pharmacy and food science (Törnqvist 2002). Today, the Greater Copenhagen region in many respects resembles a functional region. Regional planning in Greater Copenhagen has the region's advancement in science and technology as one of its cornerstones, taking advantage of its combination of universities, research facilities and technologically advanced industry (GCSC 2016).

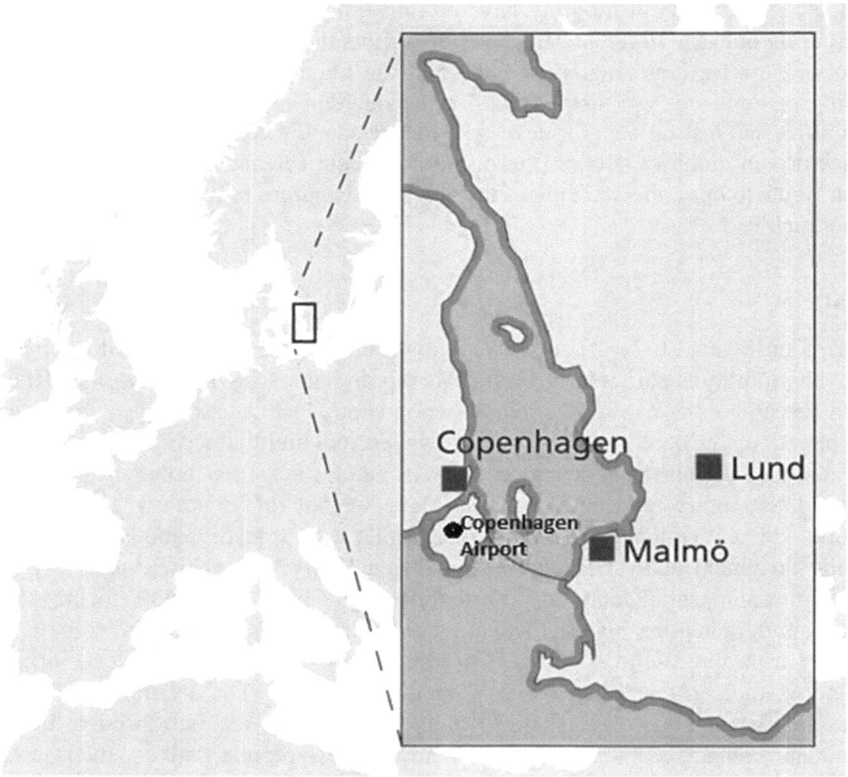

Figure 16.1 The Greater Copenhagen region © Lönegård & Co, image and media bank of Lund University. Edited by the authors.

The recent research facilities in Lund, MAX IV and ESS (European Spallation Source), can be seen as continuations of these processes. The MAX IV Laboratory was inaugurated in 2016, and it is the largest Swedish investment in research infrastructure, providing scientists with X-rays for research. It is a highly international environment with more than 30 nationalities represented among the 230 employees, as well as researchers visiting from all over the world. The decision to build ESS in Lund was taken in 2009 and it will be completed in 2023. In 2019, ESS had just over 500 employees representing 56 nationalities (ESS 2020, MAX IV 2020). When fully operational, MAX IV and ESS together estimate welcoming around 2,000 researchers annually, mainly for shorter periods as the use is scheduled in precise slots of different lengths. For example, at MAX IV 400 individual researchers used eight beamlines during the first six months of 2019 (MAX IV 2020). MAX IV and ESS are located in a new urban area called Brunnshög, planned to house a large variety of facilities for research and education, hotels and other accommodation, restaurants and cafés, as well as an urban park (Lunds kommun 2020a). Brunnshög is connected to the city centre with a purpose-built tramway.

Founded in 1666, Lund University is the largest university in the Nordic countries, with 40,000 students and 8,000 staff members (Lund University 2020a). The internationalisation process at the university has developed at a fast rate during the past 15 years. In 2007, Swedish higher education adapted to the European standard (the Bologna Process), which facilitated European students spending time at Lund University. The Erasmus partnerships play an important part in the European internationalisation of education, facilitating student exchange and staff mobility (Hannam and Guereño-Omil 2014). Lund University prioritises efforts to increase academic contacts with other parts of the world. A specific action plan for internationalisation was decided upon in 2019 with an explicit connection between the internationalisation process at the university and the research facilities (Lund University 2018).

Lund University welcomes a higher number of international students than other large Swedish universities, even if we see the same general internationalisation process across the whole higher education sector. In 2019 Lund University had 17% international students; as a Swedish comparison Uppsala University had 11%, Stockholm 9% and Gothenburg 7%. In 2019, 78% of the master's programmes were taught in English; up from 57% in 2009. Of the second cycle students, 68% have a non-Swedish bachelor's exam. Lund University recruits most of its students from Europe, but the number of non-European students is increasing. Taken together, the number of non-Swedish students grew from 2,531 in the year 2000 to 9,325 in 2019, a 368% rise during the last 20 years. The percentage of international PhD students has also increased rapidly. In 2004, 24% had a non-Swedish background; in 2018, it reached 55%. Today, 714 persons with a foreign background do their PhD studies at Lund University (Lund University 2020b).

Furthermore, there are secondary effects of student and staff mobility which are not easy to quantify, for instance in forms of additional leisure travel and VFR. Two different studies conducted at Lund University show that master's students tend to be a highly mobile category, in terms of their air travel behaviour. They also showed that leisure and VFR were important reasons for travel (Gössling *et al.* 2019, Nilsson 2015). These studies did not investigate visits to Lund University students by their friends and relatives. However, they are in line with other studies showing that international students develop personal networks during their time of study (Frändberg 2013, Rodríguez *et al.* 2012).

The internationalisation of Lund University is also apparent in the growing number of international teaching staff. More than 30% of the teaching staff now has a non-Swedish citizenship, approximately 1,000 persons. The internationalisation process is most evident among junior teachers and researchers. The percentage of international employees at short-term/limited career positions is almost twice as high as for those with a Swedish citizenship (64%). The number of non-Swedish professors is also increasing but at a much slower rate, 24% in 2018 compared to 17% in 2004 (Lund University 2020b). The trend towards internationalisation has been strongest in technology, natural sciences and medicine. These tendencies are further emphasised by developments taking place at MAX IV and ESS. Taken together, the figures above show that Lund University in many respects has changed its character from being predominantly national to being highly internationalised. Thereby, its potential international mobility (inbound and outbound) has increased substantially.

Local spatial and social impact

We argued above that different categories of science-driven mobility visitors could be included in the discussion on actors of New Urban Tourism. University cities like Lund are the arenas where this form of New Urban Tourism takes place. So how is the science-driven mobility visible in the city of Lund? We will highlight the spatial and social aspects.

Many short-term visitors come to the city of Lund for professional reasons, to attend meetings and other activities, taking place at the university, the research facilities and in companies in the research-intensive sector. The university plays an important role in this context; of a total of 132 major conferences in Lund 2019, 74 had a direct connection to the university. There are close co-operations between the DMO and Lund University in managing events and meetings (Visit Lund). Another group of short-term visitors is the VFR segment coming as guests to different academic events and celebrations. In Sweden, the academic year finishes in early June when graduations take place. In 2019, 243 new doctors took part in the conferment ceremony in the cathedral; approximately one third had a foreign background (Lund University 2019a). Each of them was allowed to bring four guests. In the international master's graduations 1,479 students (two thirds with an international background) received their diplomas with family and friends attending. These ceremonies thus generate a substantial number of visits to the city, indicating the importance of this kind of travel. These short-term visitors all need accommodation services—mainly in the form of hotels as well as other touristic services. The availability of commercial accommodation and the

Table 16.1 Hotel development in Lund, 1980–2018

	Hotels	*Available rooms*	*Overnight stays, 1000s*	*Foreign guests, 1000s*
1980	5	230	51	21
1990	7	415	90	34
2000	9	718	196	56
2005	10	822	205	62
2010	12	1031	309	86
2015	16	1482	459	156
2018	17	1502	509	171

© Lena Eskilsson and Jan Henrik Nilsson, data from Tillväxtverket, SCB, 2020

number of overnight stays have increased significantly in Lund, since the 1980s (see Table 16.1).

The growing number of temporary city users in the form of researchers coming to the university and the research facilities for more than a few days has created a demand for a different type of accommodation than conventional hotels. We see an increase in apartment hotels targeting the medium-long stays. In these types of accommodation, the guests have access to self-catering and other home-like facilities. Already, the first purpose-built researcher hotel is in place, close to MAX IV and ESS. Another one, with 250 apartments and rooms will open at Brunnshög in January 2021 (Visit Lund). In the city centre, the Clarion hotel chain plans for a new hotel, which will be, according to the CEO, "tailor made for the Lund market. Lund is growing, and ESS will attract an enormous number of researchers. They have to have somewhere to stay, often for three or six months" (Kniivilä and Kristiansen 2019).

The above data give a reasonable quantitative view on the changing visitor flows to Lund, since the market for unofficial accommodation is small. Compared to most non-Swedish cities, the number of available Airbnbs is very limited. In April and May 2020, there were fewer than 50 individual postings on the site (Airbnb 2020). In other contexts, a growing presence of Airbnb would be an indication of New Urban Tourism (Nilsson 2020). The low number of Airbnbs is a result of Swedish housing regulations, which do not allow short-time re-letting. Almost all Swedish flats are either rented or belong to housing co-operatives (a variety of condominium); the kind of owner-occupancy that is common in continental Europe is very rare in Sweden.

Housing for students is not solved by short-time accommodation like hotels. Traditionally, student housing in Lund is a matter for the Student Union's building company; the University does not own any housing property (AFB 2020). When internationalisation grew, new capacity was needed. Today, Lund University organises the accommodation of 980 international students in properties all over the city. Additional student housing is planned at Brunnshög. Beside the student housing, a guesthouse for short-term visits and 102 apartments or studios for guest researchers who stay medium-term, between two months and a year, are managed by the same system (LU Accommodation 2020).

As in similar old university cities, university buildings and student housing in Lund are not concentrated on a specific campus. Student accommodations can be found in many areas. Students, as temporary city users, are hence very visible and integrated in the cityscape. Still, there are areas that resemble student enclaves with higher concentration of departments, student housing and other services targeting students like pubs and cheap food places. International studies show that concentrations of students could have positive economic effects on urban areas. Students are portrayed as important actors in the creation of "lively, mixed-community neighbourhoods with an attractive mix of uses, high levels of local services, and vibrant cultural activities" (Munro *et al.* 2009, p. 1808), and following this, the areas have the potential to develop into "student enclaves which embody characteristics of the 'new urbanism' that is so appealing to residents and businesses" (ibid.). The vibrant student life in the city of Lund is highlighted in marketing campaigns and student information guides (Visit Lund 2020, Lund University 2019b). The Brunnshög area is another place where the city and university work together to develop and promote Lund. Targeting especially the science-driven visitors and high-tech businesses, future science-driven mobility is expected to concentrate in Brunnshög.

Specific services that are put forward in student guides are bike rentals, second-hand bikes and furnishing (Lund University 2019, SDR 2020). As in many other university cities, bicycles are very popular modes of transport, both for locals and temporary city users. These services reflect that the temporary city users generate certain types of services. The DMO participates in arrival weeks and other orientation activities when the students first arrive in Lund. The science-driven long-term visitors have in many ways a quite different social and economic situation, being mainly professional lifestyle travellers. Both the university and other employers have relocation services to help the employees and their families settle in Lund. For Lund University these services include PhD students who, according to the Swedish system, are employed during their time of studies, i.e. defined both as students and staff.

In 2015 Lund municipality established the International Citizen Hub which functions as an extra relocation service and meeting point for internationals. The overarching goal for the hub is to help attract and retain specialised competence in the region. The decision to establish the hub was a direct response to the foundation of research facilities in Lund. The hub works closely with the university, participates in welcoming days and arranges events for researchers, PhDs and master's students. Their work is done in co-operation with Visit Lund, and the hub in many ways functions as a complementary 'tourist office' for long-term visitors. The hub also contributes to promoting the image of Lund as a cosmopolitan place to visit or to settle down in.

One spatial and social dimension of the increasing presence of international researchers and other long-term professional lifestyle travellers in Lund is the establishment of international schools. At present there are four international schools in Lund that offer English language education, and one international preschool. Together, they enrol around 1,800 pupils, approximately 10% of the total number in Lund (Skolverket 2020). One of the schools is funded by the city of Lund and is particularly targeting non-permanent residents (Lunds kommun 2020b). There is a clear connection between the long-term professional travellers and the international schools. Lund international school's (2020) webpage says it is "ideally located for people who work at Lund University, the hospital, MAX IV and ESS". The schools are promoted in guides that are produced for international PhD students and post-docs moving to Sweden (SDR 2020), as well as by the International Citizen Hub.

The increasingly international atmosphere, as in the prevalent use of English in local services, affects social interaction in the city. Public sector services are also increasingly bi-lingual. In Lund, it is possible to live outside the Swedish-speaking surroundings, at least if you work or study at the university. Language has increasingly become an issue at the university, not only for teaching. Some junior teachers, particularly doctoral students, do not speak Swedish. Many international students have part-time jobs in cafés, restaurants and shops. In most cases, their Swedish is non-existent or very weak. The city as place and lived environment has changed in a fairly short period of time.

The discussion about temporary city users, the student enclaves and 'studentification' of the city, the plans for the high-profiled 'science village' at Brunnshög and the development of international schools point at interesting spatial and social perspectives for the city. The changes in the local environment and the social fabric raise sociologically interesting issues. Parts of Lund, socially and geographically, might resemble a cosmopolitan bubble or reserve, where international students and academics live their middle-class lives largely separated from the locals (cf. Munro *et al.* 2009). This social and linguistic divide does sometimes pose a problem for residents, particularly for the old, the less educated and those having migrated from non-English speaking countries.

Concluding discussion

In this chapter, we argue that science-driven mobility can be viewed as a form of New Urban Tourism. Science-driven mobility is a relatively new and growing phenomenon, definitely urban in character, and it is an example of how the borders of what we call tourism are challenged in the face of rapidly changing patterns of mobility.

The Lund case shows that science-driven mobility may be seen as resulting from fundamental changes in the economic geography of urban areas with a high concentration of high-tech industry, research and development (R&D) and higher education. The recent increase in science-driven mobility has resulted in significant spatial change: a new urban district is under construction. The increase in short-term visitors to the city has led to extensive investments in hotel capacity. Meetings and events related to science are clearly one factor behind the growth in hotel capacity. The development of apartment hotels is particularly interesting since they mainly target temporary city users. Furthermore, increasing numbers of international students have made it necessary to increase the stock of student housing. Together, all these building projects have changed the face of the city, particularly in the districts close to the university and the research facilities. There are no signs of growth in Airbnbs, otherwise often associated with New Urban Tourism. This points at the importance of recognising national regulatory frameworks when analysing tourism and urban change. New Urban Tourism is generally believed to result in social and cultural change in the hosting city. In Lund, there are few signs of radical social change. However, we argue that there is a slow but noticeable process of cultural change in the city. The increasing prevalence of the English language in the public sphere, in schools, services and workplaces, is a sign of this. The impact of such change might be rather large in the long run, since the city is relatively small.

Students, researchers and other visitors connected to the science sector all make an impact on university cities, which differs by the length of stay. The impact can be seen in immediate change on the street level, on short- and medium-term investments in the service landscape and in the long-term urban planning. There is of course a distinct difference between business tourists and VFR who stay for a few days, and guest researchers who bring their families. In-between, we have the temporary city users who do not fit into the basic guest-host or tourist-resident dichotomies, and are more like transnational urban consumers. These science-driven mobiles represent a challenge to the categorisations typical of traditional tourist segmentation, and of rigid statistical definitions of tourism. To summarise: like other forms of New Urban Tourism, science-driven mobility may bring considerable spatial change to small university cities; it questions traditional perceptions of tourism by viewing students and transient researchers as non-permanent residents, and there are indications that science-driven mobility brings about social change. It could hence be argued that science-driven mobility represents a form of New Urban Tourism precisely by the conceptual difficulties it encompasses.

Acknowledgements

The research was supported by Formas (Swedish Research Council for Sustainable Development), project 2018-02238. It is also part of a cross-disciplinary program at Lund University called "Moving In, mobility and competence provision in a region in transition".

References

Abdullateef, A.O. and Biodun, A.B., 2014. Are international students tourists? *International Journal of Business and Globalisation*, 13 (3), 298–306.

AFB, 2020. *AF bostäder* [online]. Available from: www.afbostader.se [Accessed 12 April 2020].

Airbnb, 2020. [Online]. Available from: www.airbnb.se [Accessed 14 April 2020].

Beaverstock, J.V., et al., eds., 2010. *International business travel in the global economy*. Farnham: Ashgate.

Bell, M. and Ward, G., 2000. Comparing temporary mobility with permanent migration. *Tourism Geographies*, 2, 87–107.

Bryman, A., 2016. *Social reseach methods*. 5th ed. Oxford: University Press.

Calvo, D., 2018. Understanding international students beyond studentification: a new class of transnational urban consumers. The example of Erasmus students in Lisbon (Portugal). *Urban Studies*, 55 (10), 2142–2158.

Cohen, S., 2011. Lifestyle travellers. Backpacking as a way of life. *Annals of Tourism Research*, 38 (4), 1535–1555.

Cresswell, T., 2006. *On the move. Mobility in the modern western world*. Abingdon: Routledge.

Dicken, P., 2015. *Global shift. Mapping the changing contours of the world economy*. 7th ed. London: Guildford Press.

ESS, 2020. *European spallation source* [online]. Available from: https://europeanspallationsource.se/ [Accessed 8 April 2020].

Eurostat, 2019. *Night spent at tourist accommodation establishments by NUTS 2 regions* [online]. Available from: www.appsso.eurostat.ec.europa.eu [Accessed 18 July 2019].

Flyvbjerg, B., 2006. Five misunderstandings about case-study research. *Qualitative Inquiry*, 12 (2), 219–245.

Frändberg, L., 2013. Temporary transnational youth migration and its mobility links. *Mobilities*, 9 (1), 146–164.

GCSC, 2016. The greater Copenhagen and Skåne Committee. *Handlingsplan for Greater Copenhagen 2016* [online]. Available from: https://bit.ly/375W0aV [Accessed 10 December 2020].

Gössling, S., et al., 2019. Can we fly Less? Evaluating the 'necessity' of air travel. *Journal of Air Transport Management*, 81, 101722.

Gotham, K.F., 2005. Tourism gentrification: the case of New Orleans' Vieux Carre (French Quarter). *Urban Studies*, 42 (7), 1099–1121.

Gregory, J.J., 2020. Studentification and urban change in South Africa. *In*: R. Massey and A. Gunter, eds. *Urban geography in South Africa*. GeoJournal Library. Cham: Springer, 225–238.

Hakeem, S. and Khan, Y., 2018. Urban tourism. The perspective on tourism impacts in Cambridge, United Kingdom. *Marketing & Management of Innovations*, 3, 268–275.

Hall, C.M., 2005. *Tourism. The social science of mobility*. Harlow: Prentice Hall.

Hannam, K. and Guereño-Omil, B, 2014. Educational mobilities. Mobile students, mobile knowledge. *In*: D. Dredge, D. Airey and M.J. Gross, eds. *The Routledge handbook of tourism and hospitality education*. Abingdon: Routledge, 143–154.

Kannisto, P., 2016. Extreme mobilities: challenging the concept of 'travel'. *Annals of Tourism Research*, 57, 220–233.

Kniivilä, K. and Kristiansen, I.D., 2019. Hotellkungen Petter Stordalen älskar byggkaoset i Lund. *Sydsvenskan* [online], 6 November. Available from: https://www.sydsvenskan.se/2019-11-06/hotellkungen-petter-stordalen-alskar-byggkaoset-i-lund [Accessed 10 December].

Larsen, J., 2019. Ordinary tourism and extraordinary everyday life: re-thinking tourism and cities. *In*: T. Frisch, et al., eds. *Tourism and everyday life in the city*. London: Routledge, 24–42.

LU Accommodation, 2020. [online]. Available from: https://www.luaccommodation.lu.se [Accessed 21 April 2020].

Lund International School, 2020. [online]. Available from: https://lundinternationalschool.com/ [Accessed 20 July2020].

Lund University, 2018. *Action plan for internationalisation at Lund University 2019–2021* [online]. Available from: https://www.staff.lu.se/sites/staff.lu.se/files/action-plan-for-internationalisation-at-lund-university-2019-2021.pdf [Accessed 10 December 2020].

Lund University, 2019a. *Lunds universitets doktorspromotionen, Lunds Domkyrka, 24 Maj 2019* [online]. Available from: https://www.lu.se/sites/www.lu.se/files/program-doktorspromotionen-2019-lunds-universitet-webb.pdf [Accessed 10 December 2020].

Lund University, 2019b. *The student guide, Lund University 2019/2020* [online]. Available from: www.lth.se/fileadmin/lth/english/internationella/filer/student-guide-2019-20.pdf [Accessed 10 December 2020].

Lund University, 2020a. *Lunds Universitet* [online]. Available from: www.lu.se [Accessed 10 May 2020].

Lund University, 2020b. *Lund University's internal statistics.*

Lunds kommun, 2020a. *Homepage about the Brunnshög project* [online]. Available from: www.lund.se/brunnshög [Accessed 24 April 2020].

Lunds kommun, 2020b. *Municipality's homepage, International school of Lund, Katedralskolan* [online]. Available from: www.lund.se/islk [Accessed 26 July 2020].

Maitland, R., 2006. How can we manage the tourist-historic city? Tourism strategy in Cambridge, UK, 1978–2003. *Tourism Management*, 27 (6), 1262–1273.

Maitland, R., 2007. Tourists, the creative class and distinctive areas in major cities: the roles of vistors and residents in developing new tourism areas. *In*: G. Richards and J. Wilson, eds. *Tourism, creativity and development*. London: Routledge, 73–87.

Maitland, R. and Newman, P., 2009. *World tourism cities. Developing tourism off the beaten track.* Abingdon: Routledge.

MAX IV, 2020. [Online]. Available from: www.maxiv.lu.se [Accessed 28 July 2020].

Moretti, E., 2013. *The new geography of jobs*. Boston, MA: Mariner Books.

Munro, M., Turok, I., and Livingstone, M., 2009. Students in cities: a preliminary analysis of their patterns and effects. *Environment and Planning. Part A*, 41, 1805–1825.

Nakazawa, T., 2017. Expanding the scope of studentification studies. *Geography Compass*, 11 (1), 1–13.

Nilsson, J.H., 2015. Young hypermobiles? Master students' air travel behaviour. *In*: *Paper to the 6th nordic geographers meeting*, 15–19 June 2015, Tallinn.

Nilsson, J.H., 2020. Conceptualizing and contextualizing overtourism: the dynamics of accelerating urban tourism. *International Journal of Tourist Cities*, 6 (4), 657–671.

Novy, J., 2019. Urban tourism as a bone of contention: four explanatory hypotheses and a caveat. *International Journal of Tourism Cities*, 5 (1), 63–74.

Pasquinelli, C., 2017. Tourism connectivity and spatial complexity: a widening bi-dimensional arena of urban tourism research. *In*: N. Bellini and C. Pasquinelli, eds. *Tourism in the city*. Cham: Springer, 29–50.

Poon, A., 1993. *Tourism, technology and competitive strategies*. Wallingford: CABI Publishing.

Rodríguez, X., Martínez-Roget, F., and Pawlowska, E., 2012. Academic tourism demand in Galicia, Spain, *Tourism Management*, 33, 1583–1590.

Russo, A.P. and Arias Sans, A., 2007. Students communities as creative landscapes: evidence from Venice. *In*: G. Richards and J. Wilson, eds. *Tourism, creativity and development*. London: Routledge, 161–177.

SDR, 2020. The social science doctoral student council. *The 2020 SDR PhD and post-doc guide* [online]. Available from: https://www.soclaw.lu.se/en/sites/soclaw.lu.se.en/files/2020_sdr_phd_and_post-doc_guide.pdf [Accessed 10 December 2020].

Skolverket, 2020. [Online]. Available from: https://siris.skolverket.se [Accessed 20 April 2020].

Tillväxtverket, S.C.B., 2020. *Hotel statistics*. Lund municipality, special delivery. Received 7.6.2020.

Törnqvist, G., 2002. *Science at the cutting edge. The future of the Øresund region*. Copenhagen: Business School Press.

UNWTO, 2020. *World Tourism Organization* [online]. Available from: statistics.unwto.org [Accessed 20 February 2020].

Urry, J., 2007. *Mobilities*. Cambridge: Polity Press.

Visit Lund, 2020. *Lund's DMO* [online]. Interview 3 June. Available from: https://visitlund/ [Accessed 21 July 2020].

Yu, S., et al., 2018. "Is there a bubble to burst?"–college students' spatial perception of campus and the city, a case study of rhodes college in memphis, TN. *Urban Geography*, 39 (10), 1555–1575.

Part IV

Concluding remarks

17 So, what is new about New Urban Tourism?

Maria Gravari-Barbas

Visit "as a local": at the origins of tourism

From its inception as a contemporary social, spatial and economic phenomenon, tourism *is* urban. Cities—along with sea and mountain resorts—have been the main 'laboratories' of the invention of modern tourist cultures.

Since the end of the 18th and all through the 19th century, tourists have been attracted by both the discovery of urban historical areas and by the display of modernity. In Paris for example, tourists were fascinated by the picturesque medieval areas at the central historical districts as well as by the audaciously redesigned areas by the then liberal capitalistic state around the Opéra or the Grands Boulevards. Visitors were eager to visit both Notre-Dame de Paris as well as the vast Champ-de-Mars areas, hosting the International and Universal Exhibitions which demonstrated to the world the most daring technical progress of their times.

However, if, during those early times of tourism, centrally located historical areas or modern neighbourhoods became main tourist attractions in most cities, it would be wrong to consider that tourists' imaginations and desires were limited to these 'must-see' areas. The Baedeker guide of Paris (1878) reflects the interest of the early tourists for forms of tourism which were much less mainstream than we tend to imagine. The guide invites these tourists to Parisian open markets, not to buy everyday goods but to visit the places in which local inhabitants do so. Baedeker (1878) encourages tourists to pay attention to "the shouts of Paris" (les "*cris de Paris*"), "often discordant, almost always ineligible for those who are not used to it, but many of which do not lack originality". The shouts of vendors about their products are presented as part of the soul of Paris, as a distinctive element that an informed tourist should be aware of, beyond the sole visit to iconic monuments and sites. The guide also invites tourists to visit 'off-the-beaten-track' places such as the slaughterhouse of La Villette or the Paris morgue. Though Baedeker does not mention 'authenticity' or 'intangible heritage'—as we would probably do today—it reflects the desire of tourists to see the "real Paris", to go beyond the "front stage" prepared and organised for them and to access the mythicised "back stage" (MacCannel 2013) of the 'real' city. Indeed, from the very beginnings of tourism, the ultimate dream of tourists was to experience the visited places 'from inside', and 'as a local'. Everyday life, and the tourist's desire to be part of it, has been a common feature of tourism since its very beginnings; there is nothing new about this.

If *New* Urban Tourism is characterised by the shift from 'sightseeing' to 'life-seeing', this is therefore less in terms of tourism demand than in terms of tourism offer and possibilities. The industrial tourism system prevalent in the second part of the 20th century led to the massification and rationalisation of tourist practice. Travel, hospitality and tourist attractions were organised to cater for large numbers. Creation and development of tourist specialised areas or tourist enclaves led to tourism zoning, similar to that more generally applied, at the same time, in urban planning. Tourists were invited to areas progressively adapted to their needs, where the concentration of hotels, main touristic attractions and tourist shops and services, created local spatial tourism ecosystems that locals tended to avoid for being labelled as too 'touristic'.

The post-tourism hypothesis

Recent evolution in tourism practices challenges the existing oppositions between tourist and non-tourist destinations as well as between tourist and non-tourist actors, as ordinary places are becoming tourist destinations (Condevaux *et al.* 2019, p. 21). This back-and-forth between the ordinary and the extraordinary is sometimes called upon in support of the idea that the 21st century appears to mark the advent of the 'post-tourism' era (Condevaux *et al.* 2016). This is thought to be characterised by the end of tourism as a field offering specific social, spatial and temporal horizons and a de-differentiation between tourist and non-tourist destinations and practices. The boundaries are becoming blurred between the here and the elsewhere, and between the exotic and the everyday.

This 'dilution' of tourism into everyday practices is often interpreted as a weakening of the distinction between civil society and tourism as a commercial sector. Tourists are becoming actors in the creation of tourism products. The assistance of new technologies facilitates "prosumption" (Toffler 1980), a contraction of production and consumption: tourism practices now often rely on do-it-yourself arrangements, which use social, transport and accommodation networks.

The notion of 'post-tourism' was introduced by Feifer (1985). The author defined post-tourists as individuals who are less dependent on the tourism industry, and who had different types of tourism experiences. According to Girard (2013), post-tourism can be understood in two ways, related to the points of view developed independently in the 1990s by two sociologists, Jean Viard and John Urry—both based on the observation that the lines are blurred between tourism and non-tourism.

The first perspective, developed by Urry (1995), presupposes a rupture between modernity and post-modernity, not only in the blurring of the distinctions but also in the renewal of the values on which tourism practices are based. It provides a theoretical account of the far-reaching changes within society and of the new ways of doing tourism. It gives an account of the general criticism surrounding tourism. It also offers an account of new trends in tourism practices, such as the quest for authentic experiences or new ways of exploiting this authenticity. The second perspective, which corresponds to Viard's position (2000, 2006), describes the process of extending cultural tourism values to different social horizons. It aims, more precisely, to give an account of what happens after tourism has been developed in certain highly touristic areas, when these areas attract not only tourists but also new residents who can therefore be considered as post-tourists. We thus have, on the one hand, an approach that highlights the reinvention of tourism through practices that exploit the boundaries between the ordinary and the extraordinary and which offer alternative practices or destinations "that emphasise the diversity of new destinations made touristic" (Bourdeau 2012, p. 43). On the other, post-tourism takes on a more literal meaning. It refers to a process of transition and of residential reconversion in holiday resorts and tourist regions (Viard 2000, 2006).

Both perspectives are thought to lead to a break with "tourism in the strict sense of the word, which emerged with the arrival of modernity" (Bourdeau 2012, p. 43), that is based on a "highly polarised and polarising ad hoc planning and development of spaces where the main heritage attraction (monument, famous site) and the resort are seen as symbolic" (ibid., p. 44). It is worth remembering just how much the idea of spatial as well as temporal rupture was thought to be a constituent of tourism in the 20th century. The de-differentiation of practices is one of the elements that is thought to most clearly mark the break with this tourism model and our entry into post-tourism. This de-differentiation or dissolution of tourism in ordinary practices and vice versa is also thought to translate into a weakening of the distinction between the hosts and the visitors (Cohen and Cohen 2012). It corresponds to an infusion of tourism into the everyday space, tourism becoming "a normal method of spatially organising social realities, which not only express themselves in specific areas and practices but also, and especially, become present in situations that are not ostensibly tourism-related" (Lussault 2007, p. 199).

The post-tourism perspective also focuses on changes in the behaviour of tourists themselves. Poon (1994) uses the term 'new tourists', whom he describes as more experienced and educated. Some authors highlight the fact that the new tourists seem to be increasingly in search of an authentic experience as well as a personalised, tailor-made provision (Équipe MIT 2011). They are looking for 'off-the-beaten-track' experiences and are drawn to the mundane and the familiar (Gravari-Barbas and Delaplace 2015). Rather than just a simple one-directional evolution, we seem to be dealing with a plurality of behaviours, which some authors account for through more detailed typologies.

These approaches have the merit of drawing attention to the renewal of values that form the basis of tourism practices and allow us to shed light on the mechanisms of developing tourism around the ordinary, around the end of tourism and around hybridisation but, above all, they enable us to think about the processes as a whole.

The consequences of the blurring of tourism and everyday life

A "velvet revolution" (Gravari-Barbas *et al.* 2019) in tourism emerged at the end of the 20th century, as tourists became more experienced and mature. These mature tourists had accumulated significant tourism capital through their numerous travels and their capacity to adapt and to feel "at home" in the different contexts they visited (Maitland 2010). This led to the development of new tourism offers over the last few decades, inviting visitors to experience the city in much more segmented, differentiated and plural ways than just the usual tour of the main tourist sights (Maitland and Newman 2004, Gravari-Barbas and Delaplace 2015).

The conjunction of the evolutions in both supply and demand (Ionnides and Debbage 1998) contributed to opening up, expanding and diversifying the urban tourist ecumene. While, as mentioned above, it may have been tourists who pioneered the expansion of the urban tourism frontiers, the tourism sector demonstrated recently an unprecedented responsiveness and adapted accordingly, rapidly transforming tourists' romantic expectations of local life and authentic neighbourhoods into exclusive tourism products.

At the beginning of the 21st century, tourism visits to places formerly considered inhospitable, liminal or dangerous (such as favelas, townships and slums, metropolitan suburbs, former industrial sites, urban ruins and urban undergrounds) or simply "ordinary" and every day, became commonplace (Maitland 2010). The tourist gaze (Urry 2002) has figured in the aestheticisation (Featherstone 1991/1997) or heritagisation of ordinary places (Condevaux *et al.* 2019). It has even contributed to overcoming the associated 'geographical stigma' of sites with a difficult or traumatic heritage and to transforming them into destinations offering fun, excitement, aesthetic pleasure or affective attachment.

However, the overflow of the now experienced and mature tourists from the touristic zones or enclaves to which they have been limited for decades by the tourism sector and industries, was not free of consequences. During the first decades of the 21st century, tourists where not only numerous, but also, they were potentially *everywhere*. Indeed, New Urban Tourism is not exclusive but cumulative. It does not represent a cultural shift from iconic 'touristic' sights to local authentic experiences: the iconic touristic sites are visited more than ever before—I am referring here of course to the year 2019 and the pre-COVID era. The eclectic taste of mature tourists allows them to visit simultaneously the most iconic sightseeing sites as well as the avant-garde and off-the-beaten-tracks areas. New Urban Tourism is indeed a niche tourism, compared to massive international and organised groups still dominant today.

New tourists enter, however, into a distinction game of being different (more daring, more adventurous, more cultivated) than the 'mass'; they will not only visit the Louvre or the Tate but also neighbourhoods without major sightseeing sites, for their ambiance, street life or food, alternative commerce or youth culture. However, not all neighbourhoods are prepared to host New Urban Tourists. In many areas, live 'like a local' may cause disruptions and the feeling of being overtouristified. Paradoxically enough, New Urban Tourism, by overflowing the tourist zones and enclaves in which it used to be more or less confined, is to a huge extent responsible for the overtourism phenomena experienced in the 2010s.

New Urban Tourism and the platformisation of the cities

New tourism is very different from forms of adventurous tourism in the 1960s and 1970s, when a bohemian Western youth traced the route to the remote mythicised areas (such as for example the hippie trail). Far from being alternative, New Urban Tourism is rather the reflection of the latest stage of capitalism. It owes a lot to a neoliberal approach of seeing the city as a place where everything can be packaged, "enriched" (Boltanski and Esquerre 2017) by new meanings and storytelling, and sold to new markets.

Far from romantic analyses considering New Urban Tourism as a means of creating new bridges between the local and the global, allowing tourists to live as a local and locals to communicate and exchange with the global, it is more about a new market-oriented approach than about genuine, authentic and disinterested encounters. One of the main explanations of the development of New Urban Tourism is precisely the development of tourism-rental platforms. The latter correspond to an ultimate stage of capitalism transforming hospitality into product, and millions of apartment owners into tourism providers.

New Urban Tourism is a 'platform-led' tourism, developed thanks to digitalisation and incredible social media development. As Bernardi and Mura remind us (Chapter 3), this goes beyond accommodation, and concerns also: sharing or renting mobility solutions; experiential tours through the engagement of residents; as well as sharing of food and logistic services for tourists. The authors draw the attention to the fact that 'tourist platformisation' carries risks and challenges with it. Home sharing through Short Term Renting (STR) platforms exacerbate old urban issues since it reinforces gentrification, strengthens touristification and erodes housing units from the traditional housing market (Clancy 2020).

We should not of course blame tourists, who may be truly animated by the romantic will to visit places as a local and who are animated by the expectations of real encounters with the locals. However, the achievement of their desire passing by providers such as Airbnb has important consequences that the authors of this book have underlined in their chapters. Müller and Wellner (Chapter 4) underline the inherent risks of opening up new areas of the city to tourists because of the intensified competition for residential space and infrastructure, exacerbating the already difficult situation of affordable housing. More so than ever before, and due to its diffused and transversal character, New Urban Tourism came into the centre of several other urban concerns and, in particular, housing. Even if competition between tourism and other urban functions or services is not new per se, New Urban Tourism boosted issues of gentrification to an unprecedented level (Gravari-Barbas and Guinand 2017), producing even phenomena of supergentrification in formerly gentrified areas (Gravari-Barbas 2017). As Amore and Hall put it (Chapter 12), "this should not be described as New Urban Tourism [...] it should be called what it really is: tourism-related capitalism that serves only limited interests".

Hernandez (Chapter 2) underlines for Bushwick (Brooklyn) that New Urban Tourism relies on an "identifiable and homogenous set of consumption patterns and lifestyle tastes" that constitute the hipster subculture, seen as a lifestyle category of consumption taste practices. The urban landscape of advanced capitalist cities is reshaped by the outsider aesthetic of bohemians. Tourism commodification of eventually all urban services and spaces also has other risks, underlined by Hikido (Chapter 14) for South Africa. Framing townships as hard-working communities "minimises the problems of persistent inequalities".

As a conclusion, authors in this book open important and innovative perspectives by not only presenting the different dimensions of New Urban Tourism (which, as discussed above, are not always as new as communication invites us to believe), but also by showing the inherent risks of these evolutions.

New Urban Tourism can be more pervasive than former forms of urban tourism. It introduces disruptions for both the urban and tourism sector. It pervades the housing market and the urban services. It may generate urban conflicts while depoliticising urban fights and revendications. The authors in this book call for more critical perspectives of New Urban Tourism. This seems a particularly stimulating research agenda.

References

Baedeker, K., 1878. *Paris and its environs, with routes from London to Paris*. Leipsic.

Boltanski, L. and Esquerre, A., 2017. *Enrichissement: une critique de la marchandise*. Paris: Gallimard.

Bourdeau, P., 2012. Le tourisme réinventé par ses périphéries? *In*: F. Bourlon, et al. *Explorando las nuevas fronteras del turismo. Perspectivas de la invetigacion en turismo*. Coyhaique: Nire Negro, 31–48.

Clancy, M., 2020. Tourism, financialization, and short-term rentals: the political economy of Dublin's housing crisis. *Current Issues in Tourism*, July.

Cohen, E. and Cohen, S., 2012. Current sociological theories and issues in tourism. *Annals of Tourism Research*, 39 (4), 2177–2202.

Condevaux, A., Djament-Tran, G., and Gravari-Barbas, M., 2016. Before and after tourism(s). The trajectories of tourist destinations and the role of actors involved in "off-the-beaten-track" tourism: a literature review. *Via Toursims Review* [online], 9. Available from: http://journals.openedition.org/viatourism/413 [Accessed 21 December 2020].

Condevaux, A., Gravari-Barbas, M., and Guinand, S., 2019. Lieux ordinaires et tourisme. In: A. Condevaux, M. Gravari-Barbas, and S. Guinand, eds. *Lieux ordinaires, avant et après le tourisme*. Paris: PUCA, 9–28.

Équipe MIT, 2011. *Tourismes 3. La révolution durable*. Paris: Belin.

Featherstone, M., 1991/1997. The aestheticization of everyday life. *In:* S. Lash and J. Friedman, eds. *Modernity and identity*. Oxford: Blackwell, 265–290.

Feifer, M., 1985. *Going places*. Baedeker guide of Paris, 1878. London: Macmillan.

Girard, A., 2013. Faut-il raccorder une théorie générale de la postmodernité à une théorie à moyenne portée du post-tourisme? *In*: P. Bourdeau, F. Hugues and L. Perrin-Bensahel, eds. *Fin (?) et confins du tourisme. Interroger le statut et les pratiques de la récréation contemporaine*. Paris: L'Harmattan, 43–52.

Gravari-Barbas, M., 2017. Super-gentrification and hyper-touristification in Le Marais, Paris. *In:* M. Gravari-Barbas and S. Guinand, eds. *Tourism and gentrification in contemporary metropolises. International perspectives*. London: Routledge, 299–328.

Gravari-Barbas, M. and Delaplace, M., eds., 2015. « Le tourisme urbain « hors des sentiers battus ». *Téoros* [Online], 34 (1–2). http://journals.openedition.org/teoros/2790

Gravari-Barbas, M. and Guinand, S., 2017. *Tourism and gentrification in contemporary metropolises. International perspectives*. London: Routledge.

Gravari-Barbas, M, Jacquot, S., and Cominelli, F., 2019. New cultures of urban tourism. *International Journal of Tourism cities*, 5 (3), 301–306.

Ionnides, D. and Debbage, K., eds., 1998. *The economic geography of the tourist industry: a supply-side analysis*. London: Routledge.

Lussault, M., 2007. Le tourisme, un genre commun. *In*: Ph. Duhamel and R. Knafou, eds. *Mondes urbains du tourisme*. Paris: Belin, 333–349.

MacCannell, D., 2013. *The tourist – a new theory of the leisure class*. Berkley, CA: University of California Press.

Maitland, R., 2010. Everyday life as a creative experience in cities. *International Journal of Culture, Tourism, and Hospitality Research*, 4 (3), 176–185.

Maitland, R. and Newman P., 2004. Developing metropolitan tourism on the fringe of central London. *International Journal of Tourism Research*, 6, 339–348.

Poon, A., 1994. The 'new tourism' revolution. *Tourism Management*, 15 (2), 91–92.

Toffler, A., 1980. *The third wave*. New York: Bantam Books.

Urry, J., 1995. *Consuming places*. Milton Park: Taylor and Francis.

Urry, J., 2002. *The tourist gaze*. Thousand Oaks, CA: SAGE.

Viard, J., 2000. *Court traité sur les vacances, les voyages, et l'hospitalité des lieux*. La Tour-d' Aigues: Editions de l'Aube.

Viard, J., 2006. *Eloge de la Mobilité. Essai sur le capital temps libre et la valeur travail*. La Tour-d' Aigues: Editions de l'Aube.

Index

Taylor & Francis Group
an informa business

Taylor & Francis eBooks

www.taylorfrancis.com

A single destination for eBooks from Taylor & Francis with increased functionality and an improved user experience to meet the needs of our customers.

90,000+ eBooks of award-winning academic content in Humanities, Social Science, Science, Technology, Engineering, and Medical written by a global network of editors and authors.

TAYLOR & FRANCIS EBOOKS OFFERS:

A streamlined experience for our library customers

A single point of discovery for all of our eBook content

Improved search and discovery of content at both book and chapter level

REQUEST A FREE TRIAL

support@taylorfrancis.com

Routledge
Taylor & Francis Group

CRC Press
Taylor & Francis Group